U.S. Marine Corps Wilderness Medicine Survival Course

Plus

Mountain Operations

Department of Defense

Pentagon
Publishing

UNITED STATES MARINE CORPS
Mountain Warfare Training Center
Bridgeport, California 93517-5001

WILDERNESS MEDICINE COURSE
STUDENT HANDBOOK

TABLE OF CONTENTS

TOC

TABLE OF CONTENTS (CONT.)

UNITED STATES MARINE CORPS
Mountain Warfare Training Center
Bridgeport, California 93517-5001

FMST.07.18
04/02/02

STUDENT HANDOUT

MOUNTAIN SAFETY (SUMMER)

TERMINAL LEARNING OBJECTIVE: Given a unit in a wilderness environment and necessary equipment and supplies, apply the principles of mountain safety to prevent death or injury per the reference. (FMST.07.18)

ENABLING LEARNING OBJECTIVE: Without the aid of references and given the acronym "BE SAFE MARINE", list in writing the 12 principles of mountain safety, in accordance with the references. (FMST.07.18a)

OUTLINE

1. **PLANNING AND PREPARATION:** (FMST.07.18a) As in any military operation, planning and preparation constitute the keys to success. The following principles will help the leader conduct a safe and efficient operation in any type of mountainous environment. We find this principle in the acronym "BE SAFE MARINE". Remember the key, think about what each letter means and apply this in any type of environment.

B - Be aware of the group's ability.

E - Evaluate terrain and weather constantly.

S - Stay as a group.

A - Appreciate time requirements.

F - Find shelter before storms if required.

E - Eat plenty and drink lots of liquids.

M - Maintain proper clothing and equipment.

1

A - Ask locals about conditions.

R - Remember to keep calm and think.

I - Insist on emergency rations and kits.

N - Never forget accident procedures.

E - Energy is saved when warm and dry.

a. Be Aware of the Group's Ability. It is essential that the leader evaluates the individual abilities of his men and uses this as the basis for his planning. In his evaluation, the leader must include the group's overall physical conditioning, and the consideration of change in climate and how long the unit has had to acclimatize.

(1) Mental attitude of your group. Is morale high? How much tactical training has the group had in a particular type of terrain?

(2) Technical aspect of your group. Have they been on skis, snowshoes, etc.?

(3) Individual skills. At this point, you must choose who is most proficient at the individual skills that will be required for your mission, navigation techniques, security, call for fire, rope installations, track plans, bivouac site selection, skijoring, etc.

b. Evaluate Terrain and Weather Constantly.

(1) Terrain. During the planning stages of your mission, the leader must absorb as much information as possible on the surrounding terrain and key terrain features involved in your area of operation. Considerations to any obstacles must be clearly planned for. Will you need such things as fixed ropes, rope bridges, climbing gear, etc?

(a) Stress careful movement in particularly dangerous areas, such as loose rock and steep terrain.

(b) Always know your position. Knowing where you are on your planned route is important.

(2) Weather. Mountain weather can be severe and variable. Drastic weather changes can occur in the space of a few hours with the onset of violent storms, reduced visibility, and extreme changes. In addition to obtaining current weather data, the leader must plan for the unexpected "worst case". During an operation he must diagnose weather signs continually to be able to foresee possible weather changes.

(a) Constantly evaluate the conditions. Under certain conditions it may be advisable to reevaluate your capabilities. Pushing ahead with a closed mind could spell disaster for the mission and the unit.

c. Stay as a Group. Individuals acting on their own are at a great disadvantage in this environment.

(1) Give the unit adequate rest halts based upon the terrain and elevation, physical abilities of the unit, combat load and mission requirements.

(2) Remember to use the buddy system in your group.

(3) Maintain a steady pace so that it will allow accomplishment of the mission when all members of the unit reach the objective area.

d. Appreciate Time Requirements. Efficient use of available time is vital. The leader must make an accurate estimate of the time required for his operation based on terrain, weather, unit size, abilities, and on the enemy situation. This estimate must take into account the possibility of unexpected emergencies and allow sufficient leeway to make unplanned bivouacs in severe conditions.

(1) Time-Distance Formula (TDF). This formula is designed to be a guideline and should not be considered as the exact amount of time required for your movement. Furthermore, this formula is for use in ideal conditions:

3 km/mph + 1 hour for every 300 meters ascent; and/or + 1 hour for every 800 meters descent.

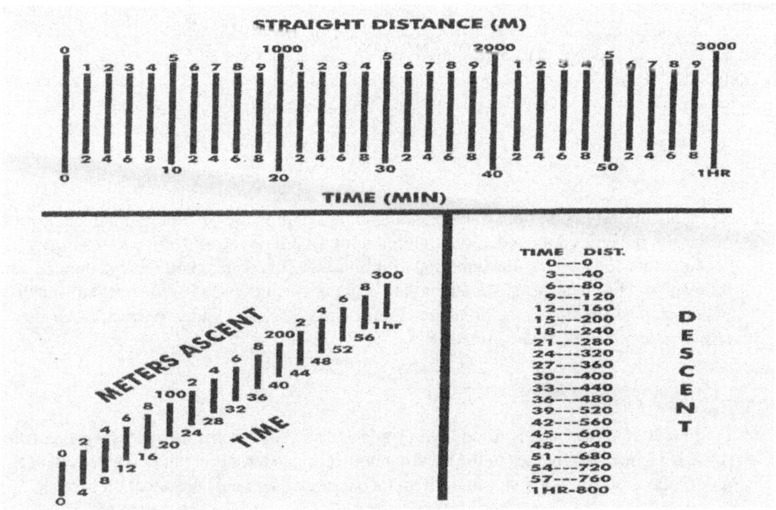

(2) Route Planning. Route cards are not to be used in place of an overlay, but as a tool to be used in route planning. Overlays/Route cards should contain the following information at the minimum:

Unit Designation:
Unit Commander.
Number of personnel.
Inclusive dates and times of movement.
Grid coordinates of each checkpoint and bivouac.
Map references.
Azimuth and distances for each leg.
Elevation gain/loss per leg.
Description of the ground.
ETA and ETD.

ROUTE CARD

UNIT I.D.		UNIT COMMANDER		NUMBER OF PERSONNEL		DATE AND TIME	MAP REFERENCE
LEG	AZM	DIST	GRID	ETA	ETD	ELEVATION	DESCRIPTION

(3) As in any military operation, route planning and execution are of vital importance. Prior to departure, the unit commander must submit a route card or patrol overlay to his higher headquarters and keeps a duplicate copy for himself. This preplanned route should be followed as closely as possible, taking into account changes based on the tactical situation. In non-tactical situations, the preplanned route should be followed to reduce search and rescue time in an emergency situation.

c. Find Shelter before Storms if required. Under certain conditions, inclement weather can provide tactical advantages to the thinking unit commander, but by the same token it can reduce the efficiency of a unit to nil if an incorrect evaluation of the situation is made.

(1) Bivouac. If the group decision is to bivouac, then it's vital that we know the principles for an unplanned bivouac.

(a) Unplanned bivouac. The principles and techniques discussed here apply both to unplanned and tolerated bivouacs. In any survival situation, especially in a mountainous environment, the most immediate danger is from exposure to the elements. Being lost will not directly kill an individual. Starvation takes time, but hypothermia can manifest itself in a matter of hours resulting in death. Adhering to the following principles will give an individual the best chance to spend a relatively safe bivouac with the prospect of continued effort toward mission accomplishment.

1. Make shelter. The requirements for expedient shelters and the building procedures will be covered in another section. The basic requirement for protection from the elements is essential.

2. Keep warm. The retention of body heat is of vital importance; any action in which body heat is lost should be avoided. The following points should be considered:

a. Adequate shelter.

b. Insulation from the ground using branches, a rucksack, etc.

c. Wear extra clothing.

d. Use extra equipment for insulation.

e. Produce external heat while trying to conserve fuel for future use.

3. Keep dry. Being wet causes the loss of body heat 24 times faster than when dry. Adequate protection from the elements is of prime importance to prevent the onset of hypothermia.

f. Eat Properly and Drink Plenty of Fluids

(1) Food. The human body can be compared to a furnace, which runs on food to produce energy (warmth). By planning the consumption of food to suit the specific situation, adequate nutrition and extra warmth can be supplied.

(2) Water. The intake of adequate amounts of water will maintain the body in proper working order. Danger from dehydration is as high in mountain regions as in hot dry areas. Loss of liquids is easily seen and felt in hot climates; whereas in the

mountains, the loss of body fluids is much less noticeable. High water intake, at least 6 quarts per day when in bivouac, 8 quarts per day when active, will help to prevent dehydration.

g. Maintain Proper Clothing and Equipment

(1) Clothing

(a) Our clothing has to perform an important function in our mission; therefore, when choosing our clothing we have to take into account some essential requirements.

1. Protection against wind and rain.

2. Layered and easily adjustable.

3. Lightweight and durable.

(b) To help us remember how to maintain and wear our clothing, we use the acronym, "COLD".

C - Keep clothing Clean.

O - Avoid Overheating.

L - Wear clothing loose and in Layers.

D - Keep clothing Dry.

NOTE: In the mountains a man should never be separated from his gear. Here are some basic and essential items that should be considered during your planning stage.

(2) Required equipment:

(a) A daypack should always be carried per squad and one Marine should carry a combat load.

(b) Map and compass. Every individual in a leadership position and his assistant should carry a map and compass. The maps should be weatherproofed and extra maps should be distributed throughout the unit.

(c) Repair kit. This kit should include those items necessary to do emergency repairs on your equipment.

(d) Survival Kit. Always carried on your person. The contents of a survival kit will be covered in another period of instruction dealing specifically with survival kits.

h. Ask Locals About Conditions. An often-overlooked source of information is the indigenous population of an area. Local weather patterns, rockslide areas, watering points, and normal routes can all be obtained by careful questioning. The leader must obtain current information of the actual conditions along his intended route. Of particular importance are recent precipitation and enemy sightings.

i. Remember to Keep Calm and Think.

(1) Emergency situation. Having recognized that you are lost and that an emergency situation exists, the following principles should be followed:

(a) Keep calm and do not panic. At this point you must make every effort to conserve body heat and energy.

(b) Think. When an individual is cold, tired, hungry or frightened he must force himself to organize his thoughts into a logical sequence.

(c) The group must try to help itself by either finding the way back to safety or by preparing shelters and procuring food.

(d) Above all else, the group must act as a tight-knit unit. In emergency situations, individual dissension can cause a total loss of control and unit strength.

(2) If the decision is reached that the group should seek its way back to safety, several possibilities exist. In most situations, the safest approach will be to retrace the route to the last known point and continue from there. The other course of action is to get a group consensus on the present location and send out a small search party to locate a known point. This party must ensure that they mark their trail adequately to return to the group. If all attempts at finding a way back to known terrain fails, a definite survival situation exists and actions discussed later in this section must be instituted.

j. Insist on Emergency Rations and Kits. Just like what was covered in the SUMMER MOUNTAIN WARFIGHTING LOAD REQUIREMENTS class, survival rations and a survival kit should always be carried.

k. Never Forget Accident/Emergency Procedures.

(1) Causes of accidents. The general procedures used to handle accidents differ little in this environment, but several distinct points should be kept in mind. The most frequent causes of accidents are as follows:

(a) Overestimation of physical and technical abilities.

(b) Carelessness.

(c) General lack of observation of one's surroundings.

(d) Lack of knowledge and experience by leaders.

(e) The failure to act as a group.

(f) Underestimation of time requirements to move through mountainous terrain and underestimation of the terrain itself.

(2) Preventive measures. The only truly effective preventive measures for the above lie in the education and experience of leaders at all levels. Too often, leaders sit by watching during training and as a result have no concept of the requirements involved in the mountainous environment. Only by active involvement can a leader gain the knowledge and experience needed to effectively lead in this environment.

(3) General procedures for handling an accident. These require only a good dose of common sense as outlined below.

(a) Perform basic first aid.

(b) Protect the patient from the elements to include insulation on top and bottom.

(c) Evacuate if necessary.

(d) Send for help if required. If possible, never send a man for help alone.

(e) Send the following information regarding the accident:

1. Time of accident.
2. Nature and location of accident.
3. Number injured.
4. Best approach route to accident scene.

(4) If one man of a two-man team is injured, the injured man must be given all available aid prior to going for help. If the injured man is unconscious, he should be placed in all available clothing and sleeping gear and anchored if on steep terrain. A note explaining the circumstances, and reassuring him, should be left in a conspicuous spot. This note must also contain the following information:

(a) When you expect to return.

(b) Where you went.

(c) What you did before you left (medication, etc.).

(5) International distress signal (whistle):

(a) Six short blasts in 1 minute from person requesting help.

(b) The return signal is three blasts in 1 minute from the respondent.

(6) Other methods if help is required:

(a) Red pyrotechnics.

(b) SOS, (... --- . . .).

(c) "Mayday" by voice communications.

1. Energy is saved when warm and Dry. With the previous 11 principles in mind this one should fall right into place. Save your heat and energy by following these steps:

(1) Dress properly.

(2) Eat properly.

(3) Drink properly.

(4) Ensure shelter meets criteria.

(5) Produce external heat (fires, stove, extra clothing, etc.) to save body heat and energy for future use.

(6) Don't lose body heat by getting wet.

UNITED STATES MARINE CORPS
Mountain Warfare Training Center
Bridgeport, California 93517-5001

FMST.07.01
03/08/02

STUDENT HANDOUT

NUTRITION

TERMINAL LEARNING OBJECTIVE: Given a unit in a mountainous environment, and the necessary equipment and supplies, apply the principles of nutrition in a mountainous environment to prevent death or injury per the reference. (FMST.07.01)

ENABLING LEARNING OBJECTIVES:

(1) Without the aid of references and from a given list, choose the correct caloric intake needed per day in a mountainous environment, in accordance with the references. (FMST.07.01a)

(2) Without the aid of references and from a given list, choose which type of diet increases your tolerance to cold, in accordance with the references. (FMST.07.01b)

(3) Without the aid of references and from a given list, choose one of the preventive measures of excessive water loss in a mountainous terrain, in accordance with the references. (FMST.07.01c)

OUTLINE.

1. **STARVATION FACTORS.** Factors that combine to result in "Near Starvation" during military operations in any mountainous environment:

a. Difficulty in carrying sufficient food.

b. Lack of time for preparing meals and beverages.

c. Loss of appetite and low food palatability resulting from hypoxia/hypothermia.

d. Fat intolerance resulting from altered metabolism.

2

2. **PHYSIOLOGICAL ASPECT TO COLD STRESS.**

a. Man's adaptation to cold is essentially behavioral. Shelters and clothing are utilized to provide a micro-environment, which will allow survival. Body composition adjustments such as increases in body weight and fat are also examples of the body's adaptation to the cold. The winter months lend themselves to greater periods of inactivity, leading to these changes.

b. In the Falkland War, constipation was a major concern. Dehydration was found to be the culprit in the majority of cases, but diet had a significant role as well. MRE's are low in fiber, containing less than 10 grams. Man needs at least 31 grams of fiber per day to ease passage of stool. Constipation can be a debilitating condition, affecting overall unit readiness.

(1) Low fiber diets can lead to diverticulosis, diverticulitis and Irritable Bowel Syndrome (IBS). IBS flare-ups tend to occur during upsetting situations and emotional stress. Educate your troops on the importance of fiber, especially in a cold weather environment.

(2) Fiber is supplied by foods derived from plants. The best sources are:

(a) Whole grain cereals.

(b) Whole wheat breads.

(c) Fresh and dried fruits.

NOTE: Vegetables also help prevent constipation, but the fiber from whole grain cereals and whole wheat bread is more effective than vegetables.

3. **CALORIC REQUIREMENTS.** (FMST.07.01a) Recommended caloric intake (Naval Medical Command Instruction 10110.1) for troops operating in a cold weather environment is 4,500 calories per day.

a. Energy Expenditure. Military operations in a cold weather environment usually involve strenuous activity. Walking on snow expends roughly twice the energy that walking on hard ground does. If troops are unfamiliar with personal over-snow mobility equipment, i.e. snowshoes and skis, awkward movement and thus wasted energy will result.

(1) The heavy clothing necessary for functioning in the cold increases energy expenditures by 5-15%. You will have the opportunity to experience this effect in the upcoming field evolution.

(2) Properly clothed, experienced individuals should have negligible heat loss from radiation, conduction, and convection. However, losses associated with respiration and evaporation can be significant.

(3) In intense cold, particularly with high winds, heat loss to the environment may be inevitable.

4. **NUTRITION IN A HIGH ALTITUDE ENVIRONMENT.** A diet high in carbohydrates and high in fat is the most beneficial, while a high protein diet is generally the least desirable.

a. The marked weight loss seen in individuals at high altitude is largely due to increased metabolic demands and anorexia secondary to hypoxia. Total caloric intake commonly does not meet energy demands.

b. Altered metabolic processes that have been reported at high altitude include delayed gastric emptying, intestinal malabsorption and decreased gastric enzyme secretion.

5. **BEST DIET FOR COLD WEATHER**. (FMST.07.01b)

a. High carbohydrate and high fat foods should be included rather than high protein foods.

b. Small meals (snacks) spaced at time intervals of two hours or less should be consumed rather than the usual three meals-a-day.

c. During a typical eight-hour day of cold exposure, three high-fat (unsaturated fat) meals providing 60% of the day's calories, coupled with two normally scheduled meals (breakfast and dinner) should provide good cold tolerance.

6. **BASIC CONSTITUENTS OF FOOD**. The three basic constituents of food are carbohydrates, fats and proteins.

a. Carbohydrates. Also known as the quick energy foods that provide energy to produce heat with the byproducts of carbon dioxide and water when oxidized by the body.

(1) Carbohydrates are present in food mainly as sugars and starches and are broken down during digestion into simple sugars that are converted into glucose.

(2) Carbohydrates are stored as glycogen in the liver and muscles and can be broken down quickly into glucose to provide a rapid source of energy. However, these stores are not large and are markedly depleted by fasting for as little as 24 hours.

(3) Eating a high carbohydrate diet for several days will double one's glycogen stores and can increase endurance by up to three times compared to the ordinary balanced diet.

(4) The main carbohydrate food sources are: CARBOHYDRATE SOURCES

(a) Fruits.

(b) Vegetables.

(c) Cereals.

(d) Sugar.

b. Fats: In the body, fat serves as the main storage form of energy. One gram of fat has the equivalent of 9.3 Cal/gm as opposed to 4.1 Cal/gm for carbohydrates and proteins. They also produce energy, heat, CO_2 and H_2O.

(1) In cases of starvation, body fat tends to be broken down into acidic compounds. If these compounds accumulate in the blood faster than they can be burned, they cause the body tissues and blood to become excessively acidic (acidosis).

(2) During brief physical activity, energy is derived equally from fat and carbohydrates, but as the duration of the activity lengthens, the percent of energy supplied by fat increases as carbohydrate energy is quickly depleted.

(a) Fat sources:

1. Butter.
2. Lard.
3. Cooking oil.
4. Mayonnaise.
5. Ice cream.

FAT SOURCES

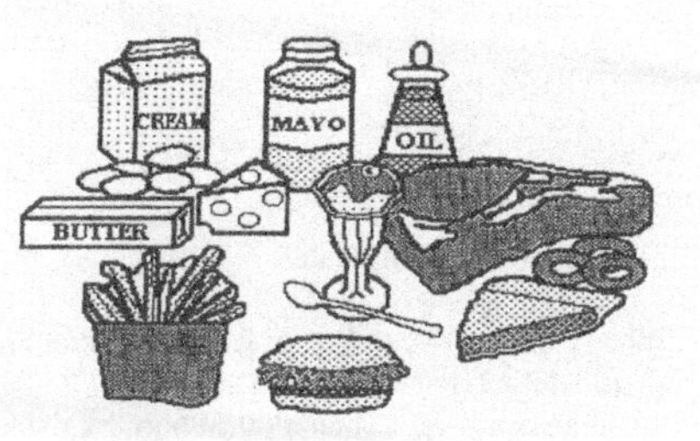

(4) Fat is also found in lesser amounts in:

(a) Dairy products. (d) Nuts.

(b) Meat. (e) Vegetables.

(c) Eggs. (f) Cereals.

c. Proteins: Protein is a reparative food of complicated molecules composed of chains of amino acids.

(1) The protein structure is made up of 25 different amino acids. Eleven of these are important for growth while eight are essential for normal tissue maintenance.

(2) Common protein food sources are:

PROTEIN SOURCES

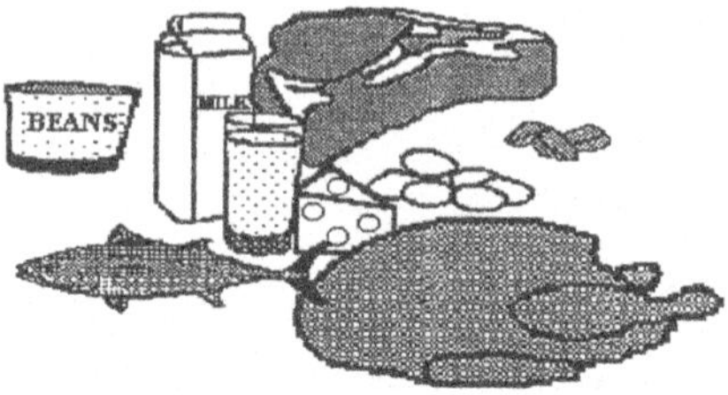

(a) Eggs.

(b) Dairy products.

(c) Meat.

(d) Poultry.

(e) Fish.

(f) Legumes (peas and beans, nuts and cereals with animal products, especially eggs, being better sources of complete protein).

(3) A pure protein diet may cause fatalities in 3-8 weeks from "rabbit starvation". This refers to those who have tried to live on relatively fat-free rabbit meat only. Fat must be a part of a normal diet.

7. **PERCENTAGE OF FATS, PROTEINS AND CARBOHYDRATES IN A SUMMER MOUNTAINOUS ENVIRONMENT**.

a. Requirements:

(1) 20% Protein.

(2) 35% Fats.

(3) 45% Carbohydrates.

b. When troops move to higher altitudes and/or are doing strenuous activity, carbohydrates should definitely be increased to at least 70 - 80% of energy intake. (Experienced mountaineers find carbohydrates more palatable at altitude).

8. **VITAMINS.** Vitamins are essential to the metabolic functioning of the body and are even more important in a cold weather environment because of an increase in metabolism and stress. Because the body cannot make vitamins, they are supplied in the food we consume. A daily multivitamin tablet supplement is recommended in a cold weather environment, but mega doses (more than 10 times the RDA) may be harmful.

a. Water-soluble vitamins are not stored in the body and need to be consumed daily. They are not stable to cooking and oxidation.

(1) B complex, Niacin, Pantothenic acid, Biotin, Choline, Folic Acid and C. (Note: Vitamin C facilitates absorption of iron from food.)

b. Fat-soluble vitamins can be stored in the body and are more stable to cooking and oxidation.

(1) A, D, E and K.

9. **FOODS THAT MAKE YOU FEEL WARM**.

a. Hot Beverages. Increases peripheral vasodilatation.

b. Adequate Dietary Iron. Mild iron deficiency increases chilling, thus decreasing cold tolerance.

c. Fat in the Diet. Fat slows gastric emptying and anecdotal evidence suggests less chilling and more restful sleep at night.

d. Carbohydrates. Keeps energy levels up. "Tired" is associated with "Cold".

e. Spices. Red pepper (Tabasco sauce) and monosodium glutamate increase peripheral vasodilatation, while capsicum and capsaicin, which are found in Cayenne, produces the same result when applied externally to the feet.

f. Water. Dehydration leads to lethargy, which tends to reduce food intake resulting in less heat production.

10. **INCREASING COLD TOLERANCE**.

a. Frequent Feedings. A high carbohydrate diet is superior to a high protein diet in increasing tolerance to cold.

(1) Diets deficient in both calories and protein reduce physical work capacity in the cold. The provision of both components is required to maintain work capacity.

11. **HIGH FAT DIETS AND SUSCEPTIBILITY TO FROSTBITE**.

a. High fat diets consumed at frequent (2 hour) intervals, maintain body temperatures better than high carbohydrate diets, but not if normal (4 hour) intervals are observed between meals.

b. Fat enriched diets lead to a propensity toward increased arterial thrombosis and increased platelet aggregation, which translates into increased potential for thrombosis.

c. Frostbite injuries are associated with disruption of the microcirculation, progressive micro vascular thrombosis, alteration of platelet function, and fibrin deposition in affected areas.

d. Therefore, diets high in (saturated) fat may predispose to frostbite.

CAUTION! A high-fat diet is important in a high-altitude, cold-weather environment. The benefits associated with the high-energy source called Fat far outweigh the risks of arteriosclerosis from the diet itself. A high-fat diet is appropriate in this specific environment during the time frame of cold, high-altitude exposure. In garrison, at sea level, a high fat diet should be avoided in order to maintain a healthy lifestyle.

12. **IRON DEFICIENCY AND THERMOREGULATION**.

a. Iron deficiency results in 9% less heat production.

b. Consuming only 1/3 of the RDA for iron results in a 29% greater heat loss during cold exposure. "The furnace is going, but the windows are open" is a phrase that helps describe this effect.

c. Iron absorption is increased by consuming iron rich foods:

(1) Lean meat.

(2) Poultry.

(3) Beets.

(4) Beans.

(5) Green leafy vegetables.

(6) Fish.

(7) Vitamin C.

d. Bottom line, IRON improves the body's response to cold.

13. **WATER REQUIREMENTS**.

a. On average, water makes up about 60% of the one's entire body weight. The average sedentary person excretes about 2.7 quarts of water each day. Of this total, 1.3 quarts are lost in the urine, 1.1 quarts through the skin and lungs and 10 ounces in the stool.

b. At high altitude, and during strenuous exercise, the amount of water lost through the skin and lungs due to the cold, dry air increases greatly.

c. Cold weather decreases the sense of thirst, which may lead to a state of chronic, mild dehydration. At temperatures below freezing and at elevations above the snow line, the lack of liquid water and the time and effort required to melt snow compound the problem.

d. Prevention of excessive water loss: (FMST.07.01c)

(1) Force fluids. Drink 6-8 quarts of water daily.

(2) Observe for dark urine and other symptoms of dehydration. (Headache, nausea/vomiting, dizziness, constipation, etc.)

(3) Limit caffeine to 1-2 cups per day.

(4) Do not rely on thirst as an indicator of hydration status.

NOTE: Eating snow or ice will cause extreme inflammation of the mucous membranes of the mouth after two days. Warm canteens of water placed in your sleeping bag will give you water to drink in the morning. They will initially act as hot water bottles. (Place them upright, as they tend to leak ...keep bag dry.)

e. Health Problems Associated with Excessive Water Loss:

(1) Dehydration.

(2) Constipation.

(3) Hypothermia.

(4) Frostbite.

(5) Heat Exhaustion.

f. The importance of hydration is essential while in a cold weather environment. If these guidelines are followed then complications arising from poor hydration should be prevented.

14. **SPECIAL PURPOSE MILITARY RATIONS**.

RATION	APPLICATION
Ration, Cold Weather, RCW.	Cold weather operations where weight and freezing are a problem.
MRE Supplement Packet.	Cold weather caloric supplements for the MRE.
Ration, Light Weight, RLW.	Long range patrols, reconnaissance, and assault.
Nutrition Sustainment Module, NSM.	Calorie dense food, bar ration for mission lasting 3-5 days where soldier must carry his own food.

Family of Operational Rations, FOR.	Modular rations that allow an optimum mix of meals for tactical operations under all climatic conditions.

15. **FIELD STRIPPING OF MRE'S.**

a. Tips on how to conserve space and weight:

(1) Remove the entire contents from the plastic wrapper.

(2) Remove the cardboard, and all materials that you will not use.

(3) Be sure to place the toilet paper and other materials from the accessory pack either back in your pocket or into the brown packaging.

(4) Carefully repack the stripped meal back into the wrapper and seal with tape.

(5) Your meal will be half its original size and two-thirds its original weight.

UNITED STATES MARINE CORPS
Mountain Warfare Training Center
Bridgeport, California 93517-5001

FMST.07.09
03/31/02

WILDERNESS PATIENT ASSESSMENT

TERMINAL LEARNING OBJECTIVE: Given a casualty in a cold weather environment and the necessary equipment and supplies, perform a patient assessment in a mountainous environment to determine the nature and extent of injury per the reference. (FMST.07.09)

ENABLING LEARNING OBJECTIVES:

(1) Without the aid of references, from a given list, choose when CPR may be legally discontinued, in accordance with the references. (FMST.07.09a)

(2) Without the aid of references and from a given list, choose which situations CPR may not be started in a remote wilderness environment, in accordance with the references. (FMST.07.09b)

(3) Without the aid of references, in the proper sequence, correctly perform a patient assessment in the proper sequence in accordance with the references. (FMST.07.09c)

OUTLINE

1. **INTRODUCTION.**

    a. An organized prioritized assessment of a wilderness trauma victim is essential to proper care. A modified Advanced Trauma Life Support (ATLS) assessment system will be introduced in this chapter.

2. **TRIAGE.**

    a. Two situations of triage may exist in a wilderness environment.

(1) The number of patients and the severity of their injuries do not exceed the ability of the care-giver and his/her equipment to render care. In this situation, treat life-threatening problems first.

3

(2) The number of patients and the severity of their injuries exceed the capacity of the care-giver and his/her equipment to render care. In this situation, those with the greatest chance of survival are treated first

3. **CPR: URBAN vs. WILDERNESS SPECIFICS.**

a. In a remote wilderness situation where advanced life support assistance may be hours or days away, CPR probably has no chance of success. In addition, administering CPR under wilderness conditions may put group members in serious danger, due to physical hazards and the effects of exhaustion.

b. Based on the above, several suggestions regarding CPR in remote wilderness areas can be made. Even though the following suggestions seem reasonable, they should be viewed as recommendations, without legal force, and not doctrine. At the present time, a rescuer who begins CPR is legally obligated to continue unless or until one of the following conditions is fulfilled:

(1) CPR may be discontinued legally when: (FMST.07.09a)

(a) The patient revives.

(b) ACLS is started by higher authority.

(c) Relieved by another rescuer.

(d) Too exhausted to continue.

(e) Pronounced dead by proper authority.

(2) In a remote wilderness environment, CPR may not be started under one of the following situations: (FMST.07.09b)

(a) The patient is in cardiac arrest caused by trauma.

(b) The patient is a drowning victim who has been immersed for over an hour.

(c) The patient is in cardiac arrest and advanced life support is more than an hour away, especially if the patient must be carried out.

(d) The patient's cardiac arrest was unwitnessed and the time of onset is unknown.

(e) The patient appears to be dead, based on rigor mortis, lethal injuries, or a body temperature below 60°F.

(f) Giving CPR would be hazardous to rescuers.

(3) After 30 minutes of CPR with no signs of life, further CPR is probably useless and may reasonably be discontinued. Administering CPR to a victim who is being evacuated by litter or sled is very difficult, if not impossible. Unless an ambulance or helicopter can be brought in rapidly, chances of survival are slim.

(4) Exceptions include patients in cardiac arrest caused by hypothermia; patients with another illness or injury complicated by hypothermia (avalanche burial and near drowning in cold water), and patients in cardiac or respiratory arrest caused by lightning injury. In these cases, the outlook is probably more favorable and CPR should be aggressively administered.

4. **PATIENT ASSESSMENT** (FMST.07.09c). There are different considerations to think about in a wilderness setting. The environment plays a major role in treatment, care and even the extraction of the patient. Not only do we have to be mindful of the enemy situation, by keeping tactical. We also have to take into consideration the elements. Snow, rain ice, and the terrain can and will pose a great danger to the patient and the rescuers. We must protect the patient from the elements. We can do this by placing an insulating barrier between the patient and the ground. Also exposing one body part at a time and recovering as you move on will help to fight hypothermia.

A) SCENE SIZE-UP.

1) Determine scene safety:

a. Rescuers: Your safety is priority. You are no good to the team as a casualty.

b. Patient: Protect the patient from further injuries, enemy, and the elements.

c. Bystanders: They can cause many accidents. Limit the personnel on scene.

d. Tactical: Have the Marines maintain security and find cover for the rescuer and the patient.

e. Fire: Forest fires can pose a formidable threat.

f. Electrical: Make sure to shut off electricity first. Rushing in injures many rescuers.

g. Special Rescues: Swift water and high angle rescues are very dangerous and only trained professionals should attempt the rescue.

2) Determine the mechanism of injury / nature of illness:

a. Obtain SAMPLE and Mechanism of Injury (MOI) history from patient, witnesses, or other caregivers.

S = Signs and/or Symptoms.
A = Allergies.
M = Medications.
P = Past Illnesses.
L = Last Meal.
E = Events of Injury.

MOI = What, when, where, and how injury occurred.

b. Environment: Survey the area to get clues. (i.e. falling rocks, next to a cliff, broken branches, etc…)

c. Bystanders: Ask them what they saw.

3) Determine number of patients:

a. Survey area for number of patients.

b. Look for the hidden patients. Patients could be buried in rock, sand or snow. They also could be spread out in bushes and trees. One of the most over looked places is in the bystanders. Be sure to have people search and identify all patients.

4) Request additional help / Organize help if needed:

a. Medical assistance: Due to the lack of medical personnel and equipment, you might need to request additional help.

b. Technical assistance: When the scope of care is over your ability or higher than the casevac corpsman can provide, you should request higher trained personnel on the casevac bird.

c. Organize help at the scene: Use the people around you to help out.

5) Consider stabilization of the spine:

a. If indicated from: Falls of 10 feet or greater, high speed collisions, unwitnessed loss of consciousness, etc…

b. Designate an assistant: Use the bystanders / marines. Tell them exactly what you want them to do and show them how to do it.

c. Improvise control: If you are alone you can use the patient's boots, sand bags, rocks, etc as forms of temporary control.

B) INITIAL ASSESSMENT.

1) Make general impressions of the patient: What can you obtain from the patient's condition as you walk up? Is the patient talking, or unconscious?

2) Determine responsiveness / level of consciousness.

a) Use AVPU.

b) Assess pupils for size, equality, and reactivity.

3) Determine chief complaint / Apparent life threats.

4) Assess Airway and breathing.

a) Assessment of airway.

- Perform chin lift or jaw thrust maneuver, with C-Spine precautions.
- Clear the airway of foreign bodies.
- Insert an oropharyngeal or nasopharyngeal airway.
- Use of other airways as needed, and as equipment and training allow.

b) Assessment of breathing.

- Expose the neck and chest, assure immobilization of the head and neck.
- Determine the rate and depth of respirations.
- Auscultate the chest bilaterally. (Ear to chest wall, if necessary)
- Check for symmetry. (Rise and fall of the chest)

c) Management.

- Ventilate. (Face-mask, bag-valve-mask, if available)
- Alleviate tension pneumothorax.
- Seal open pneumothorax.
- Administer high concentration oxygen, if available.

5) Assess Circulation.

a) Assess and control major bleeding.

- Identify and control major bleeding.
- Pulse: Quality, rate, and regularity.
- Pulse Pressure: Check radial, then carotid.
- Skin color: cyanosis, pallor and temperature.

b) Management.

- Insert two large bore IV's with saline locks. IV fluid replacement, if indicated.
- Place in shock position.
- Prevent hypothermia.

6) Identify priority patients and make transportation decisions.

a) Triage patients at this point.

b) Make transportation decisions. In this type of environment it is possible that a casevac vehicle might not gain access to your position. At this point you will have to organize how the patient will be transported out. Also, you might have to set up relay points and rewarming stations.

C) DETAILED PHYSICAL EXAM: This is the head to toe assessment. We will find and treat all secondary injuries at this time. When performing this exam in the wilderness, you will need to expose one body part at a time and cover it back up as you go. When assessing the patient, you are looking for deformities, contusions, abrasions, penetrations/punctures, bruising, tenderness, lacerations, and swelling (DCAP-BTLS).

1) Head.

a. Assessment.

- Inspect and palpate entire head and face for lacerations, contusions, and fractures.
- Re-evaluate pupils.
- Re-evaluate level of consciousness.
- Assess eyes for retinal hemorrhage, optic disc bulging, visual acuity disturbances, and contact lenses.
- Evaluate cranial nerve function.
- Inspect ears and nose for CSF leakage, hemorrhage
- Inspect mouth, tongue, and teeth.

b. Management.

- Maintain airway, continue ventilation, and oxygenate as indicated.
- Control hemorrhage.
- Prevent secondary brain injury.
- Remove Contact Lenses.

2) Assess the neck.

a. Assessment.

- Inspect for signs of blunt trauma, penetrating injuries, tracheal deviation, and use of accessory muscles during respiration, and jugular vein distention.
- Palpate for tenderness, deformity, swelling, or subcutaneous emphysema.

b. Management.

- Maintain open airway.
- Maintain immobilization and protection of cervical spine.
- Mark trachea.

3) Assess the chest.

3 a) Assessment.

- Inspect chest wall for signs of blunt trauma, penetrating injuries, and use of accessory muscles during respiration.

- Auscultate the chest wall and posterior bases for breath sounds.
- Auscultate for heart sounds.
- Palpate the entire chest wall for evidence of blunt trauma, penetrating injuries, flail chest, tenderness, crepitus, or subcutaneous emphysema.
- Percuss for hyperresonance or dullness.

c. Management.

- Needle thoracenthesis, if tension pneumothorax is suspected and proper training and equipment are present.
- Three-sided dressing for open chest wounds.
- Pericardiocentesis, if cardiac tamponade is suspected and proper training and equipment are present.

4) Abdomen.

a. Assessment.

- Inspect abdomen for signs of blunt trauma, penetrating injuries, and internal bleeding.
- Auscultate for the presence / absence of bowel sounds.
- Palpate the abdomen to elicit subtle tenderness, rebound, or guarding.

- Check pelvis for stability.

b b. Management.

c

- Correctly dress wounds.
- Keep patient NPO until definitive care is delivered.

5) Perineum / Rectum / Penis / Vagina.

d a) Assessment.

e

(1) Perineum.

- Contusions / hematomas.
- Lacerations.

(2) Rectum.

- Blood. Sphincter tone.
- High-riding prostate.

(3) Penis.

- Urethral bleeding.

(4) Vagina.

- Bleeding.
- Lacerations.

6) Musculoskeletal.

a) Assessment.

- Inspect extremities for evidence of blunt / penetrating injury. (deformities, altered range of motion).
- Palpate extremities for tenderness, crepitus, and abnormal movements.
- Palpate all pulses.
- Assess pelvis for stability. (Fractures) - Palpate spine for tenderness, crepitus, step-off's, depressions, and abnormal movements.

b) Management.

- Splint all suspected fractures in the position of function or comfort.

- Reduce dislocations as soon as possible, depending on training.
- Maintain immobilization and alignment of cervical, thoracic, and lumbar spine.

7) Reassess the vitals signs.

a) Assessment.

- Assess the vitals, pulse rate, breathing rate, pulse pressure.
- Re-evaluate the pupils and level of consciousness (GCS).
- Evaluate extremities for motor and sensory deficits.

b) Management.

- Maintain adequate immobilization of entire patient, as indicated.

5. **PATIENT PACKAGING AND EVACUATION.**

a. See the Mountain Casualty Evacuation chapter.

UNITED STATES MARINE CORPS
Mountain Warfare Training Center
Bridgeport, California 93517-5001

FMST.07.13
03/30/02

STUDENT HANDOUT

HIGH-ALTITUDE ILLNESSES

TERMINAL LEARNING OBJECTIVE: Given a simulated casualty, in a high altitude wilderness environment, treat high altitude health problems, in accordance with the references. (FMST.07.13)

ENABLING LEARNING OBJECTIVES:

(1) Without the aid of references, select from a given list the definition of acute mountain sickness, in accordance with the references. (FMST.07.13a)

(2) Without the aid of references, select from a given list the four main mechanisms of acute mountain sickness, in accordance with the references. (FMST.07.13b)

(3) Without the aid of references, select from a given list the most common and prominent symptom of acute mountain sickness, in accordance with the references. (FMST.07.13c)

(4) Without the aid of references, select from a given list the field management of acute mountain sickness, in accordance with the references. (FMST.07.13d)

(5) Without the aid of references, select from a given list the seven high altitude health preventive measures, in accordance with the references. (FMST.07.13e)

(6) Without the aid of references, select from a given list the correct definition of high altitude cerebral edema, in accordance with the references. (FMST.07.13f)

(7) Without the aid of references, select from a given list the pathophysiology of high altitude cerebral edema, in accordance with the references. (WMC.6.4g)

4

(8) Without the aid of references, select from a given list five of the symptoms of high altitude cerebral edema, in accordance with the references. (FMST.07.13h)

(9) Without the aid of references, select from a given list the field management of high altitude cerebral edema, in accordance with the references. (FMST.07.13i)

(10) Without the aid of references, select from a given list the correct definition of high altitude pulmonary edema, in accordance with the references. (FMST.07.13j)

(11) Without the aid of references, select from a given list three of the nine symptoms of high altitude pulmonary edema, in accordance with the references. (FMST.07.13k)

(12) Without the aid of references, select from a given list the field management of high altitude pulmonary edema, in accordance with the references. (FMST.07.13l)

(13) Without the aid of references, select from a given list choose the most important treatment for both high altitude cerebral edema and high altitude pulmonary edema, in accordance with the references. (FMST.07.13m)

OUTLINE

1. **ACCLIMATIZATION**.

a. Successful acclimatization depends on three factors:

(1) Degree of hypoxic stress (i.e. altitude)

(2) Rate of onset of hypoxic stress (i.e. ascent rate)

(3) Individual Physiology (i.e. genetic differences between individuals)

b. Physiologic Changes:

(1) Respiratory - Hypoxic Ventilatory Response (HVR)

- Hypobaric hypoxia
- Triggers carotid body
- Central respiratory center stimulated
- Ventilatory rate increases

HVR can be measured:

- "High" leads to successful acclimatization and superior performance.
- "Low" tends to be found in victims of altitude illnesses.
- HVR can be influenced by depressants (ETOH) and stimulants. (caffeine), and progesterone.
- **May** be negatively influenced by physical conditioning.

(2) Respiratory – Fast response: minutes to hours

- HVR leads to increased Ventilatory rate.
- Increased ventilations "blows off C02."
- Low C02 levels yield a respiratory alkalosis and rising pH.
- Elevation of pH feeds back to depress the central respiratory center.
- No further immediate rise in ventilatory rate (a brake is placed on any further increase in respiration.)

(3) Renal – Slower response: kidneys respond over the next 24 - 48 hours:

- Kidneys respond to alkalosis by inducing a bicarbonate diuresis.
- PH slowly declines - brake on central respiratory center eases.
- Respiratory rate gradually rises over 4 - 7 days to new baseline.
- Final respiratory rate is altitude dependent.

(4) Cardiovascular - changes are a result of increased sympathetic tone (due to hypoxia).

- Increased: Blood Pressure, venous tone and Cardiac output (increased heart rate and contractility.)
- Stroke Volume declines in first 1 - 3 days due to bicarbonate diuresis.
- Left ventricular pressures are normal, usually; therefore, wedge pressures are usually normal.

(5) Pulmonary Circulation - critical in understanding the pathophysiology of HAPE.

- Ascent results in global Hypoxic Pulmonary Vasoconstriction.
- Elevated Pulmonary Artery Pressure (PAP) occurs.
- Exertion at altitude worsens this Pulmonary Hypertension.
- Damage to pulmonary vascular endothelium leading to leaky capillaries.

(6) Cerebral Circulation - critical in understanding pathophysiology of HACE

- Hypoxia -----------Vasodilatation
- Hypocapnia ----------Vasoconstriction
- Hypoxia is the dominant influence and vasodilatation results.
- On average, Cerebral Blood Flow (CBF) increases 20-40% on ascent to 12.000ft.

(7) Hemopoietic Response

- Erythropoietin levels increase within hours of ascent.

- New immature Red Blood Cells (RBC) seen within days.
- New mature RBC's present within 4-10 days
- RBC mass increase seen over weeks to months (altitude dependent).
- Early changes due to hemoconcentration.
- Polycythemia is a potential danger – this can lead to a condition known as CMS, or Chronic Mountain sickness.

(8) Muscle Tissue

- Intramuscular capillary and mitochondrial density increases after three weeks exposure to altitude.

(9) Sleep Pattern

- Stage 1 increase, Stage 2 unaffected
- Stage 3 & 4 decreases
- Rapid Eye Movement (REM) decreases
- Arousals increase
- Periodic Breathing (Cheyne-Stokes).
  Increased respiratory rate leads to respiratory alkalosis
  Elevated pH depresses central respiratory center
  Apnea
  02 saturation declines, C02 levels rise
  Carotid body stimulated
  Ventilatory rate rises

(10)Summary: Physiologic changes seen at altitude result from the body's responses to hypo baric hypoxia. It has been speculated that man would be limited to an 8000-foot elevation limit without supplemental oxygen if not for the acclimatization process. As Messner and Haebler demonstrated on Mt. Everest, the process of acclimatization enables man to explore the entire planet.

2. **ACUTE MOUNTAIN SICKNESS (AMS).**

a. Definition - AMS is an acute, self-limiting illness, which results when an unacclimatized individual ascends rapidly to high altitude. (FMST.07.13a)

(1) It is rare below 8,000 feet but will occur in approximately 42% of persons rapidly exposed to altitudes greater than 10,000 feet in elevation.

(2) Aerobic fitness is no predictor for risk of AMS with exposure to altitude. There is some evidence that prior aerobic fitness leads to increased AMS incidence, most likely due to excessively rapid ascent or over-exertion at altitude. Elite athletes may also be at increased risk due to depressed cardiovascular and Ventilatory responses to exertion.

(3) Prior ascents to altitude without symptoms of AMS are no guarantee against having symptoms of AMS with future ascents. However, those with past AMS are at an increased risk of AMS with repeat exposure to high altitude. Generally, there is no correlation between the severity of the illness and increasing altitudes. The incidence is slightly increased in younger persons.

(4) There is still much we do not know about altitude-induced physiologic changes. Current explanations in the literature center around genetic differences in hypoxic ventilatory response and vascular endothelium release of inflammatory mediators and Nitric Oxide.

b. Pathophysiology of AMS: The physiologic changes associated with the symptom complex known as AMS are linked to: (FMST.07.13b).

(1) Poor ventilatory response: An individual's Hypoxic Ventilatory Response (HVR) dictates how well he/she will respond to the hypobaric hypoxia of altitude. Those with a brisk HVR will do well, whereas those with a blunted response tend to suffer the symptoms of AMS. A poor ventilatory response leads to hypoxia and hypercapnia (elevated C02 levels). These two combine to augment cerebral blood flow by inducing cerebral vasodilatation.

(2) Fluid retention: The increased sympathetic tone associated with hypobaric hypoxia explains many of the changes seen in the renal system at altitude. Low Renal Blood Flow, increased anti-diuretic hormones, low Glomerular Filtration Rates, and decreased Urine Output all contribute to a Net Fluid Gain. This retained fluid redistributes within the body's fluid spaces. As a result, intravacular and extravascular fluid shifts occur, especially in the brain, lung, and peripheral interstitial tissues.

(3) Increased Intra-cranial Pressure: Increased cerebral blood flow coupled with CNS inflammation resulting in increased vascular permeability leads to increased intra-cranial pressure and associated symptoms.

(4) Altered Pulmonary Mechanics. On ascent to altitude, there is an increase in extravascular lung water. This is thought to be mediated, in part, by a phenomenon known as hypoxic pulmonary vasoconstriction. Blood vessels in the lungs constrict when they are exposed to low oxygen levels, this global vasoconstriction leads to increased right-sided heart pressures. The result of increased right-sided heart pressures causes fluid to leak from the vascular space into the alveoli, which results in impaired gas exchange. This has been demonstrated via Pulmonary Function Testing bronchoalveolar lavage, right sided-heart catheterization and arterial blood gas analysis showing decreased Vital Capacity, decreased Peak Expiratory Flows, pulmonary exudates rich in proteins and Increased Alveolar-Arterial oxygen Gradients. One study showed a 54% increase in estimated lung mass. It is not uncommon to hear crackles in all lung fields when auscultating asymptomatic individuals. The presence of intra-alveolar fluid inhibits the diffusion of oxygen

across the alveolar membrane, thereby reducing oxygenation across the pulmonary capillary membrane.

Signs and Symptoms of AMS. Symptoms tend to occur within the first 1- 3 days of exposure to altitude. Symptoms should resolve spontaneously within 15 - 24 hours, if ascent is arrested. The three most prominent symptoms are Headache, Nausea and Vomiting. (FMST.07.13c) Although nonspecific, headache must be present to make a diagnosis of AMS. 95% of AMS victims will complain of one or more of these three. Other common complaints include dizziness, anorexia, drowsiness, malaise, weakness, and insomnia. It is possible that findings of dyspnea on exertion and mental status changes along with ataxic gait may accompany the classic AMS signs and symptoms. These are clear indications for descent, as these are the hallmark signs of progression to HACE and the development of HAPE. Again, headache is the most common and prominent symptom of AMS. The Lake Louise Consensus criteria was created to give clinicians formal guidance in diagnosing AMS. The criteria require the presence of headache plus one or more symptoms from the above list. (i.e. nausea, vomiting, anorexia, insomnia, dizziness, fatigue, and dyspnea on exertion.)

c. Field Management (FMST.07.13d)

(1) Stop any further ascent. Light duty. No tobacco.

(2) Symptomatic Treatment. Mild analgesics such as Aspirin or Tylenol for the headache, Compazine or Phenergan for nausea/vomiting.

(3) Other Meds: Acetazolamide (Diamox), Dexamethasone (Decadron), and oxygen. The effects of Diamox and Decadron are additive when used together and are a very effective means for treating AMS.

(4) Hyperventilate: It is unrealistic to try to consciously control one's respiratory rate over any appreciable period of time. It is nearly impossible to maintain conscious hyperventilation while functioning in the field.

(5) Descend. If none of these methods help, then individuals should descend 1,000-3,000 feet. This will usually result in marked relief of symptoms. Most cases, however, improve in 1-2 days with symptomatic treatment.

(6) ALL PATIENTS WITH AMS MUST BE EVALUATED FOR HAPE AND HACE.

HOW?

- Field sobriety test.
- Mini-mental status exam.
- Auscultation of lung fields.

d. Preventive Measures: (FMST.07.13e)

(1) The best approach to high-altitude travel is staged, gradual ascent, when moving at altitudes greater than 3000 meters. Gradual ascent means no faster than 3000 ft/day up to 14,000 and no greater than 1,000 ft/day over 14,000 ft.

(2) Avoid any alcohol and/or sedatives during the first two nights.

(3) Maintain adequate fluid intake.

(4) Work high - Sleep low.

(5) High Carbohydrate Diet (approximately 70% of diet).

(6) Avoid overexertion, but remember that mild exercise assists acclimatization.

(7) No smoking.

(8) Pharmacological options: Prevention of AMS is possible with Acetazolamide (Diamox). It dramatically lowers the incidence of AMS when taken prophylactically. The usual dose is 125 mg PO BID starting 24-48 hours before ascent and should be continued for 3-4 days after the start of the mission. It is also effective when used as a treatment especially when combined with Decadron. In individuals with contraindications or allergies to Diamox use Decadron (dosed PO/IM/IV 4 mg Q6).

3. **HIGH ALTITUDE CEREBRAL EDEMA (HACE).**

a. Definition. HACE is a high-altitude illness characterized by swelling of the brain. (FMST.07.13f)

(1) HACE can occur as low as 8,000 feet, but typically occurs at more than 12,000 feet.

(2) HACE, considered by most, to be a progression of AMS, occurs as a failure to heed the warnings of AMS signs and symptoms and continue ascent.

b. Pathophysiology. Increased cerebral blood flow along with the fluid shifts mentioned above lead to vasogenic edema (leaky blood vessels) inside the confined space of the skull. It is also thought that inflammation plays a key role in the development of HACE and occurs as follows. Hypoxia triggers the release of inflammatory mediators (likely Transforming Growth Factor Beta, TGF-β and Vascular Endothelial Growth Factor, VEGF.) These mediators cause increased vascular permeability that result in fluid accumulation within the brain – edema and increased intracranial pressure. Once this pressure nears arterial pressure, cerebral blood flow will be impaired. Eventually, intra-cranial pressure will rise further with a concomitant increase in the risk of brain herniation leading to marked alterations in mental status, ataxia and even cranial nerve palsies. Increased CSF production further compounds the problem. (FMST.07.13g)

c. Differential Diagnosis: Other illnesses are possible in individuals ascending to altitude. Your differential should include the following when neurologic deficits are involved:

(1) Cerebrovascular spasm and Transient Ischemic Attacks.

(2) Cerebral vascular thromboembolic events.

(3) Intra-cranial hemorrhage (aneurysm, AVM).

(3) Hypothermia.

(4) Meningitis

(5) Encephalitis

(6) "Hangover"

d. Signs and Symptoms:

(1) Early signs and symptoms are those found in AMS. (i.e. H/A, NV)

(2) When the above warning signs go unheeded the true signs and symptoms of HACE can quickly follow: (FMST.07.13h)

(a) Ataxia (loss of muscle coordination leading to difficulty maintaining balance), especially prominent in the victim's gait.

(b) Mental Status Changes, poor judgement, personality change which may progress to stupor, coma and death relatively rapidly without proper treatment.

(c) Lassitude, confusion, hallucinations, convulsions, behavioral aberrations.

(d) Paralysis of one or more extremities (see Differential Diagnosis).

(3) The most important impediment to early recognition of HACE is its insidious onset. Early signs and symptoms frequently go unrecognized or are ignored by patients, as well as their companions, who also may be suffering to some degree from the effects of altitude.

Individuals prone to HAPE may have a deficiency in the amount of Nitric Oxide produced in the vascular endothelium resulting in more profound hypoxic pulmonary vasoconstriction.

c. Pathologic Findings of HAPE: Postmortem studies of victims of HAPE have yielded some interesting results:

(1) Grossly, the lungs are congested and swollen, as seen with other causes of pulmonary edema. Average lung weight on autopsy is 2-4 times normal. Histologically, one sees pulmonary edema with a protein rich exudate filling the alveoli. It is also common to find the alveolar spaces filled with hyaline membranes.

(2) The left heart is completely normal in appearance. However, the right ventricle, right atrium, and pulmonary arteries are distended and dilated.

(3) Greater than 50% of HAPE victims are found to have evidence of HACE.

d. Signs and Symptoms. These tend to occur within 2-4 days of arrival at altitude. Usually the symptoms of AMS are present before or occur with the symptoms of HAPE. (FMST.07.13k)

(1) Early signs:

(a) Dry cough, frequently occurring at night.

(b) Dyspnea on Exertion (DOE), especially with ambulation uphill.

(c) Mild chest pain - usually perceived as an ache beneath the sternum.

(d) Decreased work performance and increased recovery time between events.

(e) Peripheral or Central Cyanosis.

(2) Later signs:

(a) Dyspnea at Rest.

(b) Productive cough which yields large amounts of pink, frothy sputum.

(c) Rapid pulse and respiratory rates.

(d) Audible crackles on auscultation.

(e) **Mental status changes**, **Ataxia**, Loss of Consciousness.

(f) CXR findings:

- Prominent pulmonary arteries.
- Normal heart size.
- Patchy infiltrates.

ECHO findings:

-Non-cardiogenic pulmonary edema -increased PAP
-Tricuspid regurgitation.
-Normal left ventricle
-Prominent pulmonary arteries -enlarged right atrium
-Enlarged right ventricle.

NOTE: Crackles heard on lung exam, as an isolated finding is NOT an indication for descent or grounds for the diagnosis of HAPE. It is common (up to 15% of climbers on Mt. Rainier) to find asymptomatic rales in climbers at altitudes of 10,000-12,000ft. Concern is heightened when crackles occur in constellation with other signs and symptoms.

e. Field Treatment (FMST.07.131)

(1) The most important emergency care measure is immediate descent to a lower altitude, since fatalities can occur within 6-12 hours in severe cases.

(a) A descent of at least 2,000-3,000 feet can be a definitive treatment if the condition is caught early enough.

(2) The patient should be placed in the most comfortable position (usually sitting) and given **high-flow O2 if available**. It is important to minimize activity, as exertion increases pulmonary artery pressures. Remember oxygen is Lasix for the lungs.

(3) Medication regimens include:

(a) Nifedipine (10mg PO initially then 30mg SR Q12) – calcium channel blocker - pulmonary artery/arteriole dilator. SR=sustained relief.

(b) Diamox and Decadron may be appropriate regimens, depending on your medication availability.

(c) Morphine (5mg IV initially decreased PAP) as a last resort.

(4) Oxygen is always appropriate, however, it is not always available.

(5) Gamow Bag can be a lifesaver if decent is not immediately possible.

f. Prevention Measures. Preventive measures are the same as those discussed for AMS.

(1) Acetazolamide (Diamox) has been proven to be successful in the prevention of HAPE. 125 - 250 mg PO bid helps prevent HAPE in individuals with recurrent episodes.

(2) Nifedipine (30 mg SRq12h) - also effective in individuals with recurrent HAPE.

(3) It should be noted that a history of HAPE is NOT a contraindication to the individual participating in high-altitude activities. Each case should be evaluated individually.

NOTE: Remember decent is the most important treatment for both HACE and HAPE. (FMST.07.13m)

5. **SYSTEMIC EDEMA AT HIGH ALTITUDES.**

Occasionally, certain individuals, especially women, will develop edema of the face, hands, and feet at altitude. This edema usually tends to occur in the absence of other symptoms. In susceptible persons, repeat episodes are common. The edema goes away upon return to lower altitude. There is no need for descent in these cases because the patient is usually able to tolerate the edema and continues to function well. However, they should be examined carefully to rule out the presence of HAPE or HACE. Diuretics are usually not warranted. It is also probably reasonable to restrict salt intake. The only indication for descent is if the edema is causing functional disability, i.e. inability to don one's boots, or facial edema severe enough to restrict vision.

6. **SICKLE CRISIS.**

Persons with sickle cell trait are at risk for developing sickling attacks when exposed to hypoxic stress, as might occur at high altitudes. However, sickling can occur at lower elevations. Individuals with Mediterranean family ties should be considered at risk for having sickle cell trait. The trait is more likely in black individuals as well. Any patient complaining of Left Upper Quadrant pain, SOB, and/or arthralgia at altitude should have sickling crisis included in the differential diagnosis regardless of race. It has been known to occur in white females as well. It has been recommended by some that individuals with sickle cell trait should be restricted to an altitude of 8,000 feet and below. However, more liberal recommendations exist, suggesting that the individual and their symptoms dictate ultimate altitude.

7. **ALTITUDE THROAT.**

At higher altitudes, mouth breathing is often increased during exertion. Inhalation of this cold, dry air can result in drying of the mucous membranes in the throat and upper airway. This condition responds to humidified air, salt-water gargles or lozenges. If exudate or edema is present, throat culture and/or antibiotics may be warranted.

8. **HIGH ALTITUDE FLATUS EXPULSION HAFE).**

Ascent to altitude can result in rapid expansion of intestinal gas, due to the decreased pressure of altitude. This produces abdominal discomfort and the passage of colonic gas. This condition is not associated with serious consequences; however, it may create tension amongst tent-team members. Treatment consists of oral administration of anti-flatulence such as simethicone, Mylanta II, or Gaviscon. Know and avoid your gas producing foods!

9. **HIGH ALTITUDE RETINAL HEMORRHAGE.**

Up to 60% of individuals ascending to altitudes greater than 18,000 feet will develop asymptomatic retinal hemorrhages. On fundoscopic exam, there will be hyperemia of the disc and dilation and tortuosity of the vessels. "Flame" hemorrhages and "cotton wool spots" can be seen on exam. The only indication for descent in an individual with retinal hemorrhages is loss of visual acuity. This usually results when a large hemorrhage involves the macular region of the retina. Remember that Carbon Monoxide poisoning can present with retinal hemorrhage as well.

10. **THROMBOEMBOLIC EVENTS.**

Thromboembolic disease occurs with increased frequency at higher altitudes. It is believed that the common circumstances of volume depletion due to dehydration and Polycythemia due to altitude exposure predispose individuals to coagulopathic events. In addition, extended periods of inactivity due to "going to ground" during storms contribute to this predisposition. There does not seem to be any direct influence of hypobaric hypoxia on the body's normal coagulation system.

11. **ACETAZOLAMIDE.**

Acetazolamide (Diamox) is a Carbonic Anhydrate Inhibitor, which is especially active in the brain, the lung, and the kidney.

a. Kidney:

(1) Uptake of bicarbonate ion is inhibited.

(2) Bicarbonate diuresis is induced.

(3) Metabolic acidosis ensues within one hour of medication.

(4) Sequence mimics and accelerates the natural acclimatization process.

(5) Enhanced HVR is the end result.

(6) Diuresis addresses problems associated with fluid retention.

b. Brain:

(1) Direct effects at the level of the Central Respiratory Center.

(2) Diminishes periodic breathing during sleep.

(3) Fewer episodes of apnea and extreme hypoxemia.

(4) CSF production is inhibited and CSF absorption is enhanced.

c. Side Effects:

(1) Polyuria.

(2) Flattens taste of carbonated beverages.

(3) Peripheral paraesthesias, myopia, impotence, nausea, and drowsiness.

(4) Contraindicated for Sulfa Allergic and G6PD-deficient individuals (reversible when medication is discontinued.)

UNITED STATES MARINE CORPS
Mountain Warfare Training Center
Bridgeport, California 93517-5001

FMST.07.35
03/13/02

STUDENT HANDOUT

HEAT ILLNESSES

TERMINAL LEARNING OBJECTIVE: Given a casualty in a wilderness environment, and necessary equipment and supplies, manage common heat related injuries to prevent death or further injury per the reference (FMST.07.35)

ENABLING LEARNING OBJECTIVES:

(1) Without the aid of references, given a list of heat illnesses match the definition to the correct condition, in accordance with the references. (FMST.07.35a)

(2) Without the aid of references, select from a given list the correct field treatment for heat cramps, in accordance with the references. (FMST.07.35b)

(3) Without the aid of references, select from a given list the correct field treatment for heat exhaustion, in accordance with the references. (FMST.07.35c)

(4) Without the aid of references, select from a given list the correct field treatment for heat stroke, in accordance with the references. (FMST.07.35d)

1. **CAUSES OF HEAT ILLNESSES**. Heat illness is caused by one of or a combination of the following factors:

a. Excessive environmental heat and humidity.

b. Faulty heat dissipation mechanisms within the body.

c. Stressed heat dissipation mechanisms.

d. Prolonged strenuous physical activity with inadequate fluid and electrolyte replacement.

2. **THERMOREGULATION**. The body is constantly adjusting itself in order to maintain its optimal temperature. This is called thermoregulation. A basic physics principle applies, in that heat is transferred from a higher to a lower temperature. What this means is that when the environmental temperature is higher than the skin temperature, the body gains heat, but when the environmental temperature is lower, the body loses heat. As a result, elevated environmental temperature adds to the heat burden of the body and can interfere with heat dissipation.

a. Heat Transfer Mechanisms. Keeping in mind the above principle we will now discuss the four methods of heat transfer.

1) Conduction: This is heat exchange between two surfaces in direct contact. An example would be lying on a hot or cold surface.

2) Convection: This is heat exchange from a surface to a gas or liquid, usually air or water. The amount of heat transferred depends on many variables, such as density, flow rate and surface area exposed.

3) Radiation: This is heat transfer between the body and its surroundings. All matter either absorbs heat or emits thermal radiation. The sun emits heat and the body absorbs it. A interesting fact is that highly pigmented skin while protective against UV radiation absorbs about 20% more heat than lightly pigmented skin.

4) Evaporation: This is the heat flow from the body surface to the environment by sweating.

b. If the body is unable to maintain thermal equilibrium by radiation, convection, and conduction and core temperature rises, sweating must occur to permit heat loss by vaporization of water. Since gastric emptying of water is about 1L/hr and sweat rates can exceed 1L/hr, a healthy diet along with constant replenishment of water lost is imperative to prevent illness.

3. **PREDISPOSING FACTORS**.

a. An underlying illness such as infection.

b. The temperature and humidity of previous days effect the likelihood of becoming a heat casualty.

c. Most heat illnesses occur during the morning hours.

d. Thirst is not an accurate indication of hydration; therefore you must drink even when you don't feel thirsty.

e. People usually do not voluntarily drink as much water as they lose.

f. All persons in the heat should be considered dehydrated, unless they have been forced to drink more than they desired.

4. **PREVENTIVE MEASURES**.

a. Drink 6-8 qts every day to replace fluids lost by sweating or from other bodily functions.

b. Consumption of a solute rich meal or a fluid replacement beverage about 2 hours before exercise can potentially counteract the sweat/salt losses.

5. **EARLY SIGNS OF HEAT RELATED ILLENSSES**. Changes in cognitive functions appear to occur frequently prior to the physical symptoms of heat illnesses. It is therefore imperative that medical personnel and leaders watch for confusion and impaired ability to function or work in the members of their group.

6. **SPECIFIC HEAT RELATED ILLNESSES**.

a. Heat Induced Syncope. (FMST.07.35a) This event is caused by the gravitational displacement of blood into the legs of an upright and stationary person combined with hemodynamic displacement of blood into peripheral circulation to support heat transfer at the body surface.

1) Prodromal symptoms.

(A) Restlessness

(D) Nausea

(B) Sighing

(E) Yawning

(C) Dysphoria

2) Treatment.

(A) The fainting is usually brief and responds to horizontal positioning.

(B) Rest in a cool, shaded area.

(C) Oral fluids.

(D) Persons should be cautioned against standing in hot environments and advised to flex leg muscles repeatedly while standing still to enhance venous return and warned to sit or lie down if warning signs reappear.

b. Heat Cramps. These are painful spasms of muscles, caused by sweating resulting in excessive loss of body salt. Generally, the spasms involve voluntary skeletal muscle subjected to prior exertion. This can occur even if a person is replacing water loss, but not salt losses.

1) Precipitating Factors: Cramps are preceded by several hours of sustained effort, combining heavy sweating in hot surroundings, with the ingestion of large volumes of water.

2) Symptoms: Painful muscle cramps in the arms, legs, and abdomen.

3) Treatment: (FMST.07.35b)

(A) Rest in a cool, shaded area.

(B) Gentle massage of affected muscles (may relieve spasms) or place the affected muscle in a stretch.

(C) Generally, heat cramps respond quickly to salt solutions.

1.) 1/4 to 1/2 teaspoon of salt in a quart of water (sip).

2.) In severe cases IV isotonic saline 0.9% NaCL.

NOTE: Salt tablets are not recommended due to the gastric irritation they cause.

c. Heat Exhaustion. This occurs when body salt losses and dehydration from sweating are so severe that a person can no longer maintain adequate blood pressure. It tends to develop over several days, that present ample opportunity exists for electrolyte and water imbalances to occur. Some 20% of cases are found to also have some form of gastroenteritis. Additionally, 60% of heat exhaustion casualties will also have symptoms of heat cramps.

1) Problems in recognition.

(A) Early signs may go unrecognized or be mistaken for malingering.

(B) Headache and visual disturbances may mimic prodromal of migraine headache.

(C) Dizziness and syncope may lead to an evaluation for arrythmia.

(D) Chills, ^ temperature, and sweating may lead to an evaluation for infection.

(E) Altered mental status or ataxia may lead to an evaluation for Meningitis or high altitude medical problem.

2. Symptoms:

(A) Headache

(B) Fatigue

(C) Piloerection

(D) Nausea/Vomiting

(E) Profuse sweating

(F) Dizziness

(G) Hyperirritability, Anxiety

(H) Chills

(I) Heat cramps

3. Clinical Findings:

(A) Tachycardia

(B) Hypotension

(D) Hyperventilation

(E) Syncope

(C) Core temperature may be normal or elevated. Spontaneous cooling may occur, usually with profuse sweating. This will not happen with Heat stroke.

(4) Treatment is based on symptoms and focuses on reduction of body temperature and rehydration. (FMST.07.35c)

(A) Lay down in a cool, shaded area.

(B) If body temperature is elevated, lower by any means available.

(C) Sprinkling with water and fanning.

(D) Ice bags over superficial great vessels and fanning.

(E) Periodic toweling of the skin renews the evaporative process.

(F) If patient is alert, orally rehydrate with cool water or commercially available rehydration mixture. Goal is 2L over 1st 2 hours.

(G) If patient is not alert, rehydrate via IV with LR.

(H) Victim must rest and continue to be rehydrated over next 24 to 36 hours.

d. Heatstroke (Hyperthermia). " A true medical emergency". This is a failure of the body's cooling mechanisms that rid the body of excessive heat build-up. Prior to 1950 the mortality rate was 40 to 75%. Long term survival of victims is directly related to the speed of effective treatment measures. Traditionally, the diagnosis of Heatstroke required the following 3 signs: a core temperature > 104°F(41°C), altered mental status, and cessation of sweating. Although, this symptom complex does represent full-blown Heatstroke adherence to this strict criteria may delay critical interventions. Unless an alternative etiology is obvious, the previously healthy person who collapses after physical exertion in hot weather should be considered to have exertional Heatstroke. Treat for Heatstroke whenever you suspect it and always suspect it in the heat.

1. Symptoms:

(A) Headache
(B) Shortness of breath
(C) Nausea/ vomiting
(D) Low BP
(E) Skin may be wet or dry
(F) Constricted pupils
(G) Seizures
(H) Dry mouth
(I) Core temp > 104°F (41°C)
(J) Weakness/ dizziness
(K) Hot, red skin
(L) Tachycardia
(M) Confusion, Aggression, Combative
(N) Coma

2. Clinical Findings.

(A) CNS: altered mental status, agitation, ataxia, delirium, hallucinations, convulsions, seizures, coma, pupillary constriction.

(B) Cardiovascular: hypotension (a late and ominous finding), shock, sinus tachycardia with ST segment & T wave abnormalities.

(C) Pulmonary: hyperventilation, pulmonary edema.

(D) Renal: acute renal failure, hematuria, pyuria, proteinuria, UA casts.

(E) Gastrointestinal: nausea, vomiting, +/- disseminated intravascular coagulation, hematemesis, melena.

(F) Hematologic: WBC 20-30 thousand, < platelets, < clotting factors.

(G) Hepatic: increased liver function tests.

3. Emergency Treatment. (FMST.07.35d)

(A) Rapid reduction of body core temperature by any means available until core temperature is 102°F.

1) Ice water bath (controversial).
2) Cool water bath.
3) Pouring or hosing cool water over victim and fanning.
4) Placing ice bags at neck, armpits, groin, forehead, chest

etc..

(B) Maintain ABC's.

(C) Monitor core temperature.

(D) IV access infuses RL or normal saline 500cc/hr x 4hrs. (watch for pulmonary edema).

(E) Control seizures with valium.

(F) Transport STAT. Cool enroute.

UNITED STATES MARINE CORPS
Mountain Warfare Training Center
Bridgeport, California 93517-5001

FMST.07.15
04/01/02

STUDENT HANDOUT

COMBAT CASUALTY CARE

TERMINAL LEARNING OBJECTIVE: Given a simulated casualty in a simulated combat environment, treat the casualty in accordance with the references. (FMST.07.15)

ENABLING LEARNING OBJECTIVES:

(1) Without the aid of references, choose from a given list the three correct phases of care of tactical casualty management, in accordance with the references. (FMST.07.15a)

(2) Without the aid of references, given a simulated casualty and the necessary equipment, conduct combat casualty care, in accordance with the references. (FMST.07.15b)

OUTLINE

1. Summary of Findings and Recommendations.

a. Pre-hospital care of combat casualties is divided into three phases: (FMST.07.15a)
   1. Care under Fire.
   2. Tactical Field Care
   3. Combat Casualty Evacuation (CASEVAC) Care.

b. Suppression of hostile fire may take temporary precedence over the rendering of care in the Care under Fire phase.

c. Cervical spine immobilization is not required for victims of penetrating head or neck trauma.

d. CPR should not be attempted on the battlefield for victims of blast or penetrating trauma who have no pulse, respirations, or other signs of life.

e. The use of tourniquets is encouraged in the Care under Fire phase. The tourniquet is removed and bleeding controlled with direct pressure in subsequent phases when feasible.

f. The nasopharyngeal airway is the airway of first choice for unconscious patients until the CASEVAC phase.

g. For patients with airway obstruction in the Tactical Field Care phase unrelieved by a nasopharyneal airway, the next airway maneuver is cricothyroidotomy.

h. Endotracheal intubation is not the standard of care until the CASEVAC phase.

i. Evaluation of the laryngeal mask airway and the Combitube for use by SOF corpsmen and medics is recommended.

j. Progressive, severe respiratory distress in the setting of unilateral blunt or penetrating chest trauma on the battlefield should result in a presumptive diagnosis of tension pneumothorax and decompression of that hemithorax with needle thoracostomy.

k. Chest tube insertion is not indicated until the CASEVAC phase of care.

l. Oxygen is not expected to be carried onto the battlefield, but CASEVAC assets should have it.

m. No IV fluids are to be given until the CASEVAC phase unless the patient is in hypovolemic shock from a bleeding site which has now been controlled. Fluid resuscitation is not necessary for patients who are not in shock and not appropriate for patients in hypovolemic shock from uncontrolled hemorrhage (penetrating wounds of the chest and abdomen).

n. Saline locks may be used instead of IV's if fluid resuscitation is not required.

o. 18 gauge catheters may be used instead of 14 or 16 gauge.

p. Hespan (1000cc) is the fluid of choice for initial resuscitation of patients with controlled hemorrhagic shock.

q. Morphine is to be used IV (5mg) instead of IM.

r. IV cefoxitin is to be used as soon as possible for patients with penetrating abdominal trauma, grossly contaminated wounds, massive soft tissue trauma, open fractures, or any patient in whom a long delay until definitive treatment is expected.

s. Casualties should not be completely undressed for a secondary survey in the field. Removal of clothing should be limited to that necessary to expose known or suspected wounds.

t. More specific combat casualty care planning should be carried out and based on specific mission-oriented scenarios.

u. The establishment of Combat Casualty Transport Teams and their use on CASEVAC assets is recommended.

v. Electronic monitoring should be routinely used on CASEVAC assets unless conditions prohibit.

w. MAST trousers are not recommended as a standard of care in any phase.

2. Phases of Care:

a. Phase One: Care Under Fire

1. Return fire as directed or required.

2. Keep yourself from getting shot.

3. Try to keep the casualty from sustaining additional wounds.

4. Stop any massive external hemorrhage with a tourniquet.

5. Take the casualty with you when you leave.

b. Phase Two: Tactical Field Care

1. Airway management

- Unconscious casualty without airway obstruction: > Nasopharyngeal airway.

- Unconscious casualty with airway obstruction: > Cricothyroidotomy.

- Cervical spine immobilization is not necessary for casualties with penetrating head or neck trauma.

2. Breathing

- Consider tension pneumothorax and decompress with needle thoracostomy if a casualty has unilateral penetrating chest trauma and progressive respiratory disstress.

3. Bleeding

- Control any remaining bleeding with a tourniquet or direct pressure.

4. IV

- Start an 18 gauge IV or saline lock.

5. Fluid Resuscitation

- Controlled hemorrhage without shock: > No fluids necessary.

- Controlled hemorrhage with shock: > Hespan 1000cc

- Uncontrolled (infra-abdominal or thoracic) hemorrhage: > No IV fluid resuscitation.

6. Inspect and dress wound.

7. Check for additional wounds.

8. Analgesia as necessary: > Morphine (5mg) IV, wait 5minutes, repeat as necessary.

9. Splint fractures and recheck pulses.

10. Antibiotics

Cefoxitin 2gm slow IV push (over 3-5 minutes) for penetrating abdominal trauma, grossly contaminated wounds, massive soft tissue trauma, open fractures, or any patient in whom a long delay until CASEVAC or definitive treatment is expected.

11. Cardiopulmonary Resuscitation

CPR on the battlefield for victims of blast or penetrating trauma who have no pulse, no respirations, and no other signs of life will not be successful if attempted.

c. Phase Three: Combat Casualty Evacuation (CASEVAC) Care

1. Airway management

Unconscious casualty without airway obstruction: > Nasopharyngeal airway, endotracheal intubation, Combitube or laryngeal mask airway.

Unconscious casualty with airway obstruction: > Cricothyroidotomy if endotracheal intubation and/or other airway devices are unsuccessful.

2. Breathing

Consider tension pneumothorax or hemothorax and decompress with needle thoracostomy if a casualty has unilateral penetrating chest trauma and progressive respiratory distress.

Consider chest tube insertion if a suspected tension pneumothorax is not relieved by needle thoracostomy

Oxygen.

3. Bleeding

Consider removing tourniquets and using direct pressure to control bleeding if possible.

4. IV

Start an 18 gauge IV or saline lock if not already done.

5. Fluid resuscitation

No hemorrhage or controlled hemorrhage without shock: > Lactated Ringer's at 250cc/hr.

Controlled hemorrhage with shock: > Hespan 1000cc initially.

Uncontrolled (infra-abdominal or thoracic) hemorrhage: > No IV fluid resuscitation.

Head Wound patient: > Hespan at KVO unless there is concurrent controlled hemorrhagic shock.

UNITED STATES MARINE CORPS
Mountain Warfare Training Center
Bridgeport, California 93517-5001

FMST.07.37
04/01/02

STUDENT HANDOUT

BURN MANAGEMENT

TERMINAL LEARNING OBJECTIVE. Given a casualty in a wilderness environment, and necessary equipment and supplies, treat burn injuries to prevent death or further injury per the reference. (FMST.07.37)

ENABLING LEARNING OBJECTIVES

1. Without the aid of references, given a list of the depths and appearances of burns, match them to the type of burn, in accordance with the references. (FMST.07.37a)

2. Without the aid of references, given a list choose the major burns that require rapid casualty evacuation, in accordance with the references. (FMST.07.37b)

3. Without the aid of references, given a list choose the first step in the treatment of burn injuries, in accordance with the references. (FMST.07.37c)

OUTLINE

1. **GENERAL:** Each year there are about 2 million individuals burned severely enough to seek medical attention. Around 70,000 require hospitalization. Deaths are on the decrease largely due to the increase in the use of smoke detectors. It has been estimated that 90% of all burn injuries are preventable. Children and young adults are frequently affected by burn injuries.

2. **PHYSIOLOGY:** For non-chemical burns, the primary events of injury occur during the time of heat contact. Coagulation necrosis takes place within cells and denaturation of collagen in the dermis. Blood vessels are completely destroyed or endothelium damage is severe enough to cause clotting which leads to ischemic necrosis of remaining viable cells. Burn wounds are not static. Surrounding the "coagulation zone" is a zone of capillary and small vessel stasis. Circulation becomes stagnant from a strip formed by red blood cells, and aggregation of platelets and white blood cells. The fate of the burn wound depends on the progression of this zone of stasis.

7

3. In patients with burns less than 10% of the total body surface area (TBSA), the inflammatory process is generally limited to the burn site itself. As burns approach 20% TBSA, the local

response becomes systemic. Burns of 30% TBSA or greater, often lead to multiple organ system failure, which in a patient with a severe burn almost invariably leads to a fatal outcome, often due to acute renal failure.

Cardiovascular changes begin immediately after a burn. The degree of these changes depends initially on the size of the burn and to a lesser extent the depth of the burn. Burns less than 15% TBSA can be fluid resuscitated with crystalloid fluids. As burns pass 20% TBSA, massive shifts of fluid and electrolytes occurs. This results in massive fluid shift from the intravascular to the extravascular space. Unless intravascular volume is repleted, classic hypovolemic shock occurs.

3. TYPES OF BURNS

a. Scald Burns: Scalding is the most common cause of burns, usually resulting from hot water. Water at 140°F creates a deep partial-thickness or full-thickness burn in three seconds. At 156°F the same burn occurs in one second. Freshly brewed coffee is generally about 180°F. Boiling water always causes a deep burn. Soups and sauces are thicker, remaining in contact with the skin longer. Exposed areas tend to burn less deeply than areas covered with a thin layer of clothes. Clothing retains heat and keeps liquid in contact with the skin longer. Immersion scalds are deep and cause severe burns resulting from longer contact time. These burns are more common in children and elderly patients because of their thin skin. Burns from hot oils are generally deep partial-thickness or full-thickness. Temperatures for cooking oils reach around 400°F, and tar can be up to 500°F.

b. Flame Burns: The next most common burn injuries are flame burns. Although household injuries have gone down with the advent of smoke detectors, a high number of burn injuries still occur from careless smoking, improper use of flammable liquids, automobile accidents, and clothing ignited from stoves or space heaters.

c. Flash Burns: Flash burns result from intense heat for a brief time caused by explosions of natural gas, propane, gasoline, and other flammable liquids.

d. Contact Burns: These burns usually result from direct contact of hot metals, plastics, glass, or coals. Contact burns are usually limited in size, but are deep wounds. With the increase of wood stove usage, this type of burn injury is on the rise in children. Contact burns, dealing with an unconscious patient or involving molten materials, are usually fourth degree. The most common contact burn in the wilderness is from hot coals.

e. Electrical Burns: Electrical burns are actually thermal burns from very high-intensity heat. When electricity encounters resistance from the body it turns into heat. Severity of the burn depends on the portion of the body affected relative to the amperage. The smaller the body part, the more intense the heat is, and the part is less able to dissipate the heat. Consequently the fingers and toes are almost always destroyed with severe damage to the foreams. Larger areas can dissipate the current enough to prevent extensive damage. Arc burns take the most direct route, not the path of least resistance. These are deep and destuctive burns which mostly occur at the joints.

Electrical burns may have other associated injuries that should not be overlooked on the initial assessment. Fall related injuries are common. The intense muscle contraction resulting from electrocution can cause fractures of the lumbar vertebrae, humerus, or femur, and also may cause dislocation of the hips and shoulders. The casualty may have cardiac symptoms of a myocardial contusion or infarction. There can be a conduction system failure or an actual rupture of the heart wall. If there are no signs present of cardiac problems intially after a shock of 110 or 220 volts there is a minute chance they will appear later. Nervous system damage is possible whenever a current passes from one side of the body to the other more than likely affecting the spinal cord.

f. Chemical Burns: Usually caused by strong acids or alkalis and are most often the result of industrial accidents, home use of drain cleaners, assault, and improper use of harsh solvents. Chemical burns continue to burn until they are inactivated by reaction with tissue or dilution by flushing with water. Acids tend to tan the skin, creating an impermeable barrier that limits futher penetration. Alkalis combine with cutaneous lipids and saponify[1] the skin until they are neutralized. The appearance of the wound can be misleading. Unless the observer can be 100% sure, all chemical burns should be considered full thickness.

g. NBC Burns: Nuclear weapons cause burns in two ways: by direct absorption of thermal radiation through exposed surfaces (flash burns); or by the indirect action of fires caused by the weapon (flash burns). The first step in treatment is to prevent contamination to the rescuers and other patients.Vesicants such as mustard gas cause skin burns with edema and blister formation. Burns resulting from vesicants should be flushed with copious amounts of water. Absorbent powder (flour, talcum powder, Fuller's earth) can be used if water is scarce. The powders should be wiped away with a moist towel. The military uses M258A1 kits for skin decontamination. White Phosphorus ignites spontaneously when exposed to air. Remove the air with an air-tight seal; mud is optional. A brief rinse with 1% copper sulfate[2] will impead further oxidation.

4. **CLINICAL PRESENTATION:** The severity of the burn injury is related to the size of the burn, the depth of the burn, and the part of the body that is burned.

a. Burn Size: Burns are the only quantifiable form of trauma. The burn size is the single most important factor in predicting mortality, need for specialized care, and the complications expected from the burn. We can measure burn size with the "rule of nines". The "rule of nines" is where areas of the body represent approximately 9% of a person's total body surface area (TBSA), each leg 18% (the front of the leg 9%, and the back of the leg 9%), the front of the torso represents 18%, the back of the trunk is 18%, the head represents 9%, and the perineum 1%. For infants and small children under four years of age, the head represents a larger percentage 18%, and the legs a smaller percentage 13%, each arm 9%, front troso 13%, back troso 13%, and the perineum 1%. The percentage of the burn area and that of the unburned area need to add up to equal 100%. For smaller burn areas, use the Rule of Palmar Surface: the patient's palmar surface equals about 1% TBSA.

---

[1] Saponify: to convert fat into soap by treating it with an alkali.
[2] Navy uses sodium bicarbonate instead of copper sulfate

b. Burn Depth: The depth of the burn is either described in degree: first, second, third or fourth, or by the depth of the injury. The following describes burns in terms of partial-thickness or full-thickness. While these descriptions appear to separate burns into defined categories, many burns have a mixture of characteristics making a precise diagnosis difficult.

(1) First-Degree: These burns involve the epidermis only and do not blister. They are most commonly caused by ultraviolet light. An example would be a sunburn. The burned skin is painful and red. It should heal in 7 days without scarring.

(2) Superficial Second-Degree, or Superficial Partial-Thickness: These burns include the epidermis and upper layers of the dermis. They characteristically form blisters, under which the skin is red and moist, these burns are painful to the touch. Blisters can take up to two days to appear. These burns are usually caused by hot liquids. Wounds should heal in 14 to 21 days; they may scar depending on the extent of the burn.

(3) Deep Second-Degree, or Deep Partial-Thickness: Burns that involve deeper layers of the dermis. There is damage to hair follicles and sweat glands. They are characterized with a molted pink and white color with blisters forming immediately. These wounds may be less sensitive to touch than the surrounding normal skin, or maybe tender to touch. The patient often complains of discomfort rather than pain. Capillary refill may be slow or absent, when pressure is applied to the wound. These burns are caused by hot liquids, oil, steam, or flame. They may be difficult to distinguish from Third-Degree/Full-Thickness burns. Healing takes 3-9 weeks. Scarring is probable, the degree is related to the amount and the depth of dermal injury. Surgical grafting may be required.

(4) Third-Degree, or Full-Thickness: Involves the entire thickness of the skin, epidermis through the dermis, down to subcutaneous fat. All structures of the epidermis and dermis are destroyed. The wound is classically described as leathery, firm, and depressed when compared to normal skin. The tissue is charred, pale, and insensitive to light and touch. It is often misdiagnosed as a Deep Second-Degree/Deep Partial-Thickness due to the similar clinical findings. This burn will not spontaneously heal. Surgical repair and/or skin grafts are necessary. There is significant scarring.

(5) Fourth-Degree: These burns involve all layers of the skin, subcutaneous fat, muscle, and bone. These are devastating, life threatening injuries. They almost always have a charred appearance, and often only the cause of the burn gives a clue to the amount of underlying tissue destruction. (FMST.07.37a)

c. Burn Classification: The American Burn Association has devised a classification of burns, dividing them into major, moderate, and minor burns. Patients are placed in groups related to their risk. Low-risk patients are between the ages of 10 and 50 years old. High-risk patients are less than 10 and greater than 50 years old. Poor-risk are patients with underlying medical illnesses such as heart disease, diabetes, or chronic pulmonary problems.

(1) Minor Burns: Imply outpatient treatment. Minor burns involve TBSA of less than 15% in the low-risk group, or less than 10% TBSA in the high-risk group; full-thickness burns of less than 2% TBSA in anyone; and no other injuries.

(2) Moderate Burns: Are partial-thickness burns of 15-25% TBSA in the low-risk group, 10-20% TBSA in the high-risk group; full-thickness burns of equal to or less than 10% TBSA in anyone; burns not involving the hands, face, feet, perineum, or circumferential limbs.

(3) Major Burns: Defined as partial-thickness burns greater than 25% TBSA in the low-risk group, or greater than 20% TBSA in the high risk group; full-thickness burns of greater than 10% TBSA in anyone; burns involving the hands, face, feet, or perineum; burns crossing major joints; circumferential limb burns; burns complicated by inhalation injury; electrical burns; burns complicated by fractures or other forms of trauma; and burns in poor-risk patients. These burns require rapid casualty evacuation. (FMST.07.37b)

5. **TREATMENT:**

a. Initial Care:

(1) The first step in the treatment of burn injuries is to stop the burning process. (FMST.07.37c) Remove patients from heat, remove all their clothes that are affected by hot liquids, shut off electricity (very important), remove chemicals by flushing the affected area for 10 to 15 minutes, and decontamination of NBC. Heat can continue to injure tissue for a surprisingly long time. No first aid will be effective until the burning process has stopped.

(2) Manage the ABC's. Once an airway has been established give oxygen. If qualified, consider intubation earlier rather than later.

(3) Assess for associated injuries such as fractures or lacerations and inhalation injury.

(4) Remove clothing and jewelry from the burn area. This will help prevent constriction from swelling of the burn tissue. Do not try to remove anything that has adhered to the wound.

(5) Evaluate the Burn (depth, extent, pain).

b. General Treatment for the Patient:

(1) Gently wash the burn with tepid water and mild soap to remove any debris and to clean the skin surface around the burn site. Pat dry. Remove the skin from blisters that have popped open (do not open blisters unless necessary for function of hands or feet).

(2) Dress burn with a thin layer of antibiotic ointment.

(3) Cover the burn with Second Skin (if the burn is small enough) or cover with a thin layer of gauze, or clean, dry clothing. Burns on the face, neck and hands may be left open to the air after applying silvadene ointment. Covering the wound reduces pain and evaporative losses, but do not use an occlusive dressing.

(4) When evacuation is imminent, do not redress or reexamine the injury. If evacuation is prolonged, redress once a day. Remove old dressings, reclean (removing the old ointment), and apply fresh ointment and a clean dry covering. (Note: if stuck, soak off old dressings with clean, tepid water.)

(5) Do not pack wounds, which are larger than 20% of the body surface area, in ice. Do not leave wet coverings on burns for more than two hours at a time to reduce the risk of hypothermia.

(6) Stabilize the body temperature. When skin is lost, so is the patient's ability to thermoregulate.

(7) Have the patient drink as much fluid as he or she can tolerate without vomiting. Include some salt in the oral fluids, but do not make these solutions stronger than . 9%. This is the equivalent of a pinch of salt per 8 ounce glass.

(8) IV therapy as follows, Baxters (Parkland) Formula: In the first 24 hours give Ringers Lactate, 4ml times the weight in kilograms times the percent TBSA burned. Half of the solution is given in the first eight hours. The second half is given over the next 16 hours. The second 24 hours should be given fluids to maintain blood pressure, colloids (usually in the form of albumin) are the preferred choice.

4 x wt (kg) x TBSA= amount LR given in the first 24 hours

Example a 70kg patient with 50% TBSA this is the formula for IV therapy for the first 24 hours.

4 x 70kg x 50% = 14,000ml in 1ST 24 hours
7,000ml hours 1-8
3,500ml hours 9-16
3,500ml hours 17-24

c. General Management of the Burn: Caring for the wound itself is often the least important aspect of burn care. All burn wounds are sterile for the first 24 to 48 hours. Burn management is aimed at keeping the wound clean and reducing the pain.

(1) Elevate burned extremities to minimize swelling. Swelling retards healing and encourages infection. Have the patient gently and regularly move burned areas as much as possible.

(2) Burns are tetanus-prone wounds. Check on the last known tetanus shot. The patient may need a renewal.

(3) Ibuprofen is probably the best over the counter analgesic for burn pain (including sunburn).

(4) If you have no ointment or dressings, leave the burn alone. The burn's surface will dry into a scab like covering that provides a significant amount of protection.

6. **INHALATION INJURIES:** Half of all fire-related deaths are from smoke inhalation. Smoke inhalation doubles the mortality rate for any burn. Smoke inhalation is a general term for carbon monoxide poisoning, thermal airway injury, and smoke poisoning.

a. Carbon Monoxide Poisoning: Carbon monoxide (CO) is a colorless, odorless, tasteless gas that has an affinity for hemoglobin 200 times greater than that of oxygen. Carboxyhemoglobin (COHb) levels are measured as the percentage of hemoglobin bound to carbon monoxide. The following COHb levels correlate with clinical symptoms:

<10%	No symptoms
20%	Headache, nausea, vomiting, loss of dexterity
30%	Confusion, lethargy, ECG ST depression
40%	Coma
50%	Death

The half-life of COHb is 4 to 5 hours at room air, 90 minutes with 100% oxygen, and 20 to 25 minutes at 3 ata in a hyperbaric chamber. Treatment for less than 40% COHb is a 100% oxygen. At 40% or greater the hyperbaric chamber should be used. If patient is still in a coma after COHb levels are normal, prognosis is poor, and it has been experience that they rarely awaken.

b. Thermal Airway Injury: Injury to the respiratory tract from inhalation of hot gas or steam. The lower respiratory tract is usually not damaged, unless there was an inhalation of flame or steam. Any patients who were in an explosion, with burns of the hands, face, and upper torso are at risk. Signs of carbon such as soot in the pharnyx should alert the caregiver to airway injury. Maintaining the airway is the important issue here.

c. Smoke Poisoning: Is the inhalation of noxious gases that are the products of combustion. Cyanide poisoning is known to occur in victims of smoke inhalation. Other toxic gases released are sulfur dioxide, hydrogen chloride, phosgene, and ammonia. These gases cause damage to the respiratory tract similar to chemical burns. Physical signs include facial burns, intraoral or pharyngeal burns, singed nasal hairs, soot in mouth or nose, hoarseness, carbonaceous sputum, and expiratory wheezing.

UNITED STATES MARINE CORPS
Mountain Warfare Training Center
Bridgeport, California 93517-5001

WMC
04/09/02

LESSON PLAN

NOMENCLATURE AND CARE OF MOUNTAINEERING EQUIPMENT

LESSON PURPOSE. The purpose of this period of instruction is to introduce you to the types of equipment used here and how to care for it so as to prevent its untimely failure. This lesson relates to all climbing and installation work that you perform here.

1. **ROPES.** All ropes used in the military must meet UIAA standards or U. S. Federal Test Standard 191A. Most ropes have a 5-year shelf life and maximum 2-year service life.

a. Static. Black in color.

(1) Construction. Kernmantle

(2) Minimum tensile strength. 7500 lbs.

(3) Maximum elongation. 1.5%

(4) Diameter. 11mm

(5) Sizes.

(a) 165 ft ± 5 ft

(b) 300 ft ± 10 ft

(6) Usage. Rescue operations and bridging where a low amount of elongation is desirable under a working load.

b. Dynamic. Olive Drab in color.

(1) Construction. Water-resistant treated Kernmantle to reduce friction.

(2) Minimum tensile strength. 6500 lbs.

(3) Maximum elongation. 6%

(4) Diameter. 10.5mm and 11mm

(5) Size. 165 ft ± 5 ft.

(6) Usage. For lead climbing/party climbing.

c. Maxim Dry Rope. Olive Drab or Multi-Colored.

(1) Construction. Water-repellent treated Kernmantle.

(2) Minimum tensile strength. 3472 lbs.

(3) Maximum elongation. 6%

(4) Diameter. 9mm

(5) Size. 150 ft ± 5 ft

(6) Usage. For glacier travel/ice climbing

d. Gold Line II.

(1) Construction. Eight strand braided nylon plymor.

(2) Minimum tensile strength. 4500 lbs.

(3) Maximum elongation. 20%

(4) Diameter. 11 mm

(5) Size. 300 ft or 600 ft spools

(6) Usage. Sling Ropes and litters only.

Three strand twisted or laid rope.

Static, kernmantle rope showing core (kern) and sheath (mantle).

Dynamic, kernmantle rope used for rock climbing.

Double braid rope showing the braided sheath and braided core.

INSIDES AND OUTSIDE SHEATHS OF VARIOUS TYPES OF ROPES

NOTE: Sling ropes are made from 15 foot lengths of plymor or dynamic rope ONLY. Twenty-five foot practice coils should be constructed with static rope, but dynamic rope can be used.

2. (5 Min) **ADVANTAGES/DISADVANTAGES**

a. Advantages of Nylon Rope

(1) High strength to weight ratio.

(2) Good energy absorption in dynamic ropes.

(3) Flexible.

(4) Rot resistant, not affected by frost.

b. Disadvantages of Nylon Rope

(1) Low melting point. Nylon fuses at 400°F and melts at 480°F.

(2) Susceptible to abrasions and cuts.

(3) Affected by chemicals and light.

c. Advantages of Manila Rope

(1) Easily gripped.

(2) Hard wearing.

(3) Does not deteriorate in heat.

d. Disadvantages of Manila Rode

(1) Heavy, kinks, especially when wet. Absorbs water and swells.

(2) Burns at +300°F.

(3) Edible by rodents.

3. (3 Min) **GENERAL INFORMATION**

a. Nylon rope stretches under tension and will rupture at between 30% and 70% elongation, depending on construction.

b. Nylon rope loses as much as 30% strength when wet.

c. Temperatures as low as 250°F will damage a nylon rope.

4. (5 Min) **NYLON WEBBING**

a. The type of nylon webbing available is tubular. Tubular nylon webbing is very strong and flexible. All rules that apply to nylon rope apply to tubular nylon webbing. The size of nylon webbing used is:

(1) 1 inch tubular nylon. Tensile strength approximately 4,000 - 4,500 lbs., depending on the manufacturer.

b. Pre-sewn Spectra Runners. Tensile strength approximately 5,500 lbs.

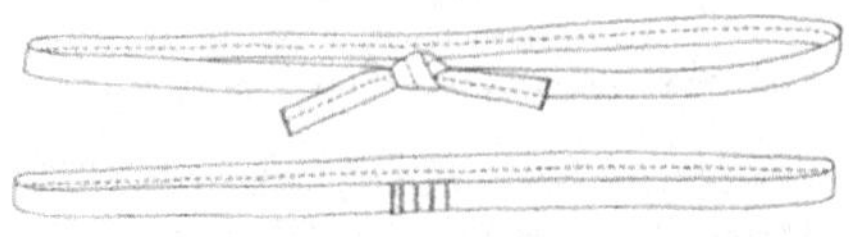

NYLON WEBBING

NOTE: These are minimum strengths. Some manufactures make even stronger webbing.

TRANSITION: We have just discussed general information in nylon webbing, are there any questions? Not only do we use ropes, but we also use carabiners in our installations, we will discuss the types of carabiners used:

5. (5 Min) **CARABINERS**. Also commonly known as snaplinks. Both locking and non-locking are used.

a. Purpose. Carabiners are used for the following purposes:

(1) To attach ropes or runners to pieces of protection.

(2) To attach the rappel rope to the rappel seat for seat-hip rappels or for crossing rope bridges.

(3) To attach the individuals safety rope to a safety line on a rope installation.

(4) To form field expedient pulley systems.

b. Nomenclature of a non-locking carabiner

(1) Gate

(2) Gate pivot pin.

(3) Locking pin.

(4) Body.

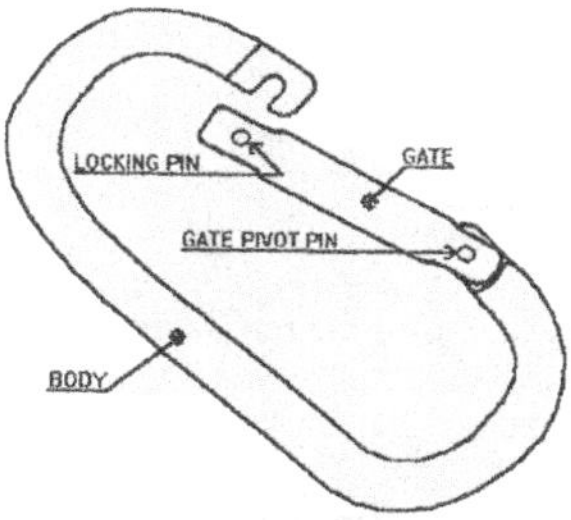

c. Nomenclature of a locking carabiner

(1) Gate.

(2) Gate pivot pin.

(3) Locking notch.

(4) Locking nut.

(5) Body.

NON LOCKI... UM CARABINER

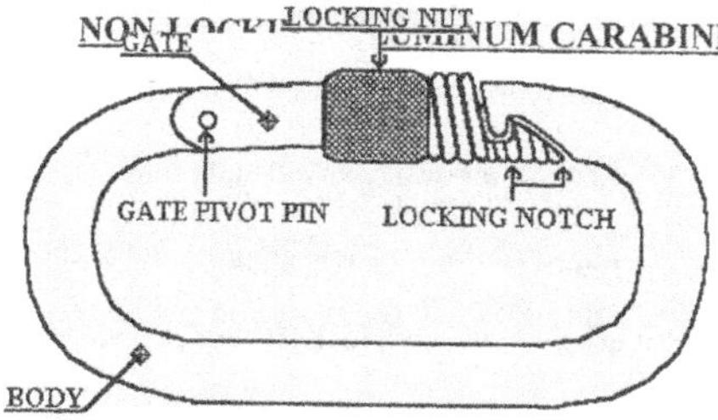

STEEL LOCKING CARABINER

d. There are two types of carabiners used. The two types and their characteristics are:

(1) Steel locking carabiners

(a) Large steel locking "D" (various manufacturers): Minimum tensile strength of 5,500 lbs.

(b) Steel-locking oval Stubai 82 is not in the MAC Kit and obsolete. However, it is being used at MWTC to save money, even though they are beyond the service life. Tensile strength of only 3,300 lbs.

(2) Aluminum non-locking carabiners

(a) Aluminum non-locking oval (various manufacturers): minimum tensile strength of 4,200 lbs.

e. Serviceability Check for a Carabiner. The following steps are used for you to check a carabiner for serviceability:

(1) The gate snaps shut with no friction and with no gap between the locking pin and locking notch.

(2) There is no excessive side to side movement of the gate.

(3) The pivot pin is tight.

(4) The locking pin is tight.

(5) The locking nut travels freely and locks securely.

(6) There are no cracks or flaws in the metal.

NOTE: The weakest part of a carabiner is the gate. If an engraver is used to mark a carabiner, it should be applied to the gate and not the load bearing side.

f. Preventive Maintenance for a Carabiner.

(1) Remove all dirt, moisture and grime.

(2) Lubricate with tri-flow graphite and clean off thoroughly.

NOTE: Whenever you use a locking carabiner ensure that the locking nut is always locked down (tightened).

6. (2 Min) **CARE OF THE CARABINER**. Do not drop the carabiner as this may result in either actual damage to the carabiner or in dirt getting into the workings of the carabiner and damaging it.

UNITED STATES MARINE CORPS
Mountain Warfare Training Center
Bridgeport, California 93517-5001

FMST.07.11
05/01/01

STUDENT HANDOUT

HYPOTHERMIA/REWARMING

TERMINAL LEARNING OBJECTIVE: Given a hypothermia casualty in a cold weather environment and necessary equipment and supplies, manage a hypothermia casualty to prevent death or further injury per the reference. (FMST.07.11)

ENABLING LEARNING OBJECTIVES:

(1) Without the aid of references, select from a given list the definition of hypothermia, in accordance with the references. (FMST.07.11a)

(2) Without the aid of references, select from a given list three of the five mechanisms by which the body loses heat, in accordance with the references. (FMST.07.11b)

(3) Without the aid of references, select from a given list the reason why gentle handling of a hypothermia victim is important, in accordance with the references. (FMST.07.11c)

(4) Without the aid of references, select from a given list the four rewarming techniques used to rewarm a hypothermia victim, in accordance with the references. (FMST.07.11d)

(5) Without the aid of references, select from a given list the proper steps of the sleeping bag rewarming method of rewarming a hypothermia victim, in accordance with the references. (FMST.07.11e)

(6) Without the aid of references, select from a given list the temperature of the water when rewarming a hypothermia casualty in a warm water bath, in accordance with the references. (FMST.07.11f)

8

1. **HYPOTHERMIA**:

a. Definition: A fall in core body temperature to 95°F (35°C) or less. (FMST.07.11a)

b. Hypothermia is potentially lethal. The mortality rate is greater than 50% in severe cases and cases complicated by trauma where the mortality can approach 100%. The most common cause of death in hypothermia is ventricular fibrillation.

c. When speaking of hypothermia, core temperature values must be used. In a controlled environment of an MTF or ER the gold standard is a core temperature taken by rectal or esophageal thermometry. The term "exposure" has often been used to speak of hypothermia but this term is inexact and has numerous other connotations. It should not be used as a synonym for hypothermia.

d. Incidence.

(1) Classically, hypothermia has been thought to occur in extremely cold environments. However, it can occur in temperate climates and at Southern latitudes. In fact, greater than 50% of cases of hypothermia occur at ambient temperatures of 50°F or greater. Contrary to popular belief, the highest incidence of hypothermia is seen in urban areas.

(2) Hypothermia tends tends to occur more frequently in males with a male to female ratio of 2.8:1.0. It's also more common in the elderly, those with drug and alcohol dependency and those with psychiatric diagnoses.

Many standard clinical thermometers don't register a low enough temperature to accurately diagnose hypothermia. Currently, a low-reading rectal thermometer is available through the Federal Stock System (NSN 6515-00-139-4593). All medical personnel attached to cold weather units or deploying to cold weather environments should carry this thermometer. In a high altitude cold weather environment medical providers should be adept at correlating signs and symptoms with specific temperature ranges so as not to unduly expose the patients to the elements. The more accurate, and invasive temperature determination should only be performed when the patient can be adequately protected from the elements.

2. **PATHOPHYSIOLOGY**:

a. Heat Conservation/Loss.

(1) The body is a furnace that generates heat. The ability of the body to thermo regulate depends on the ratio of heat production to heat loss. An imbalance of these two processes can lead to hypothermia. About 75% of food energy consumed goes toward generating body heat while the other 25% goes toward maintaining basic metabolic function.

(2) Thermogenesis can be divided into two main categories:

(A) Shivering thermogenesis creates heat by muscle activity with subsequent breakdown of glycogen. Shivering usually begins at a core temperature of around 97°F (36°C) and finally ceases at core temperatures of 85°- 90°F (30°- 34°C).

(B) Non-shivering thermogenesis creates heat by hormonal interaction with target organs such as the heart. Epinephrine, norepinephrine, growth hormone and thyroid hormone have all been implicated in playing a role in non-shivering thermogenesis.

(3) Changes in hormone secretion patterns are extremely important in acclimatization to cold weather.

(4) Once heat is generated it is retained by three main mechanisms: Peripheral vasoconstriction, abolition of sweating and piloerection. Peripheral vasoconstriction is by far the most important and efficient mechanism of heat conservation. The main internal temperature sensors in the body are located in the hypothalamus in the brain. When these areas sense a drop in core temperature they initiate peripheral vasoconstriction at the surface of the skin to prevent a loss of core body heat.

(5) The body loses heat by five main mechanisms: (FMST.07.11b)

(A) **Radiation** is heat transferred to the surrounding environment via infrared radiation.

(B) **Convection** is the loss of heat via air/water current.

(C) **Conduction** is the direct contact-transfer of heat from a hot to a cold object.

(D) **Evaporation** is the conversion of water from a liquid phase to a gaseous phase (perspiration).

(E) **Respiratory heat loss** takes place when warm humidified air from the lungs is expelled into the atmosphere.

(F) The mechanism accounting for the greatest percentage of heat loss is highly variable and depends on ambient conditions, i.e., a nude man standing in a cold room will lose approximately 65% of his heat by radiation. The same man standing in calm cold water will lose most of his heat by conduction and at a rate 25 times faster.

**NOTE: The liver and brain are the most thermogenic organs on a per weight basis. However, due to their greater mass, skeletal muscle and skin produce the largest percentage of body heat.

(6) Numerous factors predispose to the development of hypothermia. These include but are not limited to:

(A) Malnutrition.
(B) Lack of adaptation.
(C) Inactivity.
(D) Sedatives.
(E) Physical exhaustion.
(F) Burns.
(G) Extremes of age.
(G) Dehydration.
(H) Comorbid illness.

3. **SIGNS AND SYMPTOMS**:

a. Clinical manifestations of hypothermia reflect CNS, and cardiorespiratory involvement.

(1) 98°-95°F (37°-35.6°C): Mild shivering, cold sensation and impaired fine motor coordination.

(2) 95°- 92°F (35°-32.8°C): Violent shivering, difficulty speaking, sluggish thinking, and amnesia. Large muscle in coordination.

(3) 92°-86°F (32.2°-30°C): Shivering is replaced by muscular rigidity. Exposed skin is blue or puffy. Movements are jerky. Dulled sensorium, but victim may be able to maintain posture and the appearance of being in contact with surroundings. Possible atrial arrhythmias.

(4) 85°-81°F (29.4°-27.2°C): Coma, lack of reflexes, possible ventricular arrhythmias. Cessation of shivering below 86oF.

(5) Below 78°F (25.6°C): Failure of cardiac and respiratory centers, pulmonary edema, ventricular fibrillation, and asystole.

b. Some patients with hypothermia will exhibit a phenomenon that is called, "paradoxical undressing."

(1) In these instances the hypothermia victim is found nude with their cloths beside them. If the victims are women they are often misdiagnosed as sexual assault cases. This phenomena results from loss of hypothalamic control over peripheral vasoconstriction which leads to flushing of the periphery with relatively warm blood causing the individual to feel hot. Subsequently the individual undresses to cool down.

c. Again it must be stated that the only way to correctly diagnose hypothermia is with a core temperature reading.

4. **FIELD TREATMENT**:

a. ABC's, C-Spine precautions, remove wet clothing, replace with dry clothes and insulate with a vapor barrier system. (Wrapping the victim in two plastic garbage bags or bubble wrap is ideal. Further insulation may be necessary). Handle the victim carefully to prevent ventricular fibrillation caused by unnecessary jarring and rough handling. (FMST.07.11c) CASEVAC.

b. Victims with altered consciousness should be NPO. Under no circumstances should ETOH be given. This is a peripheral vasodilator and will only enhance heat loss.

c. If CASEVAC is not feasible for a long period of time, or if the victim is only mildly hypothermic, it may be necessary to rewarm in the field by using a proper sleeping bag rewarming technique. Additionally, insulated hot packs placed at the high heat loss areas will help.

d. The hypothermic victim often has no discernible vital signs.

(1) Before beginning CPR thoroughly assess pulses for 30-45 seconds to detect viable cardiovascular activity.
(2) Severely hypothermic victims in ventricular fibrillation and asystole have been successfully resuscitated even after periods as long as 4 hours.

5. **REWARMING**: The re-warming of a hypothermia casualty in a BAS setting is less than optimal since the AMAL for a BAS does not include monitoring devices, laboratory facilities, etc., but rewarming of a casualty can be successfully conducted even in these fairly primitive conditions. The following are three basic techniques of re-warming:

- Passive external re-warming: a sleeping bag or blanket that is not pre warmed.
- Active external re-warming: a pre-warmed sleeping bag, bear hugger or warm water bath.
- Active internal or core re-warming: IV fluids, warmed/humidified O2, lavage, extracorporal re-warming.

a. Four ideal re-warming strategies that may be available in the field include: (FMST.07.11d)

(1) Sleeping bag rewarming.

(2) Water bath/counter Current rewarming.

(3) Warm/humidified airway rewarming.

(3) Warmed I.V. solutions. NS is preferable as a cold liver does not like to metabolize Lactate.

6. **REWARMING TECHNIQUES (BAS)**:

a. Sleeping Bag Rewarming. (FMST.07.11e) This is the simplest, but least effective method.

(1) Make the diagnosis. Take a rectal temperature to determine if the casualty is clinically hypothermic (95°F or less). Use this temperature as a baseline to determine stabilization.

(2) Warm sleeping bag (pre-warm sleeping bag with two volunteers). Placing a hypothermic casualty into a cold sleeping bag will cause further heat loss by conduction.

(3) Strip casualty. Remove all wet clothing avoiding unnecessary handling. Muscular movement will pump cold blood to the core.

(a) Place the casualty between the two volunteers, if space permits. Their body heat is transferred from the two volunteers to the cold casualty.

(4) Monitor core temperature frequently throughout transport to ensure that the casualty's core temperature is not continuing to drop.

(5) Adequate insulation is required to prevent further heat loss to the environment.

(6) Augment heat by placing insulated heating pads in the high heat loss areas: head, axilla, groin, poplitcal region and antecubital fossa. (Be careful not to cause burns)

**NOTE: It should be noted that an evacuation bag can be used; this will provide more room for the casualty and volunteers.

b. Water Bath Re-warming. (FMST.07.11f) This method has a long history of success in rewarming even severely hypothermic casualties. The casualty is placed in a portable/field expedient bathtub or a life raft. The extremities and head are kept out of the water to avoid significant afterdrop. The water bath temperature should be 104°-108°F. An anesthesia temperature probe should be used to constantly monitor the core temperature. Since the casualty will cool the bath water around him, it will be necessary to continue to add warm water to maintain the proper water bath temperature. Keep in mind however that this technique has definite drawbacks. In the event of ventricular fibrillation caused by re-warming shock, unsynchronized cardioversion can be very dangerous in a wet environment. Furthermore, if the casualty in the water is not on a firm surface CPR, if needed, can be very difficult to perform ineffective.

**NOTE: Individual in heat distress can be rapidly cooled using this method with "cool" water instead.

c. Warm Airway Re-warming. A warmed and humidified air/oxygen mixture is used to provide a warming media within the lungs. These devices consist of a method of generating warm humidified air and are usually portable. The warm humidified air modestly increases the amount of heat that can be delivered to core. There should be a thermometer in the airway tubing to monitor the temperature. To prevent injury to the bronchi, the temperature should not exceed 115°F (108° to 115°F is ideal). Warm airway rewarming should be used in conjunction with other rewarming methods. This method, by itself, will probably not deliver sufficient heat to rewarm a severely hypothermic casualty by itself and may in fact blunt the shivering response robbing the body of an important endogenous mechanism for producing heat.

Caveat: When possible, active external re-warming should always be combined with active core re-warming techniques so as to minimize the effects of afterdrop. This occurs as a result of peripheral vasodilatation causing warm core blood to move to the surface while relatively cold surface blood moves to the core and may lower the fibrillation threshold.

**NOTE: Hypothermic submersion incident casualties cannot tolerate humidified air for any length of time.

d. Warmed I.V. Solutions. Hypothermia is a common response to I.V. therapy. Solutions that have been prewarmed have been shown to prevent this complication. There are various methods for warming I.V. solutions:

(1) Crystalloid solutions can be warmed by warm water bath and microwaved with no adverse changes to its integrity.

(2) Fresh Frozen Plasma (FFP; -20°C): Studies have shown that microwaving for 30 seconds followed by gentle manipulation of the bag for 10 seconds, repeated five times will result in only 2.6% cell destruction.

(3) Packed Red Blood Cells (PRBC 4°C): Using conventional rewarming methods, studies have shown that the outer five millimeters of the bag can have isolated areas of hemolysis. Until a device is made that can uniformly shake up PRBC during warming, we do not recommend microwaving of blood. Rewarming of blood by dilution with warm, calcium-free crystalloid solution is useful.

WIND CHILL CHART

WIND	AMBIENT TEMPERATURE (FAHRENHEIT/CENTIGRADE)													
CALM F	35	30	25	20	15	10	5	0	-5	-10	-15	-20	-25	-30
CALM C	27	-1	-4	-7	-9	-12	-15	-18	-21	-23	-26	-29	-32	-34
EQUIVALENT WIND CHILL FACTOR (FAHRENHEIT/CENTIGRADE)														
5 MPH F	33	27	21	16	12	7	1	-6	-11	-15	-20	-26	-31	-35
8 KPH C	1	-3	-6	-9	-11	-14	-17	-21	-24	-26	-29	-32	-35	-37
10 MPH F	21	16	9	2	-2	-9	-15	-22	-27	-31	-38	-45	-52	-58
16 KPH C	-6	-9	-13	-17	-19	-23	-26	-30	-33	-35	-39	-43	-47	-50
15 MPH F	16	11	1	-6	-11	-18	-25	-33	-40	-45	-51	-60	-65	-70
23 KPH C	-9	-12	-17	-21	-24	-28	-32	-36	-40	-43	-46	-51	-54	-57
20 MPH F	12	3	-4	-9	-17	-24	-32	-40	-46	-52	-60	-68	-76	-81

32 KPH C	-11	-16	-20	-23	-27	-31	-36	-40	-43	-47	-51	-56	-60	-63
25 MPH F	7	0	-7	-15	-22	-29	-37	-45	-52	-58	-67	-75	-83	-89
40 KPH C	-14	-18	-22	-26	-30	-34	-38	-43	-47	-50	-55	-59	-64	-67
30 MPH F	5	-2	-11	-18	-26	-33	-41	-49	-56	-63	-70	-78	-87	-94
48 KPH C	-15	-19	-24	-28	-32	-36	-41	-45	-51	-53	-57	-61	-66	-70
35 MPH F	3	-4	-13	-20	-27	-35	-43	-52	-60	-67	-72	-83	-90	-98
56 KPH C	-16	-20	-25	-31	-33	-37	-42	-47	-51	-55	-58	-64	-68	-72
40 MPH F	1	-4	-15	-22	-29	-36	-45	-54	-62	-69	-76	-87	-94	-101
64 KPH C	-17	-20	-26	-30	-34	-38	-43	-48	-52	-56	-60	-66	-70	-74

7 REWARMING TECHNIQUES (HOSPITAL):

a. Optimal re-warming techniques for higher echelon care facilities still tend to employ rather simple therapeutic modalities. The three safest and most effective are listed below:

- Warmed IVF.
- Warmed/humidified oxygen.
- Bear Hugger.

b. Many texts advocate the use of extracorporal blood re-warming. However, this technique has some disadvantages. It requires a high level of technical and medical support, which are often not available. Since patients undergoing bypass must be anti-coagulated, patients with multiple traumatic injuries may not be good candidates. Lastly, research indicates that anticoagulation with heparin may actually worsen the severity of frostbite that may occur in patients with hypothermia.

c. Lavage: peritoneal, thoracic and mediastinal lavage have all been suggested for use in severe hypothermia. However, the use of these techniques in inexperienced hands tends to carry a much higher complication rate. If performed correctly, peritoneal lavage can be very beneficial with a lower complication rate than thoracic lavage. Gastric, colonic and bladder lavage are indeed safer but are of limited efficacy, as they do not offer access to areas with significant surface area.

d. ACLS: At core temperatures below 86°F attempt three progressive unsynchronized shocks at 200J, 300J, and 360Js. If no response continue CPR and REWARM THE VICTIM. Re-attempt counter-shocks when core temperature rises above 86°F. Above 86°F, perform counter shocks in 2° increments above 86°F, i.e.88°F, 90°F and 92°F. Employ ACLS drug protocols above 93°F along with conventional normothermic ACLS protocol.

UNITED STATES MARINE CORPS
Mountain Warfare Training Center
Bridgeport, California 93517-5001

FMST.07.12
03/21/02

STUDENT HANDOUT

SUBMERSION INCIDENTS

TERMINAL LEARNING OBJECTIVE: Given a casualty in a cold weather environment and the necessary equipment and supplies, manage a submersion incident casualty to prevent death or further injury per the reference. (FMST.07.12)

ENABLING LEARNING OBJECTIVES:

(1) Without the aid of references, given a list of the events, which take place to a submersion victim, select the correct sequence of events, in accordance with the references. (FMST.07.12a)

(2) Without the aid of references, select from a given list, three of the five methods that can be used to increase survival time during cold water immersion, in accordance with the references. (FMST.07.12b)

(3) Without the aid of references, select from a given list, three of the six favorable prognostic signs post submersion incident, in accordance with the references. (FMST.07.12c)

(4) Without the aid of references, select from a given list, three of the four unfavorable prognostic signs post submersion incident, in accordance with the references. (FMST.07.12d)

(5) Without the aid of references, select from a given list the correct field management priorities of a submersion victim, in accordance with the references. (FMST.07.12e)

10

OUTLINE.

1. **TERMINOLOGY**.

a. Definition: A submersion incident is that medical problem that occurs after a casualty has been submersed under water.

(1) Drowning refers to death by suffocation following submersion in water.

(2) Near drowning refers to submersion in water with at least temporary survival (greater than 24 hours).

NOTE: An effort is being made to standardize the terminology used regarding submersion incidents, in order to clarify communication between care-givers. Terms like secondary drowning and other definitions are being discouraged.

2. **PATHOPHYSIOLOGY**.

a. Sequence of events: (FMST.07.12a)

(1) Panic initially, followed by a violent struggle.

(2) Gulping and swallowing air and water to avoid aspiration.

(3) Breath holding until hypoxia leads to unconsciousness.

(4) Once consciousness is lost, the gag reflex relaxes and passive influx of water into the lungs occurs. A small percentage of victims (10-15%) have significant laryngospasm, which prevents any appreciable volume of water to enter the lungs. This is referred to as "dry drowning".

(5) Aspiration is more likely in submersion victims who enter very cold water (10°C). An involuntary gasp reflex occurs, with a marked increase in respiratory rate and depth. This individual is at increased risk of aspiration, and subsequent drowning.

b. Submersion in very cold water will quickly results in hypothermia.

(1) Rapid induction of total body hypothermia (core temperatures of 89.6°F or less) may be protective in submersion incidents. Significant decreases in oxygen requirement result, especially in the case of the brain itself. Marked decreases in cerebral blood flow and oxygen demand have been demonstrated (see Hypothermia lesson). This protective effect of hypothermia has been used to explain the many cases of complete recovery after extended periods of submersion, which have only occurred in cold water.

(2) The "Mammalian Diving Reflex" slows the heart rate, shunts blood to the brain and closes the airway when more primitive marine mammals are exposed to extremely cold water. General consensus holds that this reflex is not a significant response in human beings. However, it has been found to be more prominent in new-born human infants. It has also been speculated that this "Reflex" may be more significant in those individuals who have survived extended submersion times.

c. Fatalities in submersion victims associated with hypothermia are primarily due to three known mechanisms:

(1) Immersion syndrome. Sudden death due to cardiac arrest secondary to massive vagal stimulation. Usually associated with sudden total body exposure to frigid water.

(2) Excessive fatigue and confusion leading to fatal errors of judgment.

(3) Direct hypothermic effects. Thermal conductivity in water is approximately 25 times that of air. Despite the evidence showing hypothermia's protective effects, the complications associated with hypothermia (cardiac arrhythmias, CNS depression, loss of motor control) can lead to submersion and drowning.

d. Some authorities emphasize distinguishing between "wet" and "dry" drowning. In wet drowning, aspiration of water into the lung occurs. (85-90% of casualties). In dry drowning, laryngospasm prevents aspiration of water (10-15% of cases). Ultimately, the end result is the same: obstruction to respiration, oxygen depletion, and asphyxiation. The morbidity and mortality associated with the two types of drowning are not significantly different.

e. Fresh-water and salt-water drowning are somewhat different.

(1) In fresh-water aspiration, the osmotic difference between hypotonic fresh-water and the relatively hypertonic blood leads to net water movement into the vasculature.

This flow across the alveolar membrane also leads to "washout" of surfactant, which leads to collapse of the alveolar spaces. Hypoxemia follows.

(2) In sea-water aspiration, the osmotic shift is reversed, leading to fluid outpouring from the blood stream flooding the alveoli and the interstitium. This leads to decreased lung compliance as well as direct damage to pulmonary capillaries. Again, hypoxemia results.

(3) Hypoxemia is the end result in both types of drowning. No significant difference in morbidity in survivors has been shown between fresh-water and salt-water drowning. The only bodies of water shown to have a worsening effect on prognosis are the hyper-saline bodies, like the Dead Sea and the Great Salt Lake. Otherwise, prognosis is similar.

f. Hypoxemia is the primary insult after submersion. Efforts to correct this problem should be vigorous and sustained. All clinically dead submersion casualties should receive CPR, or ACLS if available, immediately upon removal from the water, regardless of water temperature or duration of submersion. Since the overwhelming majority of naturally occurring water bodies in the world have temperatures below 92°F, all submersion victims should be treated aggressively as hypothermic casualties. Survivals after submersion times up to 66 minutes have been documented. Remember: No one is dead until they are warm and dead.

g. In the case of prolonged submersion or significant aspiration, pulmonary derangements occur rapidly. In lesser insults, the onset of symptoms may be delayed for as long as 24 hours. Physiological changes include:

(1) Cardiovascular. Casualties develop respiratory acidosis and hypoxemia, which may trigger dysrhythmias. Electrolyte changes are not usually significant in survivors of submersion incidents and should correct once spontaneous respirations are re-established.

(2) CNS. C-Spine injuries are common in submersion incidents, most commonly a result of shallow-water diving. Injury patterns include wedge, tear-drop and burst fractures of the upper C-Spine, frequently resulting in tetraplegia. Concussive head injuries also occur. Drug and alcohol use should always be considered.

(3) Metabolic. Respiratory acidosis is the initial problem. Acid/base derangement should correct after resuscitation of the casualty without specific intervention.

3. **SURVIVAL TECHNIQUES IN COLD WATER.**

a. Swimming motion increases heat loss more than it increases heat production. The direct hypothermic effects of cold-water immersion make it impossible to swim for long periods of time. The average swimmer is unlikely to swim more than a kilometer in 50°F water. Swimming is advised only when a rescue vehicle is nearby, or the distance to shore is less than ½ mile.

b. Treading Water and Drown Proofing.

(1) Treading water results in reduced survival time due to increased heat loss secondary to increased activity.

(2) "Drown Proofing" is a technique whereby the swimmer relaxes in the water, floating facedown, arms spread, allowing his face to submerge between breaths. "Drown proofing" is the least desirable survival technique in cold water since the head and neck (high heat loss areas) are maximally exposed.

c. Personal Floatation Device (PFD). Without the aid of the PFD, the swimmer caught in cold water is forced to use less desirable methods to stay afloat.

(1) If wearing a PFD, the individual can utilize the "Heat Escape Lessening Posture" (HELP). This position minimizes the exposure of the swimmer's high heat loss areas because the arms are folded across the chest and pressed to the sides, the knees are drawn up to the chest and the legs are crossed at the ankles. Unfortunately, this position is fatiguing and becomes difficult to hold

(2) If there is more than one person in the water, the "Huddle" position should be used. Similar to a football huddle, the chest, abdomen and groin area should be pressed together. Children or injured individuals should be placed in the middle of the huddle, since they become hypothermic more quickly.

d. Increasing Survival Time: (FMST.07.12b)

(1) Try to enter the water in a lifeboat or raft.

(2) Locate and don a PFD as quickly as possible.

(3) Wear several layers of clothes, especially to protect high heat loss areas.

(4) If you must enter the water, enter gradually rather than jumping or diving.

(5) If afloat in the water, avoid unnecessary movement. Assume the HELP or Huddle position.

(6) Don't try to swim.

(7) Remember the situation is not hopeless. Have the will to SURVIVE!

4. **RESCUE TECHNIQUES**.

a. Reaching the Casualty. The basic order for water rescue is: Reach and Pull, Throw, Tow and, as a last resort, Go.

(1) Reach and pull. When the casualty is responsive and close to the shore, begin by holding out an object for him to grab then pull him from the water. Floating line is considered the best choice, but if it is not available, a tree branch, fishing pole, oar or other object will do as well.

(2) Throw. Should the casualty be alert, but too far away for you to reach and pull, throw an object that will float to them. A PFD or ring buoy is best, if available. The primary course of action is to throw anything that will float, and to do it as quickly and as accurately as possible.

(3) Tow. Once the conscious casualty has an object to hold onto, try to find a way to tow him to shore. From a secure position, tow the casualty ashore.

(4) Go. Swim to the casualty as a last resort. You should be a strong swimmer, trained in water rescue, in order to avoid the dangers associated with this form of rescue. If the casualty is unconscious or unresponsive, you must go to the casualty.

5. **SIGNS AND SYMPTOMS**.

a. History of Incident. Several questions need to be answered, as they will help determine the approach to the casualty.

(1) When a person is found floating face down on the water surface or is brought to the surface by a rescuer, there usually is little question about the cause of unconsciousness.

(2) When near-drowning is unwitnessed or occurs after a dive or other water activity (surfing), a head injury and a C-spine injury should be presumed, and C-Spine precautions should be instituted.

(3) Determine how long the casualty was submerged. This helps determine prognosis.

(4) How cold was the water? Very cold water (less than 70°F/21°C) may be protective.

(5) What was the casualty's health prior to submersion? Drowning could have been secondary to alcohol ingestion, cardiac arrest or other illness/injury.

b. Casualties may be symptomatic or asymptomatic at the scene:

(1) Respiratory signs and symptoms may include dyspnea, tachypnea (SOB), sore throat, substernal burning, pleuritic chest pain, persistent cough, frothy sputum, cyanosis, rales, rhonchi and wheezes.

(2) Cardiovascular signs may include tachycardia and hypotension.

(3) Neurological findings may include anxiety, restlessness, lethargy, confusion, convulsions, incontinence, hyporeflexia and coma.

(4) Abdominal distention is common as is fever and hypothermia.

c. Favorable prognostic signs: (FMST.07.12c)

(1) Casualty is alert.

(2) Basic or advanced life support is received at the scene.

(3) Good response to initial resuscitation.

(4) Hypothermic.

(5) Older child or adult.

(6) Brief submersion time.

d. Unfavorable prognostic signs: (FMST.07.12d)

(1) Age under three years.

(2) Fixed dilated pupils.

(3) Submersed five minutes or longer.

(4) No resuscitation attempts for the first ten minutes.

(5) PH less than 7.10.

(6) Cardio-pulmonary arrest as initial presentation.

(7) Resuscitation time greater than 30 minutes.

6. **FIELD MANAGEMENT**.

a. The scene of a submersion incident is frequently chaotic and first consideration should be the safety of the rescuers, then retrieval of the casualty from the water.

b. Priorities: (FMST.07.12c)

(1) Establish an airway and check breathing and circulation (ABC's). Rescue breathing, can be performed by a trained rescuer, while the casualty is still in the water.

(2) Observe C-Spine precautions, since cervical spine and head injuries are common, especially in diving accidents.

(3) Begin CPR as soon as possible, if necessary, when the casualty reaches the shore, or suitable rescue platform.

(4) The symptomatic patient should be given high-flow 02 and IV access should be established, if available.

(5) Medevac ASAP.

(6) It is recommended that all victims of submersion be observed in a medical facility for a minimum of 6-8 hours. It is not uncommon for asymptomatic near-drowning victims to refuse further care. All that can be done is attempt to get a contact number in order to follow-up with the victim via telephone 8 hours later.

c. Vomiting is common and should be anticipated by turning the patient's head and having suction ready, if available. Frequently, the cause of vomiting is gastric distention.

(1) The American Heart Association recommends that gastric distention be relieved only when it interferes with CPR.

(2) Gastric decompression can be achieved by "breaking" the patient. This maneuver is performed by placing the patient in the prone position and lifting upwards at the abdomen.

(3) Another method is to turn the patient to a lateral recumbent position and compress the abdomen with the palm of one hand.

(4) The preferred method of gastric drainage is to pass an NG tube and relieve distention with gentle suction.

(5) Gastric drainage should not be attempted without suction equipment available, due to the risk of aspirating vomitus.

7. **PROGNOSTIC PREDICTORS**.

a. 0 - 2 "Unfavorable Prognostic Indicators" yield a greater than 90% survival rate (see above).

b. 3 or more "Unfavorable Prognostic Indicators" yields a less than 5% survival rate.

c. Corm's Criteria: Based on the status of the patient 1 hour after initial resuscitation.

(1) A - Awake	All survive (99+%)
(2) B - Blunted	Most survive (95+%)
(3) C – Comatose	
C1 - Flexor Response	Good survival with ICU care (80+%)
C2 - Extensor Response	Good survival with ICU care (80+%)
03 - Flaccid	Poor survival
C4 - Arrest	Dismal survival

UNITED STATES MARINE CORPS
Mountain Warfare Training Center
Bridgeport, California 93517-5001

FMST.07.14

04/02/02

STUDENT HANDOUT

WILDERNESS ORTHOPAEDIC/ TRAUMA INJURIES

TERMINAL LEARNING OBJECTIVE: Given a casualty in a mountainous enviornment and, necessary equipment and supplies, manage common orthopedic injuries in a mountainous enviornment to prevent death or further injury per the reference. (FMST.07.14)

EABLING LEARNING OBJECTIVES:

(1) Without the aid of references, select from a given list the phases of clearing a cervical spine in the field in accordance with the references. (FMST.07.14a)

(2) Without the aid of references, select from a given list, the correct treatment for an open book pelvic fracture in accordance with the references. (FMST.07.14b)

(3) Without the aid of references, select from a given list, six considerations for orthopedic patient evacuation in accordance with the references. (FMST.07.14c)

OUTLINE

1. INTRODUCTION.

Orthopedic injuries occuring in the wilderness are not significantly different from those suffered in garrison. Bones break, ligaments are torn and strained, and joints are dislocated in both settings. The key to preparing for and managing wilderness orthopedic injuries is understanding that the provider will have significant limitations with regard to diagnostic tools and initial management resources at his/her disposal. The personal risks each expedition member is exposed to can rise dramatically if the team is needed to transport a victim out of the back-country. Determination of a casualty's mobility becomes a major consideration for the wilderness medical provider. Assessment of the terrain, creative use of available natural resources, and effective leadership are all necessary components of caring for orthopedic injuries in the wilderness.

10

←

2. GENERAL: Making the diagnosis.

The wilderness "doc", whatever his/her training background, needs to have good physical exam skills. NO X-RAY and NO LABS will be available in the back-country. These basic tools are not available in the field. The provider will need to rely on his/her observation, palpation, and auscultation skills to make a diagnosis. This person will also need to use the patient, if possible, to uncover the likely injury by taking a thorough history which includes careful review of the mechanism of injury.

A. HISTORY: Critical information can be gained from talking to your patient.

- Mechanism of Injury - Determine:
    1. High vs. low-velocity accident
    2. Direction of force
    3. Nature of activity
    4. Loss of consciousness

- Ask the casualty:
    1. Did they hear a pop, snap, crack, or breaking sound?
    2. How long ago did the injury occur?
    3. Can they move the injured body part?
    4. Was there swelling within 15 minutes of injury or did it take longer?
    5. Was there immediate disability or was there a gradual loss of function?
    6. Was there immediate discoloration of a joint?
    7. Can they bear weight on the injured extremity?

- Other important information:
    1. Evaluate the environment. (i.e. Hot/Cold, Wet/Dry)
    2. Allergies.
    3. Medications currently being taken.
    4. Previous injuries/surgeries to the site in question.

B. PHYSICAL EXAMINATION

Your powers of observation become crucial.

- Look for:
    1. Swelling.
    2. Discoloration and bruising.
    3. Obvious deformity/angulation.
    4. Open wounds. (with or without protruding bone fragments)
    5. Differences right vs. left.

6. Muscle spasm surrounding injured site.
7. Point tenderness.
8. Crepitus.

- Beware:
  Consider the environment while assessing your patient
  - Don't undress the casualty in the cold!
  - Protect your patient from the elements. (Sun, Rain, Cold)
  - Talor your physical exam to meet the constraints of the environment.
    * palpate under clothing.
    * visualize one region at a time, then re-dress.
    * set up temporary shelter from the elements.

3. Head Injuries:

a. A blow to the head may lead to increased intracranial pressure (ICP) or intracranial bleeding neither of which are managable in the wilderness. The job of the clinician is to diferentiate a serious life threatening injury from a minor one.

   1. Minor injury: No loss of consiousness (LOC), or LOC of less than 15 seconds with immediate return to full alertness. The casualty can not be on medications which increase risk of bleeding or have a history of bleeding disorders. Patient may be monitored every two hours for mental status changes, lethargy, irritibility, persistent nausea and vomiting, changes in speech or visual changes.

   2. Serious Injury: LOC greater than 15 seconds, and/or persisitent confusion or memory loss; signs or symptoms of increased ICP: Debilitating headache, mental status changes, persistent nausea and vomiting, appearance of clear fluid in external auditory canal, Battle sign, raccoon eyes, or seizures.

   3. Field treatment of serious head injury: Suspect injury of C-spine, manage airway-be able to clear vomit, elevate head 30 degrees. Evacuate to treatment facility ASAP.

4. Spinal Injuries.

A. Cervical Spine

1. High Risk Activities:

C-spine injuries in the wilderness usually occur after either a fall from a significant height or high-velocity accident from skiing. Common winter activities predisposing participants to C-spine injuries are skiing, snow-boarding, and snow-mobiling.

2. Anatomy:

The cervical spine consists of seven cervical vertebrae interposed between the base of the skull and the thoracic spine. The cervical spine has a great deal of mobility to allow maximal range of motion for the skull. This increased mobility comes at a high price; this portion of the spine is less stable than the rest of the vertebral column, and hence more subject to injury. The C-Spine is stabilized by three longitudinal ligaments: The anterior longitudinal ligament (ALL) runs longitudinally along the anterior surface of the vertebral bodies. It is broad and very strong and helps to prevent hyper-extension of the head and neck. The posterior longitudinal ligament (PLL) also runs longitudinally, this time along the posterior surface of the vertebral bodies, within the vertebral canal itself. This ligament is relatively narrow and somewhat weaker than the ALL, and helps to prevent hyper-flexion at the neck. Third is the supraspinous ligament, connecting the spinous processes. This, too, is a strong ligament, acting to inhibit hyper-flexion.

The peripheral nerves innervating the muscles and the sensory nerves of the upper extremities originate from the cervical portion of the spinal cord and these are some of the nerves likely to be affected in the event of a C-Spine injury. Hence, it is important to assess the neurologic status of the upper extremities if a cervical injury is suspected.

3. C-Spine Injury Statistics:

-Most common injury: Flexion injury at C5/6
(so look for a deficit in the C6 distribution)

-28% of C-Spine injuries have another spinal fracture associated with it.
(so examine entire spine)

-10% of Head injuries/Facial fractures also have a C-Spine injury, especially if there was LOC.

4. Remember:

a. Head injury -> Assume C-Spine injury.
b. Perform complete Neurovascular exam
   -Motor, Sensory, Reflexes, Pulses, Babinski reflex.

B. Clear a C-Spine. (FMST.07.14a)

When faced with a possible C-Spine injury in the Wilderness, clearing it without X-ray should be considered only in extreme situations. However, there are situations which demand that every effort be made to make a casualty ambulatory, because the movement of the individual would demand herculean effort from the rest of the team, potentially placing more individuals at risk of injury or prolonging exposure to extreme environmental conditions.

Clearing a C-Spine in the field should follow a three-phased assessment. If the patient fails the assessment at any one of the phases, then the team is obligated to maintain spinal precautions and transport the patient. The phases are outlined below, and expanded upon in the lecture.

PHASE ONE:

-ensure no alteration of mental status.
-maintain in-line traction.
-perform complete neurovascular exam (focusing on the upper extremities)
-palpate spinous processes one-by-one while asking casualty whether palpation causes pain (supraspinous ligament)
-distinguish between pain over spinous process and muscle soreness associated with paraspinal musculature.
-only if palpation was completely pain-free can you move on to ...

PHASE TWO:

-continue to maintain in-line traction
-palpate anteriorly to the left and right of trachea. along the anterior vertebral bodies (anterior longitudinal ligament).
-Assess for bony tenderness under the angle of the jaw to the clavicle.
-Distinguish between tenderness of the sternocleidomastoid muscle and actual vertebral pain.
-only if palpation was pain-free can you progress to...

PHASE THREE:

-while loosely maintaining in-line traction, have patient move head through active range of motion

*first flexion/extension
*followed by rotational movement

-stop test at the first indication of pain with movement

Note: If you are able to move the patient's head through a full range of motion without pain, you have effectively cleared his C-Spine. Have the patient resume activity slowly and cautiously; evaluate any other complaints noted.

B. Thoracic and Lumbar Spine:

-Look for a T-L Spine injury with calcaneal fractures. Approximtely 10% of patients with a calcaneal fracture will also have an associated lumbar fracture.

C. Pelvis:

-Think HEMORRHAGE/SHOCK if faced with a Pelvic Fracture
Up to 6 liters of blood loss possible – (internal iliac arteries)

Place gentle constricting wrap around pelvis if open book fracture is suspected. (FMST.07.14b)

4. FRACTURES.

Some common wilderness fractures:

a. Metacarpal/Phalangeal	-the hands are your means of interfacing with the environment, sometimes it's a forceful meeting.
b. Distal Radius	-Very common wilderness fracture.
c. Scaphoid	-check for snuffbox tenderness, pain with axial load to thumb.
d. Lunate	-also common, pain dorsally at base of 3rd MC.
f. Clavicle	-need good NV exam, check for pneumothorax.
g. Long bone	-Radius/Ulna: if proximal, check elbow -Humerus: radial nerve runs in spiral groove -Tib/Fib: think compartment syndrome- Fib fx may ambulate with cane -Femur: think Hemorrhage, think traction.
f. Ankle	-commonly a fracture/dislocation, check NV
g. Hip	- Leg will typically be externally rotated

REMEMBER:

- Always think HEMORRHAGE with Long Bone fractures
- Always perform NEUROVASCULAR Exam before and after treatment
- Always SPLINT AND PAD for stability, comfort, function and transport.
- Indications for REDUCTION are: NV deficit , severe angulation, severe pain, if angulation prediposes to open fx or makes transport difficult.
- OPEN FRACTURES: Have 6-8 hours to get to surgery. Field Tx- Gently wash off with betadine/ iodine solution to get off obvious dirt- don't scrub. Wrap in sterile gauze. Don't place exposed bone back under skin. Give antibiotics: Augmentin, 2nd or 3rd generation cephalosporin, a quinolone or tetracycline.

THE REASONS WE PLACE FEMUR FRACTURES UNDER TRACTION.

1. Re-establishing length tightens fascia and tamponades bleeding.
2. Dramatic pain relief.
3. Helps prevent open fractures.

4. Helps reduce secondary soft tissue damage.
5. Increases ease and safety of transport.

SOME BASIC RULES OF EXTREMITY SPLINTING:

- Splint all fractures before moving the casualty, unless in imminent danger.
- Splint all fractures as they are found, unless severe angulation complicates transport or is causing neurovascular deficits.
- Splint to include the joint above and below.
- Construct splints using uninjured extremity, then splint in the position of function, comfort or stability.

5. DISLOCATIONS.

a. Rapid Diagnosis and Reduction is imperative, if evacuation time is > 1-2 hours.

- Easier to reduce immediately after injury. (muscle spasm)
- Makes transport much easier. (increased patient comfort)
- Dramatic pain relief.
- Early reduction reduces risk of Neurovascular injury.
- Reduction could make difference between an ambulatory vs. litter patient
  the safety of the entire expedition could be placed at risk

b. Signs and Symptoms: Remember, NO X-RAYS!

- Decreased Range of Motion.
- Obvious Deformity. (compare right vs. left)
- Usually NO Crepitus.
- Typical Posture.

c. Reduction Techniques:

- Phalanges	Linear traction, buddy tape. Can not reduce volar displacement of 1st phalanyx.
- Shoulder	Bedsheet method- counter traction. Prone with weights method. External rotation. Snowbird technique.
- Patella	Extension with medially directed pressure.
- Ankle	Linear traction with knee flexion.

- Hip — Anchor pelvis, hip flexed at 30 degrees, upward traction.

- Sternoclavicular — Roll between scapulae, and apply forceful downward pressure.

-Knee — Should only be reduced as an absolute last resort. Apply anterior traction gently.

REMEMBER: Analgesia and Sedation should be utilized when available before attempting joint or fracture reduction, unless multiple traumatic injuries are present.

6. EVACUATION CRITERIA:

Conditions demanding expeditious evacuation: (FMST.07.14c)

1. Suspected cervical, thoracic, and/or lumbar spine injuries.
2. Pelvic injury with instability and/or significant blood loss.
3. Any open fractures. (6-8 hour window, antibiotics early)
4. Compartment syndromes. (pain with passive motion, pallor, pulseless, paresthesia)
5. Hip or Knee dislocations.
6. Any other injury the medical provider feels he/she is not prepared to manage.

7. OTHER TRAUMA.

Heavy bleeding: Usually can be treated by direct pressure, if a tourniquet is required,release pressure every five minutes to reassess.

Contusion: Apply ice 1st 48 hours which will provide pain relief and limit expansion of hematoma.

Lacerations and avulsions: High pressure irrigation with clean water, may remove debris with flame treated forceps. Use steri-strips for minor clean lacs. Keep dirty or puncture wounds open and give antibiotics.

Animal Bites: Copious irrigation, and antibiotic, do not close. Consider rabies if bite was unprovoked and animal was acting strange. (bats, raccoons, skunks, canines, and felines) Has never been seen in livestock, rabbits, squirrels, rats or mice. Most common zoonotic source in North America is the skunk.

Severed part:	Can be reattached up to 24 hours after injury if the cut is clean. Gently clean part, wrap in gauze and keep cool without freezing.

UNITED STATES MARINE CORPS
Mountain Warfare Training Center
Bridgeport, California 93517-5001

FMST.07.38
04/02/02

STUDENT HANDOUT

REPTILE AND ARTHROPOD ENVENOMATION BITES AND STINGS

TERMINAL LEARNING OBJECTIVE: Given a simulated casualty, in a wilderness environment, treat reptile and arthropod envenomation, in accordance with the references. (FMST.07.38)

ENABLING LEARNING OBJECTIVES:

1) Without the aid of references, select from a given list the first aid measures used for pit viper envenomation, in accordance with the references. (FMST.07.38a)

2) Without the aid of references, select from a given list the first aid measures used for coral snake envenomation, in accordance with the references. (FMST.07.38b)

3) Without the aid of references, select from a given list the first aid measures used for widow spiders (lactordectus species) bites, in accordance with the references. (FMST.07.38c)

4) Without the aid of references, select from a given list the first aid measures used for brown recluse spider (loxosceles species) bites, in accordance with the references. (FMST.07.38d)

5) Without the aid of references, select from a given list choose the first aid measures used for bee, wasp, and hornet stings, in accordance with the references. (FMST.07.38e)

12

OUTLINE.

1. **PIT VIPERS**. Crotalidae have a triangular head, cat-like vertical pupils, hinged fangs and a heat-sensitive "pit" on each side of the head between the tip of the nose and the eye. Rattlesnakes have a variable number of rattles depending upon age and number of molts. They sometimes strike without rattling. About 60% of this country's venomous bites are attributed to rattlesnakes. Cottonmouths (water moccasins) and copperheads are the other two commonly encountered North American pit vipers. Copperhead and cottonmouth venom are quite similar but weaker than most rattlesnake venom.

a. Anatomy of Pit Vipers.

(1) Venom apparatus: Venom gland; Compressor glandular muscle; Primary duct; Accessory gland; Secondary duct; Fang sheath; and fangs.

(2) Facial pit organs: Highly sensitive paired receptors of infrared radiation that can detect temperature changes of less than 0.2°C; used to detect warm-blooded prey/predators and aim strike; also pit organs may have a role in determining volume of venom injected.

(3) Rattles or buttons: Interlocking keratin rings at tip of tail in rattlesnakes; a new rattle is added with each molting cycle (every 50-400 days); some are lost due to trauma.

(4) The strike: Aimed primarily by facial pits; rarely strike farther than 1/2 their body length; speed - approximately 8 feet/second.

(a) Rattlesnakes generally inject 25-75% of their venom when they bite humans.

(b) Can bite without injecting any venom ("dry bite").

1. 20% - no envenomation.
2. 20-30% - mild envenomation.
3. 30-50% - moderate to severe envenomation.

b. Signs and Symptoms of Snake Venom Poisoning:

(1) Pit Vipers:

(a) Local:

1. Puncture wounds/scratches:

A. Pattern can be misleading (venomous vs. nonvenomous).

B. Must differentiate from other animal/insect bites or plant puncture wounds- usually venomous snakebite wounds are larger, with more bleeding.

C. May be one or many fang marks present; range from a few mm to 4cm apart.

2. Pain, edema, erythema:

A. Usually presents within 30 minutes if envenomation.

B. More severe envenomation generally yields more rapid progression and more severe pain.

C. Pain is usually characterized as burning and immediate in onset.

D. Pain usually confined to bite site.

E. Edema limited to subcutaneous tissues (with no increase in intra-compartmental pressures).

F. Usually starts within 5 minutes (if none in 10 minutes, probably no significant envenomation unless dealing with Mojave rattlesnake).

G. Gradually spreads up extremity for 36 hours.

H. Danger of airway obstruction in bites to head and/or neck.

3. Ecchymosis:

A. Ecchymosis starts within several hours.

B. May involve entire extremity.

4. Lymphangitis:

A. Venom absorbed rapidly through lymphatic system results in lymphatic inflammation.

5. Petechiae, vesicles, hemorrhagic bullae:

A. Will occur in untreated rattlesnake envenomation.

B. Rarely seen when treated early with adequate antivenin.

C. Usually occurs 6-36 hr after bite.

6. Necrosis, tissue destruction:

A. Due to direct venom effects.

B, Can be prevented if adequate antivenin is given within 2 hours of bite.

(b) Systemic:

1. Nausea, vomiting:

A. Common (early onset may indicate severe envenomation).

2. Weakness, diaphoresis, chills, dizziness/vertigo.

A. Syncope is common with all pit viper bites.

B. Frequency is proportional to severity of bite.

3. Change in taste:

A. May complain within minutes of rubbery or metallic taste.

4. Increased salivation.

5. Fever.

6. Tingling, numbness in scalp/face/fingers/toes:

A. Can occur within 10 minutes and indicates moderate to severe envenomation.

7. Fasciculations in face/neck/back/ or other involved extremity:

A. Can occur early and indicates severe envenomation.

8. Visual disturbances:

A. Blurred vision, yellowing of vision and blindness.

9. Tachycardia, bradycardia.

10. Hemorrhage, hemolysis, thrombosis and DIC:

A. Bleeding from wounds, gingival bleeding, epistaxis, Hematuria, hematemesis, Melena, lower GI bleeding, hemoptysis, peritoneal hemorrhage and cerebral hemorrhage.

B. Can occur as early as 6 hours; more common 12-72 hours after bite.

C. Systemic bleeding only occurs in moderate to severe envenomation.

D. Most coagulopathic effects secondary to proteases acting at various sites in the coagulation cascade.

11. Hypotension, shock:

A. Can occur rapidly in severe envenomation.

B. Early, due to pooling of blood.

C. Late may be due to loss of volume.

D. Generally little effect from decreased cardiac contractility.

12. Pulmonary edema:

A. Common in severe envenomation.

B. Secondary to toxin and inflammatory injury of pulmonary capillary membranes and pooling of blood in major vessels and capillary beds of the lung.

C. Compounded by direct cardiodepressant factor in some venoms.

13. Oliguria, anuria:

A. Hypotension with resultant decreased GFR. (#1).

B. Hemoglobinuria and myoglobinuria with renal tubular obstruction.

C. Direct toxic effect of venom on kidneys.

D. Acute tubular necrosis from rhabdomyolysis.

14. Paresis, Paralysis:

A. Seen with Eastern Diamondback and Mojave rattlesnakes.

B. Uncommon with other rattlesnakes.

15. Respiratory failure:

A. Uncommon with pit viper bites (except Mojave).

B. Complicated by cardiovascular failure.

16. Seizures:

A. Probably due to hypertension and hypoxia.

17. Coma:

A. Usually sensorium is clear.

B. When coma occurs, it is secondary to cerebral anoxia.

18. Death:

A. Generally occurs in 6 to 48 hours in untreated severe envenomation.

B. Can occur sooner with IV envenomation.

c. First Aid Measures for Pit Viper Envenomation. (FMST.07.38a)

(1) Do not try to kill the snake and bring it in for identification.

(a) Only vital to identify the snake when Mojave rattlesnake or coral snake bites are a possibility, as management is altered.

(b) Risk of a second bite to the victim or rescuer.

(c) Should never delay transportation of the victim.

(2) Remove any jewelry, which could become a tourniquet with progression of edema.

(3) Calm and reassure patient to decrease heart rate and circulation of venom.

(4) Constriction band.

(a) Most widely accepted first aid measure in the literature.

(b) Should be more than 1/2 inch wide.

(c) Apply to proximal end of extremity affected and wrap to approximately 5 cm above swelling (site of envenomation). Then wrap back towards proximal end of extremity from where the wrap started.

(d) Apply only tight enough to occlude lymphatic and superficial venous return.

(e) Probably no benefit when applied 30 min or more after bite.

(f) Maintain until antivenin is started.

(5) Incision and suction.

(a) Very controversial; should only be used by experienced medical personnel.

(b) Do not incise wound.

(c) Suction best applied by mechanical means (Sawyer extractor).

(d) Do not use mouth (unknown oral lesions may expose rescuer to risk).

(e) Should be started within 30 minutes of envenomation.

(f) Continue suction for 30 to 60 min.

(6) Rest and immobilization.

(a) Put victim at rest and splint extremity in position of function, at or just below heart level.

(b) Allow room for swelling in splint.

(7) Ice.

(a) Avoid any method of cooling.

(8) Watch for adverse reactions.

(a) Rare documented cases of anaphylaxis caused by snake venom in patients previously bitten or otherwise exposed to snake venom.

(b) Some evidence that venom can activate the alternate-complement system directly and cause a similar reaction.

(9) Rapid transport to a medical facility.

(a) The most important measure in first aid.

(b) No measures should delay this; antivenin is the only proven definitive therapy.

2. **CORAL SNAKES**. The elapidae species of coral snakes are brightly colored, with black noses and alternating red-yellow-red-black bands around their bodies (remember "red on yellow, kill a fellow"). They have relatively small mouths with fixed fangs. From southern Mexico through tropical South America, the rules for distinguishing coral snakes are highly unreliable. Unless you are a knowledgeable herpetologist, it is best not to pick up colorful snakes in the tropics.

a. Signs and Symptoms of Coral Snake Envenomation:

(1) Fang marks may be hard to see.

(2) Frequent delay in onset of symptoms, followed by extremely rapid progression.

(3) Little or no pain at bite site.

(4) No local necrosis.

(5) Earliest evidence may be drowsiness or euphoria.

(6) N/V, increased salivation, paraesthesias at bite site.

(7) Bulbar paralysis progressing to peripheral paralysis.

(8) Paraesthesias, and fasciculation.

(9) Occasional seizures.

(10) Mild Hypotension.

(11) Death is due to respiratory and cardiovascular failure.

b. First Aid Measures For Coral Snake Envenomation. (FMST.07.38b)

(1) None of proven benefit except rest, reassurance and rapid transportation to medical facility for antivenom administration.

3. **GILA MONSTERS**. No direct injection mechanism for their venom, but have powerful jaws that allow them to chew and tear at their victims. Drooling venom produces a substantial amount of pain and possible complications resembling pit viper envenomation. Follow the recommendations for pit viper envenomation. Fatal Gila monster encounters are extremely rare.

4. **VENOMOUS ARTHROPODS**. Arthropods represent the largest phylum in the animal kingdom. It contains spiders, scorpions, insects, ticks, kissing bugs, water

bugs, caterpillars, moths, butterflies, grasshoppers, centipedes, and millipedes, among others. Some arthropods sting (bees, ants, scorpions, etc.), others bite (spiders, centipedes, kissing bugs, etc.), while still others discharge a secretion that is toxic (millipedes, caterpillars, etc.). Although the number of arthropods that are sufficiently venomous to endanger humans is not known, these animals are implicated in far more bites and stings than all other phyla combined. Almost all of the 20,000 species of spiders are venomous, but luckily for, man, only relatively small numbers have fangs long and strong enough to penetrate the human skin. There are some 500 species of scorpions and all are venomous, although only a small number pose a significant danger to humans.

a. The number of deaths from arthropod stings and bites is not known, nor do most countries keep records of the incidence of such injuries. In Mexico, parts of Central and South America, and in North Africa, deaths from scorpion stings may exceed several thousand per year. Spider bites probably account for no more than 200 deaths per year, worldwide. The number of deaths from arthropod bites or stings in the temperate countries is far greater than the number of deaths from snakebite. Almost all of these deaths, however, are from anaphylactic reactions.

5. **<u>SPIDERS</u>**. -There are at least 200 species of spiders that have been implicated in significant bites to humans (Russell, 1988a). Spiders got the misplaced blame for the bites seen in one large series of case studies. Of some 600 suspected spider bites, 80% were found to be caused by other arthropods. The arthropods most frequently involved in these mis-diagnoses were ticks (including their embedded mouth parts), kissing bugs, mites, bedbugs, fleas, flies, beetles, water bugs and various other stinging arthropods.

a. Lactordectus species (Widow spiders): It is most commonly referred to as the Black Widow, Brown Widow or Frog-legged spider depending on the species. There are many other commons names such as: hourglass, poison lady, deadly spider, red-bottom spider, T-spider, Gray Lady spider, or shoe-button spider. Although both male and female widow spiders are venomous, only the latter have fangs large and strong enough to penetrate the human skin. Mature females range in body length from 10 to 18mm, whereas males are from 3 to 5mm.

(1) Signs and Symptoms of spider envenomation.

(a) Sharp, pinprick-like bite, but in some cases the incident is so minor that it goes unnoticed.

(b) Initial bite-pain is often followed by dull-numbing pain in the affected extremity, and cramps in one of the large muscle masses.

(c) Muscle fasciculations can be seen within 30 minutes of the bite.

(d) Sweating is common.

(e) Complaints of weakness and pain in the regional lymph nodes.

(f) Pain in the lower back, thighs, or abdomen is common and rigidity of the abdominal muscles is seen in severe envenomations.

(g) Severe paroxysmal muscle cramps may occur accompanied by arthralgias.

(h) Tissue reaction around the bite may vary, depending on the species involved. In most cases in the United States, there is no reaction at the bite site and the puncture wounds may be impossible to find. There may be an immediate 2 to 4 mm blanched area around the puncture wounds, surrounded by a slightly erythematous area, which in time becomes pallid. This may persist for 20 to 30 minutes.

(i) Hypertension is a characteristic finding in moderate to severe poisonings. It usually appears 2 to 3 hours post envenomation.

(j) Elevation of body temperature.

(k) Increased salivation.

(1) Anorexia.

(2) Treatment for Widow Spider envenomation.
(FMST.07.38c)

(a) First aid measures are supportive in nature. If the pain is intense, ice can be placed over the wound until the patient arrives at a medical facility.

(b) Antivenin is given in severe cases at medical facilities.

b. Loxosceles species (Violin spiders): Are known in the United States as fiddle-back, or brown recluse spiders. There are over 100 species of Loxosceles. The abdomen of these spiders varies in color from gray to orange or reddish-brown to dark brown. The violin on the cephalothorax is brown to blackish and distinct from the pale yellow to reddish-brown background of the cephalothorax. This spider has three pairs of eyes, forming a curved row. Females average 8 to 15 mm in body length whereas males average 6 to 10 mm. Both are venomous.

(1) Signs and Symptoms of Violin spider envenomation.

(a) Pain; about the same degree of a bee sting but patients may be completely unaware of the bite.

(b) Localized burning sensation around the site of injury.

(c) Pruritus is often present and the area begins to appear red, with a small-blanched area around the immediate bite site.

(d) The reddened area enlarges during the subsequent 1- 8 hours. It often becomes irregular in shape and, as time passes, hemorrhages may develop throughout the area.

(e) A small bleb or vesicle forms at the bite site and increases in size.

(f) Vesicle subsequently ruptures and a pustule forms.

(g) The whole area becomes swollen and painful.

(h) Lymphadenopathy may develop.

(i) Necrosis can develop from superficial to deep layers.

(j) Rare systemic signs and symptoms include:

1. Malaise.
2. Stomach cramps.
3. Nausea and vomiting.
4. Jaundice.
5. Spleen enlargement.
6. Hemolysis.
7. Hematuria.
8. Thrombocytopenia.

(2) Treatment for Loxosceles spider envenomation: (FMST.07.38d)

(a) There are no first-aid measures of value. In fact, all first-aid measures should be avoided, because the natural appearance of the lesion is most important in determining the diagnosis.

(b) An ice pack may be placed over the wound if the pain is severe.

(c) Excision of the bite site is no longer recommended as this has been demonstrated to cause worsening of necrosis and tissue damage. Definitive therapy includes use of dapsone, antibiotics, tetanus prophylaxis and possible steroid use.

c. There are many other species of spiders that include Steatoda species (cobweb spiders), Phidippus species (jumping spiders), Chiracanthium species (running spiders), Lycosa species (wolf spiders), and the newest member to the United States is the Hobo spider. There have been several reported deaths in the Northwestern United States. Originally a European species of brown spider, the Hobo spider was introduced into the U.S. in 1936, and little research has been done on it to date.

6. **SCORPIONS**. Approximately 75 of the 650 species of scorpions can be considered of sufficient importance to warrant medical attention. In the United States, members of the genera Hadrurus and Vejovis are capable of inflicting a painful and erythematous, but non-lethal lesion. However, species of the genus Centruroides are sufficiently dangerous to warrant definitive medical care. Most stings are inflicted by the Vejovis species.

a. Centruroides species. There are approximately 30 species of this genus confined to the New World. Of these, eight are of considerable medical importance, and most of these species are found in Mexico. In the United States, they are commonly referred to as "bark scorpions", because of the preference for hiding under the loose bark of trees, in dead trees, or in logs. Their general color is straw to yellowish-brown or reddish-brown, and they are often easily distinguishable from other scorpions in the same habitat by their long thin tail with its "telson" on the end, the pedipalps, or pincer-like claws. They can reach lengths up to 55 mm.

(1) Signs and symptoms of envenomation.

(a) Initial pain, rarely severe.

(b) Area becomes sensitive to touch (tap test).

(c) Tachycardia may ensue after one hour.

(d) Respiratory rate increases.

(e) Fasciculations may be seen over the face or large muscle masses.

(f) Complaints of generalized muscle weakness.

(g) Respiratory distress may progress to paralysis.

(h) Slurring of speech may be present.

(i) Convulsions have been reported.

(2) Treatment for scorpion envenomation.

(a) There is no first-aid measure of value for scorpion stings.

(b) Ice may be placed over the wound to reduce pain.

(c) In severe cases, antivenin may be recommended.

b. Vejovis species is closely related to the Urectonus. Both genera are ground scorpions. Whereas Urectonus usually inhabits mountain habitats from southern California to Oregon, Bejovis has a wide distribution from the southern portions of Canada, south through Wyoming and Colorado to Texas and west to California. Some species measure up to 85mm, but most are from 30 to 35mm. Vejovis spinigerus, a common western U.S. species, is often called the striped-tail scorpion.

c. Hadrurus species is native to North America. These are the longest and most stout-bodied of our scorpions, generally referred to as the giant hairy scorpions because of their size and the conspicuous bristles on their legs, pedipalps and caudal segments. Adults may can measure up to 135 mm. These are burrowing species and may be found as deep as 2 feet in sandy soil. They are native to Arizona, California and parts of Utah, Nevada and Idaho, as well as Mexico.

7. **BEES, WASPS, HORNETS AND ANTS**. While it may take over 100 bees to inflict a lethal dose of venom in most adults, one sting can cause a fatal anaphylactic reaction in a hypersensitive person. There are 3 to 4 times more deaths in the United States from bees than from snakebites. In the few fatalities that have resulted from multiple bee stings, death has been attributed to acute cardiovascular collapse.

a. Treatment for bee stings: (FMST.07.38e)

(1) The stingers of many bees may remain in the skin and should be removed by teasing or scraping rather than pulling, to avoid even more venom being injected into the patient.

(2) Ice may be applied to the sting site.

(3) Persons with known hypersensitivity to stings should carry a Bee-Sting Kit, containing an antihistamine and epinephrine when in endemic areas.

(4) Desensitization can be carried out using whole-body antigens or, preferably, whole-venom antigens.

8. **TICKS AND MITES**. Carry many diseases. In North America, some species of Dermacentor and Amblyomma cause tick paralysis.

a. Signs and symptoms of Tick Paralysis.

(1) Restlessness and irritability.

(2) Lethargy and anorexia.

(3) Generalized muscle weakness.

(4) Incoordination and ataxia.

(5) Nystagmus.

(6) Ascending flaccid paralysis with loss of deep tendon reflexes.

(7) Bulbar or respiratory paralysis may develop.

b. Treatment for Tick Paralysis:

(1) Remove the tick.

(2) Treat symptomatically (ventilator support may be necessary).

(3) Lesions should be cleaned.

(4) Antivenin and/or corticosteroids are used in severe cases.

9. **LEPIDOPTERA (CATERPILLARS)**. Caterpillars commonly cause envenomation in humans. There are at least ten families of caterpillars that are venomous.

a. MEGALOPYGE OPERCULARIS (Puss Caterpillar or Woolly Slug) The most commonly seen venomous caterpillar in the United States.

(1) Description: Hairy, flat, ovoid shaped: 30-35 millimeters in length.

(2) Location: Texas north to Maryland and Missouri.

(3) Stinging apparatus: Consists of spines intermingled with the hairs of the body:

b. SIGNS AND SYMPTOMS.

(1) Needling pain (Can be intense).

(2) Redness.

(3) Swelling.

(4) Nausea.

(5) Headache.

(6) Fever.

(7) Vomiting.

(8) Lymphadenopathy.

(9) Hypotension.

(10) Shock.

(11) Bleeding of the mucus membranes.

a. Treatment

(1) Apply contact tape to affected area, and remove tape. This will effectively remove spines from affected area.

(2) Group I corticosteroid cream and ointment.

(3) Codeine or Demerol may be required for pain.

(4) Antiemitics may be needed to control vomiting.

Note: Anaphylactic reactions have been reported, but are rare.

UNITED STATES MARINE CORPS
Mountain Warfare Training Center
Bridgeport, California 93517-5001

FMST.07.10

04/02/02

STUDENT HANDOUT

FREEZING AND NEAR FREEZING TISSUE INJURIES

TERMINAL LEARNING OBJECTIVE: Given a casualty in a cold weather environment, and necessary equipment and supplies manage common cold weather injuries in a cold weather environment to prevent death or further injury per the reference. (FMST.07.10)

ENABLING LEARNING OBJECTIVES.

1) Without the aid of references, from a given list select the correct definition of frostbite, in accordance with the references. (FMST.07.10a)

2) Without the aid of references, from a given list select the correct two mechanisms of injury for frostbite, in accordance with the references. (FMST.07.10b)

3) Without the aid of references, from a given list select the correct signs and symptoms of frostbite, in accordance with the references. (FMST.07.10c)

4) Without the aid of references, from a given list select the correct favorable post thaw signs of a frostbite injury, in accordance with the references. (FMST.07.10d)

5) Without the aid of references, from a given list select the correct unfavorable post thaw signs of a frostbite injury, in accordance with the references. (FMST.07.10e)

6) Without the aid of references, from a given list select the correct field treatment for frostbite, in accordance with the references. (FMST.07.10f)

12

7) Without the aid of references, from a given list select the correct primary field treatment consideration when treating frostbite in the field, in accordance with the references. (FMST.07.10g)

8) Without the aid of references, from a given list select the correct four methods to reduce the chance of becoming a frostbite casualty, in accordance with the references. (FMST.07.10h)

9) Without the aid of references, from a given list select the correct definition of immersion foot, in accordance with the references. (FMST.07.10i)

10) Without the aid of references, from a given list select the correct mechanism of injury for immersion foot, in accordance with the references. (FMST.07.10j)

11) Without the aid of references, given a list of categories and characteristics match the category of immersion foot to its characteristics, in accordance with the references. (FMST.07.10k)

12) Without the aid of references, from a given list select the correct field treatment of immersion foot, in accordance with the references. (FMST.07.10l)

OUTLINE.

1. **FROSTBITE**

a. Definition: Frostbite is the actual freezing of tissue. (FMST.07.10a)

(1) Frostbite usually occurs as a result of exposure to subfreezing temperatures for a long period of time. However, brief exposure to high wind-chill or very cold temperatures can also cause frostbite. Frostbite usually occurs after more than twelve hours of exposure to subfreezing temperatures. Skin temperature of the extremities is dependent on heat production and is more a function of underlying muscle mass than of circulation. Thus, the areas of the body most prone to frostbite (hands, feet, ears, nose, chin and molar regions) have small amounts of underlying muscle tissue.

(2) Incidence. Frostbite, unlike hypothermia tends to spare the extremes of age. Thus, it is most prevalent in the middle years of life and like hypothermia, the greatest incidence occurs in urban areas. Other factors predisposing a person to frostbite include: poor nutritional status, dehydration, prior cold injury, individuals from the southeastern part of the United States, diabetics, individuals with atherosclerotic disease, and use of tobacco products.

(3) Although the actual numbers are not known, persons born in the south are 3.7 times more susceptible to frostbite. Persons with a history of prior cold injury are 1.7 times more prone to have a second injury. This includes parts of the body other than the originally frostbitten region.

b. Mechanisms of Tissue Injury. Tissue injury resulting from frostbite is produced in two ways: (FMST.07.10b)

(1) *Actual freezing of the tissues.* Ice crystal formation in the extra-cellular space leads to an osmotic gradient across the cell membrane drawing water out of the cell. This process, known as crenation, causes a marked change in intra-cellular electrolyte and enzyme concentrations, which can lead to cell death or recovery depending on the duration of exposure to freezing conditions and post thaw care of the tissue. Intra-cellular ice crystal formation is much more destructive, however, this requires much faster freezing rates.

(2) *Obstruction of blood supply to the tissues.* Cell injury and death caused by ice crystal formation leads to the release of inflammatory mediators, which damage vascular endothelial cells. The net result is increased vascular permeability. Fluid leaks out of the vascular space and into the intercellular space causing tissue edema and increased blood viscosity. Elevated viscosity combined with platelet aggregation, to the walls of damaged blood vessels, lead to vascular obstruction with resultant tissue damage.

c. Signs/Symptoms. (FMST.07.10c) Initially there is discomfort and pain in the affected tissues, which can progress to loss of sensation. The skin appears white and bloodless. There is decreased or no capillary refill. The skin may feel waxy and will not glide freely over a bony prominence. With minor injury, blisters form, usually in the first 24 hours, and edema may also be present. Deep injury is not associated with substantial formation of blisters. There is also little edema seen in deep injury.

(1) Favorable prognostic signs of frostbitten tissue. These signs indicate probably little or no tissue loss, post thaw. (FMST.07.10d)

(A) Warmth of tissue.

(B) Normal tissue color.

(C) Preservation of sensation.

(D) Blister formation. *A* blister extending the full distal length of the injured digit is a good prognostic sign. However, a blister limited to the line of demarcation between normal and frostbitten tissue is a less favorable sign.

(E) Edema persisting more than 24 hours.

(F) Rapid capillary refill.

(2) Unfavorable prognostic signs. These indicate a poor prognosis with probable tissue loss, post thaw: (FMST.07.10e)

(A) Complete absence of edema.

(B) Cyanotic tissue.

(C) Continued loss of sensation.

(D) Affected areas continue to stay cold.

(3) Since it is extremely difficult to predict ultimate tissue loss from initial examination, any classification must be done retrospectively. Classically, frostbite is divided into four degrees of severity:

(A) First degree - Erythema, edema, transient tingling and/or burning.

(B) Second degree - Blisters, edema, anesthesia and/or paresthesias.

(C) Third degree - Involves the entire thickness of skin and extends into subcutaneous tissue.

(D) Fourth degree - Involves the entire thickness of the skin along with muscle and bone.

(F) Frostnip is a completely reversible injury that is technically not considered frostbite. Frostnip classically presents as a white patch and numbness and is readily warmed in the field by placing the affected extremity in one's axilla or groin. No resultant skin changes occur.

d. Field Management (FMST.07.10f)

(1) The correct treatment for frostbite is as follows:

(A) Once an extremity has suffered frostbite, the primary consideration of treatment is not to rewarm the region if there is a chance of refreezing. (FMST.07.10g) This occurs because the amount of tissue damage is greatly increased if the part is thawed then allowed to re-freeze. Generally, the sooner a frostbitten region is thawed the lower the probability and extent of damage. Rapid rewarming is the preferred method of thawing a frostbitten region. This consists of immersion of an affected part in circulating warm water of 104°-108°F until the tissues are soft and pliable, which usually takes about 40-60 minutes.

(B) If slow rewarming methods (such as allowing a part to sit at room temperature) is used, some refreezing of the melted fluid occurs and the new ice crystals are actually larger than those formed during the original freezing causing further tissue destruction.

(C) Out dated methods such as rubbing with snow or rewarming next to an open fire, etc., only enhance tissue damage.

(D) Once thawing has taken place no benefit of continued rapid rewarming can be demonstrated.

(E) Cold injuries that are old and have thawed spontaneously should not be rapidly re-warmed again.

(F) Nicotine use should be prohibited since it is a vasoconstrictor and may enhance tissue loss.

(G) There is no need for ointments, salves, etc.

(H) Blisters should be allowed to remain intact.

(I) Medevac all cases of frostbite, as soon as the tactical situation permits.

(J) One final note; in military tactical situations such as ours, the victim often must walk down below the snow line to reach a BAS or base camp where rapid rewarming can be carried out. In this situation let the victim walk if he has frostbitten feet. Once his feet have thawed, he will be unable to walk and will have to be carried out.

(K) Hospital and long-term care: After re-warming the frostbitten tissue, attention must turn to prevention of infection and functional restoration. The patient's tetanus status should be determined and meticulous care should be taken to keep the injury clean. Daily whirlpool therapy for 30-45 minutes has been shown to be beneficial. The use of post-thaw medications is controversial but there is evidence that topical aloe vera and NSAIDS to control inflammation are probably beneficial. Some studies have demonstrated benefit from the use of allopurinol as a free radical scavenger. Also, animal studies have shown a definite benefit from the use of 90% DMSO gel, however, this use is not FDA approved on humans and it may be difficult or impossible to obtain this medication. The wide scale use of heparin and dextran 40 has been suggested but a definitive recommendation that supports their use still cannot be made. Studies suggesting the use of thrombolytics are very small and not well controlled therefore there is no current recommendation for use of thrombolytics. Lastly, it has been shown that there is a clear benefit for the use of early and aggressive physical therapy to preserve function and prevent contracture.

e. Prevention (FMST.07.10h)

(1) Some general principles for the prevention of frostbite are listed below

(a) Dress to keep comfortably cool. The head radiates 30-50% or more of all body heat and so goes the old adage, "If your feet are cold put on your hat".

(b) Nutrition. Eat the right type and quantity of foods.

(c) Avoid tight fitting and constrictive clothing. This decreases blood flow and increases the chance of frostbite. Rings, jewelry, etc., conduct heat away from the skin and increase the chance of frostbite.

(d) Wear mittens instead of gloves. They are much warmer. For work involving a high level of dexterity use contact gloves.

(e) Do not touch metals with bare hands at extremely low temperatures. Your flesh will stick to the surface and this will result in subsequent frostbite and tissue loss. If you find yourself in such a predicament pour a warm fluid over the metal to thaw your flesh off the metal.

(f) Avoid nicotine.

2. **IMMERSION FOOT**.

a. Definition. (FMST.07.10i) Immersion foot (AKA trenchfoot) is a non-freezing injury of the extremities in which the tissues are damaged.

b. Usually a disease of the feet (but it can occur in the hands), it occurs when the extremity is exposed to a cold-wet environment (external temperatures above freezing and less than

50°F) for long periods of time (usually greater than 12 hours). Prolonged standing enhances the pathologic potential for injury due to wet and cold conditions.

c. Mechanism of Injury. (FMST.07.10j) Even though this injury is called immersion foot, actual immersion of the feet in water is not necessary. This injury can also occur in the hands.

(1) The circulation of blood to the affected extremity is reduced because the victim's extremity is cold, wet and the peripheral blood vessels have constricted to reduce overall heat loss.

(2) Because prolonged cooling damages sensitive tissues, nerve injuries are also involved. This nerve damage is responsible for the severe pain and paresthesias. Damaged skin and subcutaneous tissues result in erythema, edema and possible blister formation.

d. Signs/Symptoms Early signs include numbness and paresthesias followed by complete numbness and sometimes leg/arm cramps. The skin at first appears erythematous but becomes progressively pale and mottled then grayish-blue as the condition progresses. Often there is a large amount of edema and blister formation. In severe cases the arterial pulses are absent. Immersion foot is classified by severity: (FMST.07.10k)

CATEGORY	CHARACTERISTICS
Minimal	Reddening of skin; slight sensory changes
Mild	Edema; sensory changes (reversible)
Moderate	Edema, redness, blisters, intra-cutaneous hemorrhage; irreversible nerve damage
Severe	Severe edema, massive intra-cutaneous hemorrhage; necrosis, gangrene

e. Field Management (FMST.07.10l) Treatment consists of:

(1) Pat drying of extremity.

(2) Gentle rewarming.

(3) Elevation of affected extremity.

(4) Bed rest.

(5) Treatment after this is supportive. For more severe injuries, debridement or resection of a portion of the foot may be necessary. However, this is best accomplished at a higher echelon of care. Do not be fooled by this injury. These patients can get

extremely ill from wet-gangrene which may necessitate amputation of tissue. However, the initial treatment should be as conservative as possible. DO NOT TREAT THESE PATIENT'S WITH WATER IMMERSION THERAPY. If infection is present obtain surgical consult early, and plan on small sequential amputations.

3. **CHILBLAINS (PERNIO)**.

   a. Definition. Chilblains are a non-freezing injury that leads to red inflamed, pruritic skin.

   b. Mechanism of Injury. Pernio is caused by intermittent exposure to above freezing temperatures and high humidity. (More extreme cold is usually associated with lower humidity). Classically, chilblains occur in young individuals and most commonly on the anterior surface of the legs and the dorsum of the hands. The cause of pernio is not clearly known, but it is thought to reflect an abnormal vascular response to the environment.

   c. Signs/Symptoms. Skin is red, tender, and pruritic, edematous and blanches with pressure. Itching may be severe and hyperhydrosis can occur.

   d. Management. The field treatment is massage and gentle heat. This treatment is usually self-limiting and will not recur without re-exposure. Topical steroids are effective for itching and burning.

4. **CORNEAL FROSTBITE**

   a. Definition. Actual freezing of the cornea.

   b. Mechanism of injury. This occurs when large amounts of cold air pass over and cool the cornea, resulting in corneal clouding and loss of vision. This can occur as the result of even short periods of exposure to activities associated with high wind chill i.e. snowmobiling.

   c. Signs/Symptoms. On examination the cornea will be clouded over.

   d. Treatment and Prevention. Antibiotic ointment and DO NOT PATCH THE EYES. Prevention is the key.

UNITED STATES MARINE CORPS
Mountain Warfare Training
Center Bridgeport, California
93517-5001

FMST.07.22
04/03/02

STUDENT HANDOUT

LAND NAVIGATION REVIEW

TERMINAL LEARNING OBJECTIVE In a mountainous environment, navigate in mountainous terrain, in accordance with the references. (FMST.07.22)

ENABLING LEARNING OBJECTIVES

1) Given a map and compass, orientate the map, in accordance with the references. (FMST.07.22a)

2) Given a map and a protractor locate a six-digit grid coordinate, in accordance with the references. (FMST.07.22b)

3) Given a map, compass, and protractor perform a resection to locate an unknown point from three known points, in accordance with the references. (FMST.07.22c)

4) Without the aid of references, select from a given list the correct definition for true north, grid north, and magnetic north, in accordance with the references. (FMST.07.22d)

OUTLINE.

1. **THE MAP**. Before you can properly use a map, there is some basic information you will need to understand.

a. Definition. A map is a reduced or scale drawing of the ground and important things on the ground as seen from the air. Essentially it is a picture of the surface of the earth, as it would appear looking at it from an aircraft.

b. Marginal Information. As you take a quick look at the borders, you will notice that there is a large amount of information. When using a map, the instructions are placed around the outer edges of the map and are known as marginal information. All maps are not the same, some having more marginal information than others, and the information being located in different places around the border. It is important that you know how to use this information if you plan to get the most out of your map. The different information on the margins is as follows:

(1) Sheet name. The sheet name is found in two places: the center of the upper margin and lower right margin of the map. The map is usually named after it's outstanding cultural or geographical features. Whenever possible, the name of the largest city is used.

(2) The sheet number. The sheet number is found in the upper right margin of your map and is used as a reference number for the map sheet.

(3) Series name and scale.

(a) The series name is found in the upper left margin of your map. A map series usually comprises a group of similar maps on the same scale and on the same sheet lines or format designed to cover a particular geographical area. It may also be a group of maps designed to serve a common purpose. The name given a series is of the most prominent area.

(b) The scale notation represents the ratio of the maps distance to the corresponding (actual) distance on the earth's surface. For example, the scale notation of 1:50,000 indicates that one unit of measure on the map equals 50,000 of the same units of measure on the actual surface of the ground (1 map inch = 50,000 earth inches).

(4) Series number. The series number appears in the upper right margin and lower left margin. It is a comprehensive reference expressed either as a four digit number or as a letter followed by three or four numbers.

(5) Index to adjoining sheets. The index to adjoining sheets is found in the lower right margin of your map.

(6) Bar scales. The bar scale is located in the center lower margin of your map.

(7) Contour interval. The contour interval is located in the center lower margin of your map, just below the bar scales. It states the vertical distance between adjacent contour lines on the map. For example, every contour line represents an increase in elevation as indicated by the distance of the contour interval.

(8) Northern References. There are 3 different northern references located on the declination diagram. It is located in the right portion of the lower margin of your map.

(a) True North. A line from any position on the earth's surface to the North Pole, which is the exact top of the earth.

(b) Grid North. The north direction established by the vertical grid lines on the map.

(c) Magnetic North. The direction to the North Magnetic Pole.

(9) Legend. Is located in the lower left margin. It illustrates and identifies some of the symbols on the map. There is not enough room on a map sheet to show the true outlines of objects and land features so the mapmaker uses a set of standard symbols to represent them. Sometimes the symbols are not the same, or standard types of symbols typically used. This is because of the type of map, the scale, or the origin. Because of this possible difference, to prevent errors in symbol identification, the legend should be referred to every time a map is used. (The example on the legend may state, "unimproved road, hard surface road, jeep trail," etc. The right side of the symbol will have three symbols; the first being "unimproved road", the second "hard surface road", and the third being "jeep trail".)

(10) Map colors. The colors you see on your map are not there to make the map pretty or enjoyable to the user. They have important meaning. The colors vary with different types of maps, but on a standard, large scale map, there are five basic colors:

(a) Black. Represents most of the man-made features.

(b) Red. Represents major roads.

(c) Blue. Represents water.

(d) Green. Represents vegetation.

(e) Brown. Represents all terrain features, such as hills, draws, fingers, depressions, and saddles by the use of contour lines. These contour lines are shown in brown.

(f) Other Colors. Occasionally other colors are used to show special information. These, as a rule, will be indicated in the marginal information. Currently, there are changes being made to the color system, which may result in changes of the marginal information in regards to map colors.

2. **HORIZONTAL SCALES**.

a. Bar Scale. This special ruler, located at the bottom of the map sheet, enables you to measure a distance between 2 points on your map and change that map distance into the actual ground distance.

(1) The primary scale. The primary scale is the part of the scale that starts at the zero (0) and extends to the right.

(2) The extension scale. The second part of the bar scale is the extension scale. It starts at the zero (0) and extends left.

b. Measuring Straight-Line Distance. If you are given a mission that requires you to move from your present position to a different position some distance away and you plan to travel in a straight line between these two points, the following method is used to determine the distance you will have to travel:

(1) First locate your present position and your destination on your map. Then you will lay a straight edged piece of paper on the map so that the edge of the paper touches the location of your present position and the location of your destination. Make a tick mark on the edge of the paper at each point. Now move the paper down to the bar scale. Place the left tick mark on the zero (0) of the scale so that the other tick mark will fall somewhere on the primary scale.

c. Measuring a Curved Line Distance. If you are given a mission that requires you to move from your present position to a different position and this time you choose to move along a road to reach your destination, the following method is used to work out the ground distance you will have to travel:

(1) Find your present position, your destination, and the road you will follow, on your map. Position a straight edged piece of paper at your location and make a tick mark,. Now, rotate the paper so that it runs along the road until the road turns, and make a tick mark on both the paper and the map. Keeping these 2 tick marks together, rotate the paper again until the road turns, and again mark the paper and the map. Continue this procedure until you have reached your destination. Now using your bar scale as previously stated, find the actual ground distance your squad will be traveling.

3. **ELEVATION AND RELIEF**. In order to begin learning the subject of elevation and relief, let's first look at some definitions.

a. Contour lines. Of the several ways to indicate elevation and relief on maps, the most common is through the use of contour lines. A contour line represents an imaginary line on the ground along which all points on this line are the same elevation above sea level.

b. Characteristics of Contour Lines. There are a few characteristics of a contour line that you should know that will help you understand what they represent on your map. These characteristics are as follows:

(1) Contour lines never cross.

(2) Contour lines connect on themselves.

c. Types of Contour Lines. In addition to knowing the definition, purpose, and characteristics of a contour line, you need to realize that there are different types of contour lines.

(1) Index contour lines. Starting at zero elevation, every 5th contour line is drawn with a heavier line.

(2) Intermediate contour lines. These are the contour lines falling between the Index contour lines.

(3) Supplementary contour lines. These lines are shown as dashed brown lines.

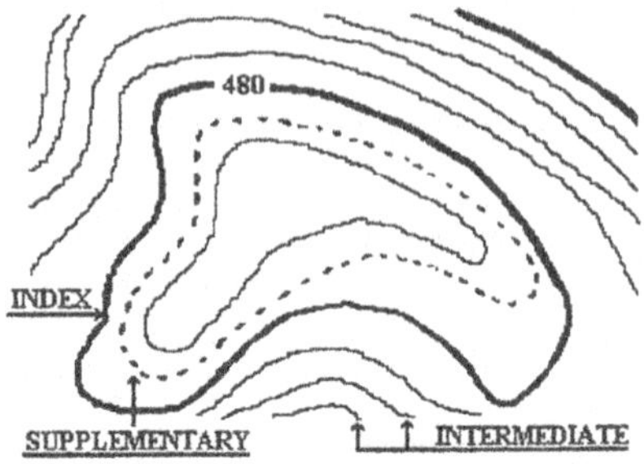

d. Determining Elevation of a Point. Now, you are ready to learn how to determine the elevation of any point of the ground represented on your map.

(1) Given elevation nearest the point, examine the area around your point until you find the nearest point that has a marked elevation. There are a few different ways that contour lines are marked, and they are:

(a) Labeled index contour line. Index contour lines are broken with the elevation of the line printed in the break.

(b) Spot elevations. At various places on your map, usually on prominent landforms such as road junctions or hilltops, a spot elevation will be given. It will be printed in either brown or black.

(c) Bench marks. Similar to Spot elevations, but more accurate. They are marked on your map by a black "x", with the letters "BM", and the elevation.

(d) A point on a contour line. If a point is located on a contour line, its elevation will be the same as that contour line.

(e) A point between contour lines. If a desired point is less than 1/4 the distance between 2 lines, its elevation is considered to be the same as the closest contour line. If the point is from 1/4 to 3/4 the distance, its elevation is considered to be in the middle of the 2 lines.

(f) A point on top of a hill. To estimate the elevation on the top of an unmarked hill, add 1/2 of the contour interval to the elevation of the highest contour line around the hill.)

4. **TERRAIN FEATURES**. There are many types of terrain features that you must be able to identify. By interpreting the arrangement of contour lines, you should be able to get a mental picture of how these terrain features would appear if you were actually out there on the ground looking at them. Some of the features you need to be able to recognize are as follows:

a. Hills. Defined as a point or small area of high ground. A hill is represented on your map by a number of contour lines each of which circles around and connects onto itself. These rough circles get smaller as they progress toward the top of the hill. You need to carefully compare the arrangement of these contour lines to the actual appearance of the hill.

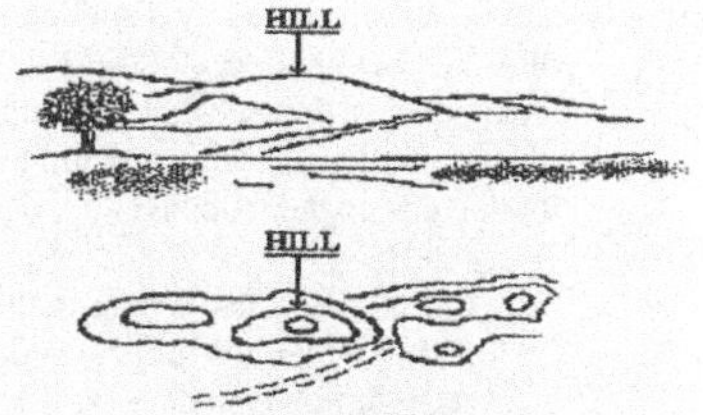

b. Spur. Defined as a line of high ground. The points along the top or crest of the spur are higher than the ground on both sides. A spur is represented on the map by a number of contour lines, each of which form a U-Shaped or V-Shaped design. The end or bottom of the U or V-Shaped lines point down to a lower elevation.

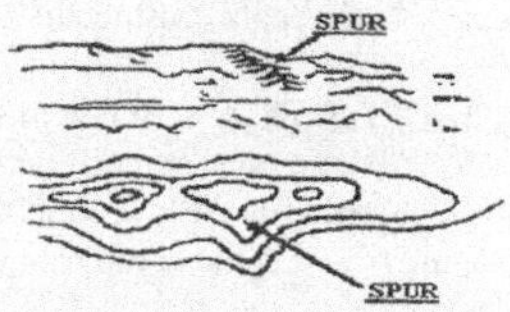

c. Draw. Defined as a line of low ground. The points located in the bottom of a draw are at a lower elevation than the ground to either side. A draw usually runs between 2 spurs. It is sometimes difficult to tell the difference between the contour lines that represent a draw from those that represent a spur because they are both U or V-Shaped lines. Remember that the curved portion representing a spur points down to ground of lower elevation, and the curved portion representing a draw points to ground of higher elevation. Now by looking closely at the area on your map, you will be able to tell which of the 2 you are dealing with. In many cases, draws will contain a stream, while spurs usually never have a stream running down their centers.

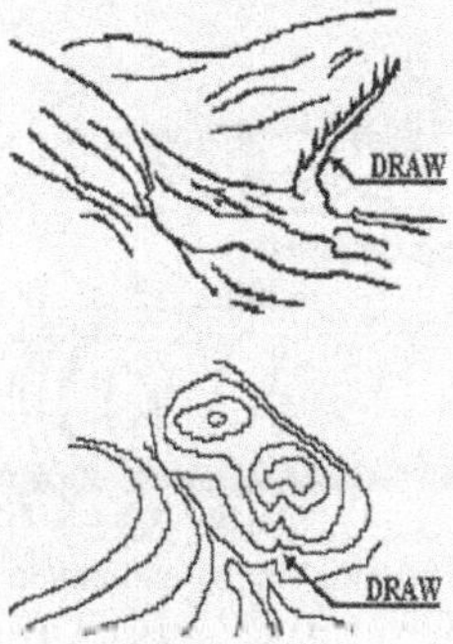

d. Saddle. Defined as the low ground between 2 hilltops, or a dip along the crest of a spur. On your map a saddle is represented by contour lines that form 2 hilltops that are located closely to each other. The area between these 2 hilltops being the location of the saddle.

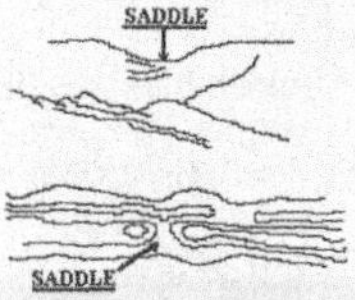

e. Depression. Defined as a point or area of low ground surrounded on all sides by higher ground. A depression is represented by contour lines forming circles within a small area on the map. These rough circles look identical to hilltops except for the presence of hachures. These are short stubby lines that are connected to the contour line at one end and point away from the contour lines at right angles in a downhill direction.

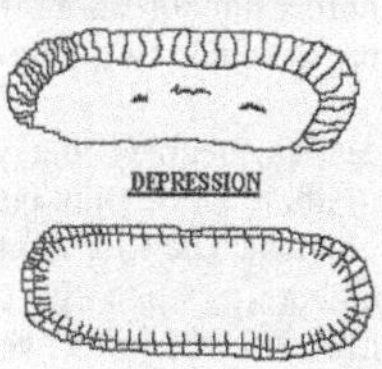

f. Cut. A cut is a man-made feature by which the bed of a road or railroad is leveled by cutting through high ground.

A cut is represented by a sudden ending of contour lines, the ends being connected by a straight line. Hachures are connected to this straight line and point downhill toward the roadbed. This sudden ending of contour lines occurs on both sides of the road bed.

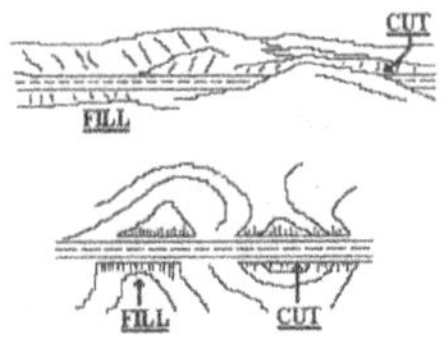

g. Fill. A fill is also a man-made feature by which the bed of a road filling in the low area. A sudden ending of contour lines can also represent a fill. A straight line connects these ends with Hachures pointing downhill away from the road bed. Again, this occurs on both sides of the road bed.

5. **SLOPES.** The rate of rise or fall of a ground form is known as its slope. In combat, this factor of slope must be considered when making plans for an operation. You must have the ability to recognize the ground forms in your area of operation by studying your map and the spacing between the contour lines on your map.

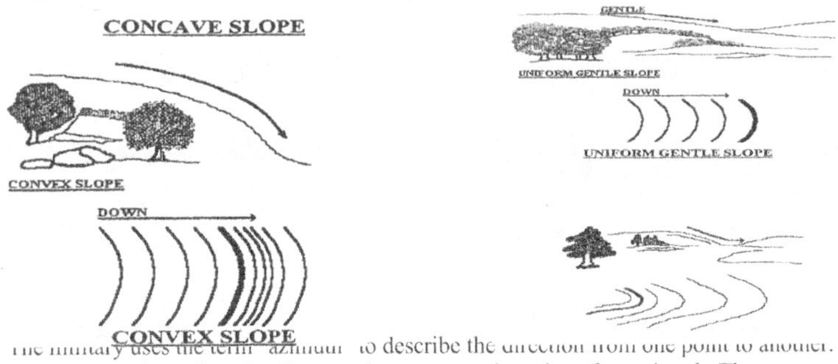

a. The military uses the term azimuth to describe the direction from one point to another. There are two base direction lines used to measure the value of an azimuth. They are as follows:

(1) Grid north. It is used when measuring direction from your map.

(2) Magnetic north. Used when measuring direction from your compass.

b. Units of Measurement. Now that you have learned about the two base direction lines that are used to measure direction, we need to learn the units of measurement that are used to measure the angle from one of these direction lines to the direction line.

(1) Degrees. The degree is the unit of measurement that is used most often. Anytime you need an azimuth to move from one point to another, you will use degrees as your unit of measurement. The base direction line, grid north or magnetic north, will be zero (0) degrees and measuring clockwise from this base direction line, the angle increases until you reach 360 degrees which brings you back around in a full circle to the base direction line.

(2) Mils. The mil is also used as a unit of measurement for direction. It is used when working with indirect fire weapons such as mortars and artillery. The azimuth you send will have to

be in mils, not in degrees. Remember, you are still measuring the angle between the base direction line and the line of sight from you to the distant object only you are using a different unit to measure the angle. Moving clockwise from the base direction line, which would be zero (0) mils, the angle back to the direction line. Now you know there are 360 degrees and also 6400 mils in the circle. This means that each degree equals approximately 17.5 mils.

c. Converting Map and Compass Directions. (WMC.5.2d) Earlier you learned that an azimuth was always measured from a base direction line. You also learned that we can use two different base direction lines. Grid north is the base direction line used when you are working with the map and getting a map direction or a grid azimuth. Magnetic north is the base direction line used when you are using the compass to get a compass direction or a magnetic azimuth. Because these two different base direction lines are used and because inmost cases there is a difference of one or more degrees between them, it is necessary for you to learn how to convert from a grid azimuth to a magnetic azimuth or from a magnetic azimuth to a grid azimuth. In order to learn this procedure we must take another look at the declination diagram.

(1) Declination diagram. A diagram that shows the relation between three different norths true north, grid north and magnetic north. It is the difference between grid north and magnetic north that we need to look at. The difference between these two north's is known as the grid-magnetic angle or G-M angle.

(a) Grid-Magnetic angle. The grid-magnetic angle is the angular difference between grid north and magnetic north. As you can see on this diagram a line that connects grid north and magnetic north represents the G-M angle. The size of the angle is stated off to the left side of the diagram in both degrees and mils. The G-M angle on your maps is 18 degrees or 320 mils. It is important to know that the G-M angle will change depending on where you are located on the earth's surface.

(2) Converting. When you receive a mission as a squad leader to move your squad from your present position to some distant point, you take out our map, a protractor and figure out the map direction or the grid azimuth to that distant point. However, before you can use your compass to direct you to your destination, you will have to convert the grid azimuth you have from your map, to a magnetic azimuth that you can set on your compass.

(a) The first step in making this conversion is to examine the declination diagram on your map. The G-M angle on your map is 18 degrees. You know that you will have to add the G-M angle to, or subtract it from the azimuth you measured on your map, or shot on your compass.

(b) The instructions as to whether you will add or subtract this angle are written on the map depending on whether you are going from grid to magnetic or vise-versa.

d. Orientation of a Map. (WMC.6.14a) In order to gain the best use of your map and to avoid confusion while navigating across terrain, you will need to know how to orient your map to the ground. *A* map is oriented when it is lying flat and level with it's north and south pointing in the same directions as north and south on the ground. There are two methods you can use to orient a map.

(1) Inspection. The first method used to orient the map is by inspection. By carefully examining your map and the ground around the area in which you are located to find linear features such as roads, railroads and powerless, you can align the feature on the map with the same feature on the ground. If there is only one linear terrain feature in the area in which you are located, you will have to be careful not to reverse your directions. Even though you have a road running north-south on your map aligned with the same road on the ground, the north end of your map may be pointed in a southerly direction down the road. To avoid this from occurring, find another terrain feature in the area that you can also identify on the map. Once you have aligned your map along the linear feature, check to insure that the second feature is positioned on the same side of the linear feature on your map as it is on the ground.

(2) Compass. This is the second method of orienting a map. You begin by placing your map on the ground so it lies flat and level. Open your compass so that the cover and eyepiece are completely open. Align the graduated straight edge of the compass along one of the north-south grid lines on your map with the cover end of the compass pointing toward the top of the map. This will place the stationary index line parallel to the grid north. Since the north seeking arrow on the dial of your compass will always point to magnetic north, you set your compass up just like a declination diagram. Grid north being the stationary index line and magnetic north being your north seeking arrow. Now rotate your map, insuring that the graduated straight edge of your compass remains aligned with the north-south grid line, until the face of the compass appears exactly as the declination diagram in the lower margin of the map. Your map is now oriented. For the Sonora Pass map sheet you would rotate the map until the north seeking arrow on the compass lies 16.5 degrees to the right of the stationary index line.

7. **GRID SYSTEM.** (WMC.6.14b) In most combat situations it is very important to be able to pinpoint your location, the location of another unit, or the location of an objective. In order to do this, it is necessary to have a location system that will allow you to accurately determine a specific location and state that location in such a way that will allow others to find that location quickly. This system is known as a grid system.

a. Grid Box. This box gives basic instructions on reading grids in determination of specific points on the map.

b. Grid Lines. The straight black lines that run across, and up and down the surface of your maps form the 1000 meter grid squares. These lines are called grid lines. If you look around the border of your map you will see that both the horizontal and vertical grid lines are identified by a two digit number.

c. Grid Square Identification. Each 1000 meter grid square on your map has it's own unique number. This number is called a grid coordinate. A four digit number is the result of combining the two digit number identification of the vertical grid line and the two digit number identification of the horizontal grid line that cross to form the lower left corner of the grid square. To find the grid coordinate of any grid square on your map you will always read your map right and up. You find the two digit number of that vertical grid line and you will have the first 2 digits of your coordinate. Next, find the two digit number of the horizontal grid line and place it after your vertical number and you now have the 4 digit grid coordinate that identifies your grid square.

d. Locating Terrain Features Within a Grid Square. If you are given a 4 digit grid coordinate to find a certain point on the earth's surface, that grid coordinate would put you within 1000 meters of that point. If you are told to find a certain feature within that grid square and you are good at reading your map, a 4-digit coordinate may get you close enough to your point to enable you to find it. However, there are many situations in which a 4 digit number will not get you close enough to a given position to allow you to accomplish your task. It is necessary to be able to locate points within a grid square and using a six-digit grid coordinate does this.

(1) Six Digit Grid Coordinate. The 1000-meter grid square will have to be broken down further into 100-meter grid squares. Because the surface of your map would become too cluttered by all these additional lines, a special instrument has been made to help you divide up a 1000-meter grid square. This instrument is called a coordinate scale or protractor.

8. **RESECTION**. (WMC.6.14c)

a. You will face a situation where you need to pinpoint the location of your position. This may not be possible by associating the terrain in your area with the map because you may have a poor map or the terrain in your immediate area is flat and has no distinguishing features. When you find yourself in this situation you will need to know how to perform a resection.

(1) Examine the terrain around your position and find two terrain features or objects that you can also identify on your map.

(2) Use your compass to find the magnetic azimuth to one of the distant terrain features you have identified.

(3) Convert the magnetic azimuth you found into a grid azimuth.

(4) Change the grid azimuth you found into a back azimuth.

(5) Drawing a line on your map from a distant point you identified in step one back toward your present position. Place a small mark on your map at this point. Using a straight edge, draw a line from where you plotted the distant object to the mark you made by the edge of your protractor. This line should be extended to insure it continues through the area where you are located.

(6) Repeat, finding the magnetic azimuth to the second distant object. That is at least 60 degrees from the first.

(7) When you have drawn your second line on the map, the point where the second line crosses the first line is the location of your position. The process that has just been covered is called a two-point resection.

(8) If your position is located on a linear line feature such as a road, power line, railroad or river, then it is only necessary to perform a one-point resection in order to pinpoint your location. Instead of identifying two distant points, you will only need to select one. Where the single line that you drew on your map crosses the linear feature on which you are located will be the location of your position.

9. **DETOURING AROUND AN OBSTACLE**.

a. There will be occasions when it will become necessary to move off your plotted route in order to avoid an obstacle such as a small lake or swamp. You must be able to move around these obstacles and return to the route you were following before you took your detour. This can be accomplished by using one of three detour methods: Far side landmark method, nearside landmark method, and the 90-degree offset method.

(1) Far side landmark. Used during hours of daylight when you are able to identify some feature on the far side of the obstacle that lies on your line of march and is obvious enough so that you can find it once you have moved around to the far side.

(2) Nearside landmark. Used during hours of daylight when you cannot find a landmark on the far side. However, you can identify a feature on the nearside that is on your line of march and you will still be able to see once you have moved around to the far side. Again, let us use the example of being on an azimuth of 198 degrees, and it brings you to a small lake. From the nearside you cannot locate any features on the far side. There is, however, a small pier on the nearside that is located on your line of march. You can now move around the lake using the easiest and safest route. Move to an area on the far side that is as close to being opposite of the nearside landmark as possible. You then figure out the back azimuth of your route azimuth, which in this case would be 18 degrees. Shoot an azimuth back across the lake to the pier. Move to the right or left until you are in a position where the pier lies at an azimuth of 18 degrees. This puts you back on your line of march and you can continue on following your original azimuth of 198 degrees.

(3) 90 degree offset. Used during hours of reduced visibility or during daylight hours when there are no visible landmarks along the line of march on either side of the obstacle. This time you are moving along a 360 degree azimuth and you encounter a large, impassable area of thick brush. The brush is too high to see any landmarks from either side of the thicket. The following procedure should be taken to detour around the thicket.

(a) Stop the pace count you have been making for movement along your assigned route.

(b) Decide which way you will detour around the thicket. Let's say in this case the right flank appears to be the easiest and safest. Add 90 degrees to your original azimuth of 360 degrees in order to move your patrol to the right.

(c) Move on this 90 degrees azimuth until you have reached a point where you can clear the right flank of the thicket. You will keep a new pace count starting from the point where you made your 90 degrees turn to the right to the point where you can clear the right flank of the thicket.

(d) Make another change of direction. This time you want to return to a 360 degrees azimuth so you will parallel your original course and move by the right flank of the thicket. In order to return to the 360 degrees azimuth you will now subtract 90 degrees.

(e) Move on the 360-degree azimuth until you have by passed the thicket. At the point of your second 90 degree turn, you will pick up the pace count you were making before you began making your detour around the obstacle. Even though you are not on your original line of march, you are moving parallel to your original line of march and therefore must include the distance you cover during this leg of the offset in your pace count for your route.

(f) Once you have bypassed the thicket, you will make another change of direction. This time, you want to move back toward your original line of march. Since that means you will be turning toward the left, you will subtract 90 degrees from the 360 degrees azimuth you have just completed. Your new direction will be 270 degrees.

(g) Pace back toward the original line of march following the 270-degree azimuth, the same number of paces as you made off the original line of march. In doing this you will have returned to your original line of march.

(h) Once you have moved back to your original line of march you will make your last change of direction. By adding 90 degrees to the original azimuth of the last leg of the offset, which in this case was 270 degrees, you will be back on the original azimuth of 360 degrees.

(i) Return to your original line of march pace count.

UNITED STATES MARINE CORPS
Mountain Warfare Training Center
Bridgeport, California 93517-5001

FMST.07.39
04/03/02

STUDENT HANDOUT

SEARCH AND RESCUE MANAGEMENT

TERMINAL LEARNING OBJECTIVE: In a mountainous environment, execute a search and rescue, in accordance with the references. (FMST.07.39)

ENABLING LEARNING OBJECTIVES:

(1) Without the aid of references and from a given list, choose the definition of a search, in accordance with the references. (FMST.07.39a)

(2) Without the aid of references and from a given list, choose the definition of a rescue, in accordance with the references. (FMST.07.39b)

(3) Without the aid of references and from a given list, choose one the four methods used to determine a search area, in accordance with the references. (FMST.07.39c)

(4) Without the aid of references and from a given list, choose one of the five methods for confining a search area, in accordance with the references. (FMST.07.39d)

(5) Given a lost individual scenario in a mountainous environment, plan a search and rescue in accordance with the references. (FMST.07.39e)

OUTLINE.

1. **GENERAL**. When discussing a search and rescue situation we must first be able to differentiate between the two facets of the operation. A "search" means that you are looking for an overdue missing person (FMST.07.39a). A "rescue" 'is the extraction of a known subject from a known location (FMST.07.39b). In either of these situations time is critical for several reasons:

14

a. Subject may need emergency medical care.

b. Size of the search area is limited.

c. Time and weather destroys clues.

d. Subject may need protection from the environment or from himself.

2. **SEARCH METHODS**. When conducting a search you will be searching for clues, instead of searching for the victim. There are two reasons for this:

a. There are more clues than victims. Anyone moving through an area will leave signs.

b. Clue detection greatly reduces search difficulty by reducing the search area.

c. This determination will assist the search manager in the use of assets, the justification for requesting outside agency assistance, and other management decisions. There are eight points that must be considered in determining the search urgency:

1. Subject profile.
2. Weather profile.
3. Equipment available to subject.
4. Subject's training, experience, and mobility.
5. Terrain hazards.
6. History of incidents in the area.
7. Time.
8. Political sensitivity.

3. **SUBJECT PROFILE**. The subject profile is based on the subject's capabilities and mobility in the terrain and conditions of the search area. The three points are age, medical condition, and number of subjects. Discussing this will assist us is rating the relative urgency:

a. Age: Very young/old, other.

b. Medical condition: Injured, ill, mental problems, healthy, or dead.

c. Number of subjects: Single or group.

d. Subject experience profile: No experience/no knowledge of area, no experience/knows area, experienced/no knowledge of area, or experienced/knows area.

e. Weather profile: Past or current bad weather, bad weather in next 8 hours, bad weather in next 24 hours, or no bad weather predicted. Temperature and visibility.

f. Equipment profile: Inadequate, questionable, or adequate for environment or weather.

g. Terrain/Hazard profile: Known hazardous terrain or, hazards or few or no hazards.

h. History of incidents in the area: Outcome of other searches for individuals with similar profiles in similar situations.

i. Time: Time since the subject went missing and the effects of time on clues.

j. Political sensitivity: External sources; VIPs, politicians, relatives, media, or higher authority in your chain of command.

4. **DETERMINING THE SEARCH AREA**. There are four methods used to determine the search area by rescue managers. The best application is to use a combination of these methods by personnel with search experience. (FMST.07.39c)

a. Theoretical: This is the distance that the victim could have traveled in the elapsed time since he/she was reported missing. (PI x R Squared)

b. Statistical: Historical data on searches for similar individuals in similar conditions.

c. Subjective: Evaluation of limiting factors for this specific incident and location.

d. Deductive reasoning: Sherlock Holmes version, based on analysis of circumstances involving the specific incident.

5. **BEHAVIOR ASPECTS OF A LOST PERSON**. General aspects of lost person behavior will give you an awareness of the actions of the individuals you are searching for by separating them into different categories. Some factors that could effect the search strategy are the general health of the subject, his/her past experiences, physiological effects of the environment, and the victims' mental state or biological cycle. Also, looking at the subject's ability to assist in his/her own survival with simple skills such as fire, shelter, signaling, and land navigation will determine strategy. The psychological impact on the victim should also be taken into account. Will the subject be overcome by Stress? Fear? Panic? Embarrassment? Or will the subject be calm and develop a positive "I'll be OK attitude? A further point to consider is the circumstances under which the subject became lost. Was the subject at a known location such as a campground? Enroute to a specific location or on a trail? Traveling in a wilderness area with no specific location or destination in mind? Or does the subject fall under one of the following categories:

a. Children age 1 to 3: Unaware of concept of being lost, no navigation skills or sense of direction, wander aimlessly, might seek out a convenient place to take a nap. i.e. hollow log, under bush/rock/picnic table.

b. Children age 3 to 6: More mobile, more inquisitive and easily distracted, will attempt to "go home" once they decide that they are lost, taught to avoid strangers and may hide from searchers.

c. Children 6 to 12: Good navigation skills and sense of direction in familiar terrain but become disoriented in unknown surroundings, may have intentionally run away and do not want to be found, may not answer when called, same fears as adults but no experiences to fall back on, once night sets in more likely to respond to searchers and except assistance.

d. Mentally retarded (all ages): Act similar to children 6 to 12, generally will not respond to name, will hide from searchers or seek shelter, once shelter found might remain in place, will do little or nothing to assist in their own survival or rescue.

e. Elderly (above 65): Possibly suffering from senility or Alzheimer's, easily confused or attracted to something that strikes their fancy, may not be oriented to the present, likely to over-extend themselves physically, and often have hearing impairments.

f. Despondent: Often are seeking solitude and will not respond to searchers, usually found near prominent terrain/scenic views, within sight and sound of civilization, rarely found in the underbrush.

g. Hikers: Rely on trails and have a set destination in mind, have poor land navigation once they loose the trail, often get confused at trail intersections or in springtime when trails are covered with snow, hiking parties of different ability levels get separated, will miss trails and head down wrong drainages. Hunters: Concentrate on game more than navigation, often pursue game into drainages/deadfalls/underbrush/deep snow, tend to over-extend themselves, typically not prepared for foul weather.

h. Cone pickers/Berry pickers/Photographers/ rockhounds/etc: Intention to stay or go to a specific location, carry little provisions or survival gear, go in good weather so they don't carry foul weather gear, often are not aware of surrounding as they are intent on pictures or the ground if pickers, these people are an extremely high risk group.

i. Fishermen: Well oriented once they arrive at lake or river, most often late due to accident, these are often recovery operations.

j. Climbers: Generally well equipped and self sufficient, tend to remain on marked routes and leave good plans with friends, primary factors are weather and hazardous conditions such as falling debris or avalanches, once located these searches often become technical rescues. 1. Marines: Subject profile will depend on individual's MOS, mission, training, and abilities learned as a civilian.

6. **BEGINNING THE SEARCH**. Once we have determined that a subject is missing we must determine whether or not he/she is in our search area. The sheriffs department will initiate a "Bastard Search", this is to ensure that the subject is not simply at the local tavern, shacked up with an old friend, or was not in an automobile accident enroute to the area. Next we will begin our search at one of the four following area:

a. Last Known Point (LKP): This is the last place that you have physical evidence to support that the subject was there. i.e. Subject's vehicle at the trailhead, name in a logbook, campsite with gear, etc.

b. Place Last Seen (PLS): The last place he was physically seen. i.e. On the trail by other hikers, at the hunting cabin, at a restaurant or gas station, or back home in San Diego.

c. Departure Point: The place where the subject was supposed to enter the search area. Usually a trailhead or specific point relating to the individual's activity.

d. Clues: As clues are discovered, and you confirm that they belong to the subject, you will update or adjust the search area.

7. **MATTSON**. Once we have determined that we do have an overdue subject and can assume that he/she is in our search area we need to break the search area into manageable sized segments. Depending upon the area to be searched, special training required to search high hazard areas, and assets available, we will divide the area by hazard or by terrain features. The lines of division should follow terrain features so that the searchers on the ground can easily navigate and stay within their zones. We will then determine the POA using the Mattson approach. This is done with each of the five or six members of the management team with the most search experience assigning a percentage to each of the segments using his individual judgment on the likelihood of the subject being in that area. Always include an extra column to represent the "rest of the world". Once all personnel have made their determinations, divide the totals and this will give you your areas to search in order of precedence:

NAME	A	B	C	D	E*	TOTAL
DUNSTON	40%	20%	10%	20%	10%	100%
KEOGH	50%	30%	5%	10%	5%	100%
KOFFEL	40%	25%	15%	10%	10%	100%
AYRES	60%	20%	10%	5%	5%	100%
TOTAL	190%	95%	40%	45%	30%	400%
AVERAGE	47.50%	23.75%	10.00%	11.25%	7.5%	
PRIORITY	1 ST	2ND	4TH	3RD	5TH	

8. **COMMAND SYSTEM**. The Office of Emergency Service (OES), has developed a command system that they use for all emergency situations. This system is called the

Incident Command System (ICS). The breakdown is very similar to the military chain of command and the duties correspond to those of the various "S" shops in the battalion. The personnel in each duty area will complete their assigned duties as you would in a military unit with the Incident Commander (IC) being overall responsible for the entire action. Normally, in California, the IC is the County Sheriff.

a. Incident Commander. CO

b. Operations-Tactics, coordination, insert resources. S-3

c. Plans-Strategy, intelligence, briefing and debriefing teams. S-2

d. Logistics-Beans, Band-Aids, beds, heads, motor-T, Special equipment. S-4

e. Communication-COC and field, landlines, frequency coordinator. S-6

f. Finance-Documentation of all man-hours, damages, legal issues.
S-1

g. Liaisons-inter agency, media, safety, family. PAO/Chaplain

9. **CONFINING A SEARCH AREA**. There are several reasons to confine the search area. The most important is that the longer the subject has free movement, the farther he/she can travel. (Remember the Theoretical Method) Additionally, this will ensure that the subject does not walk out without you being aware of it. It's real dumb to be conducting a search for a subject that is on his/her way home. The use of confinement will not interfere with the other searchers, decreases the chances of a more massive search and if promptly initiated has been shown to be highly effective. There are several methods that we can use. (FMST.07.39d)

a. Sign-cutters - Searchers that hike or ski the perimeter of the entire search area, then the perimeter of the segments. This will let you know if he/she moves from one area to another and allows you to concentrate limited assets in a specific area.

b. Trail-watchers/Road Blocks/Camp-ins - Searchers that camp at trailheads. They can alert you if the subject walks out and interview other individuals concerning possible sightings of the subject.

c. Look-outs - These personnel are already in place in fire towers. Simply notify them of the situation.

d. Track Traps - Clearing areas of the trail that are checked periodically for fresh tracks from the subject.

e. String-Lines - Spools of line with paper arrows attached that point to a safe area of a trail-watcher camp.

10. **USING CLUES**. The use of clues cannot be underestimated. They can tell you if a subject was or was not in the area, by freshness you can tell how long ago the subject was there, by the type of clue you can determine physical and mental condition, and by following clues you can determine a direction and speed of travel. Once you have located a CONCLUSIVE clue, you will re-mattson the area, using the clue as the LKP. The clues we will be looking for fall under four categories:

a. Physical - Footprints, equipment, candy wrappers, cigarette butts, etc.

b. Recorded - Summit logs or trail registers.

c. People - Witnesses who have seen the subject.

d. Event - Mirror flashes, camp fires, whistle signals, etc.

11. **ADDITIONAL ASSETS**. There are a variety of assets available to you in the search and rescue business. Some are volunteers, some charge money and some (Like the military) require higher authority to grant permission for their use. Remember that in addition to active searchers you can also use passive measures such as lights, noise, balloons, fires, smoke generators, sirens, or P.A. systems. Some examples of active searchers are:

a. Scent dogs.

b. Tracking dogs.

c. Human trackers.

d. Hasty teams.

e. Grid teams.

f. Helicopters.

g. Civil Air Patrol.

h. Snowmobile clubs.

i. Horse teams.

UNITED STATES MARINE CORPS
Mountain Warfare Training Center
Bridgeport, California 93517-5001

FMST.07.16
04/03/02

STUDENT HANDOUT

TRIAGE

TERMINAL LEARNING OBJECTIVE: Given simulated casualties, in a wilderness environment, triage casualties, in accordance with the references. (FMST.07.16)

ENABLING LEARNING OBJECTIVES:

1) Without the aid of references and from a given list, choose when actual triage most often begins, in accordance with the references. (FMST.07.16a)

2) Without the aid of references and from a given list, identify one of the four-triage categories used by NATO, in accordance with the references. (FMST.07.16b)

3) Without the aid of references and from a given list, choose the primary cause of death in a mass casualty situation, in accordance with the references. (FMST.07.16c)

OUTLINE.

1. **DEFINITION**: Triage or sorting (often used interchangeably) is the evaluation and classification of casualties for the purpose of establishing the priority for treatment and evacuation.

NOTE: Casualty sorting is one of the most important tasks for the entire medical service. No other job requires more informed judgment, hard work and courage. The Officer and the Corpsman responsible for triaging wounded Sailors and Marines must exercise sound judgment so that casualties receive proper, expedient care. Triage is a continuing process and any individual assigned should be the most capable and experienced available.

a. Decision Factors. The factors concerning casualty triage are based upon the following:

(1) The need for immediate resuscitation.

(2) The need for emergency surgery.

(3) The futility of surgery due to obviously lethal wounds.

(4) Time needed to perform surgery compared with:

(a) Probability of success.

(b) Number of other casualties needing treatment.

b. The actual triage begins most often with the casualty himself. (FMST.07.16a). This usually occurs when the injured individual determines that "Buddy-Aid" or "Self-Aid" will not be adequate.

(1) The corpsman, once presented with the casualty, must make several determinations:

(a) Is the casualty a walking-wounded or a litter case?

(b) If severely wounded, are resuscitative efforts required immediately?

(c) Does the casualty need any other immediate treatment (bleeding, asphyxia or pain control)?

(2) Once the casualty has been transported to a Battalion Aid Station (BAS), triage continues. The dental officer will commonly perform the duties of the Triage Officer at the echelon I or echelon II level, if present.

2. **<u>NATO TRIAGE CATEGORY CODES</u>**. (FMST.07.16b)

a. At each medical treatment facility (MTF) in the area of operations (AO), incoming casualties are classified by level of treatment required. Four triage category groups have been universally adopted for use by both United States and NATO Forces. These categories are listed and defined as they appear in standardization agreement (STANAG) No 2879:

(1) <u>Immediate Treatment (Group T1) (RED Tag)</u>. To include those requiring emergency life-saving surgery. These procedures should not be time consuming and should concern only those patients with high chances for survival. Some examples are:

(a) Respiratory obstruction.

(b) Accessible hemorrhage.

(c) Emergency amputation.

(2) Delayed Treatment (Group T2) (YELLOW Tag). To include those badly in need of time-consuming major surgery, but whose general condition permits delay in surgical treatment without unduly endangering life. To mitigate the often, critical effects of delay in surgery, sustaining treatment, such as stabilizing I.V. fluids, splinting, administration of antibiotics, catheterization, gastric decompression, and relief of pain will be required. Examples are:

(a) Large muscle wounds.

(b) Fractures of major bones.

(c) Intra-abdominal and/or thoracic, head or spinal injuries.

(d) Uncomplicated major burns.

(3) Minimal Treatment (Group T3) (GREEN Tag). To include those with relatively minor injuries who can effectively care for themselves or who can be helped by untrained personnel. Some examples are:

(a) Minor lacerations.

(b) Abrasions,

(c) Fractures of small bones.

(d) Minor burns.

(4) Expectant Treatment (Group T4) (BLUE Tag). This group comprises patients who have received serious and often multiple injuries, and whose treatment would be time-consuming and complicated with a low chance of survival. If fully treated, they may make heavy demands on medical manpower and supplies. Until the mass casualty situation is under control, they will receive appropriate supportive treatment. The extent of treatment depends on available supplies and manpower, and may involve the use of large doses of narcotic analgesics. These patients should not be abandoned, and every effort should be devoted to their comfort. The possibility of their survival, despite alarming injuries, must always be kept in mind. Examples are:

(a) Severe multiple injuries.

(b) Severe head or spinal injuries.

(c) Large doses of radiation.

(d) Widespread severe burns.

3. **TREATMENT CONCEPTS**.

a. The triage officer must be familiar with all aspects of combat casualty care (airway management, hemorrhage control, use of anesthesia, surgical procedures, fluid resuscitation and post-operative care, to name a few), but ever-present in his mind should be the following concepts:

(1) Asphyxia and hemorrhage are initially the primary causes of death in any mass casualty situation. (FMST.07.16c)

(2) Once at a definitive care facility, uncontrolled hemorrhage is the leading cause of death.

(3) Life has priority over limb, and preserving function overrides the correction of superficial anatomic defects.

(4) A casualty is in a constant state of change until the wound or injury has been repaired.

(5) Systemic disturbances caused by the wound continue until healing is complete.

(6) Combat conditions will dictate the modifications necessary in triage and care of mass casualties.

b. Priorities of Treatment. With the six concepts of treatment in mind, we can now address three basic priorities for treatment.

(1) First Priority:

(a) Asphyxia:

1. Mechanical obstruction of the airway.

2. Sucking chest wound.

3. Tension pneumothorax.

4. Maxillofacial trauma.

5. Massive hemo-pneumothorax.

(b) Shock:

1. Exsanguinating hemorrhage.

2. Major internal hemorrhage (thoracic, abdominal, and pelvic).

3. Visceral injuries/evisceration.

4. Cardio-pericardial injuries.

5. Massive muscle injury leading to Fat Embolization to the lung.

6. Full-thickness or deep-partial burns of 10 % or greater.

(2) Second Priority:

(a) Visceral injuries:

1. Perforation of G.I. Tract.

2. Pancreatic and Biliary system.

3. Genitourinary tract.

4. Thoracic wounds w/o asphyxia.

(b) Vascular injuries requiring repair:

1. Tourniquet cases.

(c) Closed head injuries:

1. Increasing loss of consciousness.

(d) Major burns:

1. Partial-thickness burns of greater than 25% TBSA in the low risk group, 20% TBSA in the high risk group.

2. Burns involving the poor risk group.

3. Burns involving the face, hands, feet, genitalia/perineum.

4. Circumferential limb burns.

(3) Third Priority:

(a) Brain and spinal injuries - requiring burr hole decompression.

(b) Soft tissue injuries.

(c) Debridement necessary with less than major muscle damage.

(d) Lesser fractures and dislocations.

(e) Eye injuries.

(f) Maxillofacial injuries w/o asphyxia.

(g) Burns under 20% in other locations other than priority 2.

4. **TRIAGE TOOLS**.

a. Trauma Score/Glasgow Coma Scale. In recent years, many methods have been proposed to speed up the classification and sorting of casualties. One method that has gained wide acceptance is the Trauma Score/Glasgow Coma Scale. Several studies have been performed using various parameters, i.e., respiratory rate, chest wall expansion, BP, capillary refill, etc. By assigning a number scale to each specific response, the likelihood of survival as a percentage can be calculated.

1. Glasgow Coma Scale.

a. Eye opening:

1) Spontaneous = 4.
2) To voice = 3.
3) To pain = 2.
4) None = 1.

b. Verbal Response.

1) Oriented = 5.
2) Confused = 4.
3) Inappropriate words = 3.
4) Incomprehensive words = 2.
5) None = 1.

c. Motor Response.

1) Obeys Command = 6.

2) Localized Pain = 5.
3) Withdraw Pain = 4.
4) Flexion (pain) = 3.Extension (pain) = 2.
5) None = 1.

d. Total the points up and they equal:

1) 13-15= 4.
2) 9-12 = 3.
3) 6-8 = 2.
4) 4-5 = 1.
5) <3 = 0

2. Trauma Scale.

a. Respiratory Rate:

1) 12-29 min = 4.
2) 10-12 min = 3.
3) 6-9 min = 2.
4) 1-5 min = 1.
5) None = 0.

b. Systolic BP:

1) >89 mmHg = 4.
2) 76 mmHg = 3.
3) 50-75 mmHg = 2.
4) 1-49 mmHg = 1.
5) None = 0.

c. Total the points.

3. Total trauma Scale + GCS = % survival (1-12)

12 = 99
10 = 88
8 = 67
6 = 63
4 = 33
2 = 29

1 = 25
0 = 4

UNITED STATES MARINE CORPS
Mountain Warfare Training Center
Bridgeport, California 93517-5001

FMST.07.08
04/03/02

STUDENT HANDOUT

PREVENTIVE MEDICINE AND WATER PURIFICATION

TERMINAL LEARNING OBJECTIVE. Given a unit in a mountainous environment and the necessary equipment and supplies, perform preventive medicine, in accordance with the references. (FMST.07.08)

ENABLING LEARNING OBJECTIVES.

1) Without the aid of references, from a given list select the five important areas of personal hygiene, in accordance with the references. (FMST.07.08a)

2) Without the aid of references, from a given list select the correct steps taken to ensure proper immunization prior to deployment, in accordance with the references. (FMST.07.08b)

3) Without the aid of references, from a given list select the three methods of water purification, in accordance with the references. (FMST.07.08c)

4) Without the aid of references, from a given list select the four forms of halogens used for water purification, in accordance with the references. (FMST.07.08d)

5) Without the aid of references, from a given list select the definition of giardia, in accordance with the references. (FMST.07.08e)

6) Without the aid of references, from a given list select the most common drug used to treat giardia infection in the United States, in accordance with the references. (FMST.07.08f)

7) Without the aid of references, from a given list select the three important reasons for proper waste disposal, in accordance with the references. (FMST.07.08g)

8) Without the aid of references, from a given list select the correct methods of waste disposal, in accordance with the references. (FMST.07.08h)

16

1. **PERSONAL HYGIENE** Your body's defense mechanisms break down due to the physical demands of the arduous terrain, environment, stress and metabolic changes. Personal

hygiene becomes the key to prevent the spread of communicable diseases. Five important areas are: (WMC.6.17a)

a. Body.

(1) The body should be washed frequently in order to minimize the chances of small cuts and scratches developing into full blown infections and as a defense against parasitic infections.

(2) A daily bath or shower consisting of soap and hot water is ideal. However, when this is not possible, you should:

(a) Give yourself a "sponge bath" using soap and water, making sure particular attention is given to body creases, i.e., armpits, groin area, face ears, hands and feet.

(b) If water is in extremely short supply, you should take an "air bath" by:

(1) Remove all clothing and hang it up to air.

(2) Expose the body for two hours to the sunlight, which is ideal, but the effects will basically be the same if done indoors or during an overcast day. BE CAREFUL NOT TO SUNBURN!

(3) Shaving. Shaving at night before bedtime is preferable because shaving will remove facial oils, which aid in protecting the face from a harsh cold environment during the day.

b. Hair.

(1) Clean frequently.

(2) Inspect at least once a week for parasites. Have your mountain buddy inspect the areas not within your view. This is particularly important during the summer season. Any time the area you are operating in an area that contains wild life there is the possibility of parasites like fleas and ticks.

c. Hands. Hand washing with soap and water will cut down in the transmission of viral and parasitic infections via the person-person and fecal-oral route.

(1) Keep fingernails trimmed and clean to prevent accidentally scratching yourself and to prevent harborage for bacteria.

d. Feet. Are your prime sources of transportation, so take care of them!

(1) Inspect frequently for:

(a) Blisters.

(b) Infections, bacterial and fungal.

(2) Preventive measures:

(a) Keep your feet dry, change socks frequently.

(b) Use antifungal foot powder.

(c) Antiperspirants. Pretreatment for 2 weeks prior to field with an antiperspirant containing Aluminum Chlorhydrate. This will aid in the prevention of athlete's foot.

e. Oral Hygiene. The mouth and teeth should be cleaned at least daily to prevent tooth decay and gum disease.

(1) Ideally with:

(a) Tooth brush.

(b) Tooth paste.

(c) Dental floss.

(2) If these items are not available:

(a) A "chew stick" made from a clean twig about 8" long and finger width. Chew one end until it becomes frayed and brush-like, then use as a toothbrush.

(b) Field expedient dental floss can be pulled from the center lines of paracord.

(c) The gums should be stimulated at least once daily by rubbing them vigorously with a clean index finger.

2. **IMMUNIZATIONS (NAVMEDCOMINST 6230.3)**. In addition to being required, immunizations are your best front line defense against communicable disease.

a. Eight weeks prior to deployment: (WMC.6.17b)

(1) Ensure Basic Series Completed (BSC) on required immunizations.

(2) Determine if additional immunizations/chemophylaxis will be required in that geographical area of operations.

(3) Identify and flag health risk personnel: i.e. TB converters (NAVMEDCOMINST 6224.1), hepatitis (NAVMEDCOMINST 6230.1A), mononucleosis, parasitic infections and anemic personnel.

(4) Review medical intelligence reports:

(a) EPMU, Navy Environmental Preventive Medicine Unit.

(b) G-2, USMC.

(c) NAVDISVECTOECOLCONCEN, Navy Disease Vector Ecology and Control Centers.

(d) CDC - Center For Disease Control.

(e) WHO - World Health Organization.

3. **WATER.** A satisfactory source is one where quantity and quality is enough to supply the needs of all the troops. Water or ice taken from the environment should be considered contaminated and must be purified with available means. Be aware that cold weather slows the chemical reaction time of purification measures.

a. Sources.

(1) Rivers

(2) Streams

(3) Ponds

(4) Lakes

(5) Ice

(6) Snow

b. Water Source Selection. The choice of a water source is influenced by:

(1) Not contaminated by: (Check all to one mile upstream).

(a) Sewage

(b) Enemy pollution

(c) Chemicals

(d) Biologicals

(e) Radiologicals

NOTE: Check condition of vegetation or for dead animals.

(2) Quantity.

(3) Ease of procurement.

(4) Ease of purification.

(5) Freedom from turbidity.

(6) Excessive organic or non-organic contamination.

(7) Should be easily protectable.

4. **PURIFICATION.** Simply consists of removing or destroying enough impurities to make water safe and pleasant to drink.

a. The first step in purifying is to select a treatment method. (All figures are based on one quart of water.) The three different purifying methods are: (WMC.6.17c)

(1) Boiling. The standard recommendation is to bring the water to a rolling boil for 1 minute at sea level for complete sterilization of the water. Recent research showed that additional boiling time for sterilization was not necessary and that boiling for two minutes at any altitude rendered water safe for consumption.

(2) Halogens. The number of tablets and contact time must be considered. All charts based on one-quart canteen. (WMC.6.17d)

(a) Iodine tablets.

TYPE WATER	TABLETS	CONTACT TIME
CLEAR	1	25 MIN
CLOUDY	2	25 MIN
COLD	2	25 MIN

(b) Chlorine bleach. The strength of the solution, number of drops and contact time must be considered.

PERCENT	CLEAR WATER	CLOUDY WATER	COLD WATER
1%	10GTTS/30 MIN	20GTS/30 MIN	10GTTS/60MIN
4-6%	2GTTS/30 MIN	4GTTS/30 MIN	2GTTS/60 MIN
7-10%	1 GTTS/30 MIN	2GTTS/30 MIN	1 GTTS/60 MIN

(c) Iodine solution. The strength, number of drops and contact time must be considered.

PERCENT	CLEAR WATER	CLOUDY WATER	COLD WATER
2%	5GTTS/30 MIN	10GTTS/30 MIN	8GTTS/60 MIN

(d) Betadine solution.

CLEAR WATER	CLOUDY WATER	COLD WATER
8GTTS/30 MIN	16GTTS/30 MIN	8GTTS/60 MIN

(1) Filtration Most filters are not small enough to completely disinfect water. They remove particulate debris, which allows lower halogen dose and improved appearance and taste of "dirty" water. Filters clog quickly if the water is dirty or has a lot of suspended particles. NEVER USE FILTERS AS YOUR SOLE METHOD OF WATER PURIFICATION.

(a) Maximum Filter Pore Size:

1) Parasitic eggs and larvae: 20 micrometers (um).

2) Giardia, E. histolytica: 5 um.

3) Cryptosporidium: 1 um

4) Enteric bacteria: 0.2 um.

5) Viruses: 0.01 um (too small for field water filtration).

5. **GIARDIA**

a. Definition. (WMC.6.17c) An intestinal flagellate protozoa that can cause clinical illness in humans.

(1) Found worldwide in tropical and temperate areas, the infectious cysts are passed by humans and animals and can be transmitted by direct fecal-oral route or waterborne contamination (cysts maintain viability for months in cold water). Due to these factors, Giardia is both a common cause of diarrhea in travelers to foreign countries, and campers in the United States where it has become endemic and is now considered the most common human parasite.

(2) It is found with increasing frequency in formerly "pristine" mountain streams, rivers and lakes and outbreaks have occurred in remote recreational areas and resorts throughout the western mountain states.

(3) Giardiasis is currently a risk for anyone drinking untreated surface water in the western mountains and for communities with water treatment techniques inadequate for eliminating the cysts.

b. Pathophysiology. The organism has two forms: Trophozoite and cyst.

(1) These may be passed in the stool of an infected person. The trophozoites die within one hour outside the body, however the cysts are very hardy, retaining infectivity more than two months in cold water. If the cysts are ingested, excystation is encouraged by stomach acid, and the motile trophozoite attaches by suction disc to the wall of the duodenum and proximal jejunum, where they may cause symptoms of illness.

(2) Water, ice, or food prepared with contaminated water, are the major sources of infection, but direct person-to-person transmission can occur from sexual contact or hand-to-mouth contamination.

(3) Incubation time: 5 - 25 days (7 - 10 medium). Most infections are thought to be asymptomatic carrier states, however, a wide array of clinical syndromes may occur.

c. Symptoms.

(1) A number of people have abrupt onset of explosive diarrhea accompanied by abdominal cramps, foul gas, vomiting, fever and malaise. This commonly lasts three to four days before developing the more common subacute syndrome. In most individuals, the onset is more insidious, and symptoms are persistent or recurrent.

(2) Stools become mushy, greasy and malodorous. Over a period of weeks, watery diarrhea may alternate with soft stools and even constipation. Upper GI symptoms accompany stool changes, mid and upper abdominal cramping, substernal burning, acid indigestion, sulfurous belching, nausea and foul flatus, distention with an early "full" feeling are all typically increased after a meal.

(3) Symptoms of anorexia, fatigue and weight loss are common, but fever and vomiting are infrequent except during an acute onset of illness.

d. Laboratory Parameters.

(1) The "typical" victim should suggest a diagnosis of giardiasis. This individual may have a history of camping, vacationing at a mountain resort, or foreign travel. Symptoms of upper abdominal discomfort, foul gas and soft stools arose one or two weeks after a short trip or during a lengthy one and have fluctuated for weeks. Sometimes improving, the individual held off visiting the doctor, only to have the symptoms recur. Work or school is not interrupted, however, physical discomfort and embarrassment from the gas are concerns.

(2) Laboratory confirmation of giardiasis can be difficult. Stool samples are the first means of diagnosis. Trophozoites may be found in fresh, watery stools, but disintegrate rapidly. Cysts remain in fresh stools for at least 24 hours. Cysts passage, however, is extremely variable and not related to clinical symptoms. Three samples, taken every other day, should be examined for a definite diagnosis.

(3) Average duration of symptoms has varied from three to ten weeks. Acquired resistance has been shown. Immunity is suspected, but not proven and human reinfection definitely occurs.

e. Treatment. (WMC.6.17f) Complete cure of most Giardia infections can be achieved 90% of the time with *Metronidazole* (flagyl) 250 mg three time s a day for 5 days.

f. Prevention There is no safe or effective chemoprophylactic drug and prevention should be practiced by travelers and campers. The purity of American wild waters can no longer be assumed and in developing countries, even tap water should be suspected. In addition to choosing a technique for individual field water disinfection, individuals must become more responsible for prevention of water contamination. This requires attention to personal latrines and camper trailer waste disposal, a subject that most people do not want to touch.

6. **WASTE**. The importance of proper waste disposal cannot be over emphasized.

a. It serves to: (WMC.6.17g)

(1) Eliminate harborage for rodents and vermin.

(2) Prevents attraction of rodents and vermin.

(3) Prevents a pathogenic contamination source.

b. Two basic types of waste are organic and non-organic. The basic ways to dispose of them are burning, burying, or hauling it away. (WMC.6.17h)

(1) Organic wastes:

(a) Human waste - Burn or haul away to a designated waste pit area.

(b) Urine - Use only assigned marked areas.

(c) Edible garbage - Burn. It should not be left exposed for animals, vermin or the enemy.

(d) Paper -Burn.

(e) Contaminated bandages - Sterilize/burn.

(2) Non-organic waste:

(a) Metals - Haul away or bury.

(b) Liquids - Burn or bury.

UNITED STATES MARINE CORPS
Mountain Warfare Training Center
Bridgeport, California 93517-5001

FMST.07.19
04/03/02

STUDENT OUTLINE

REQUIREMENTS FOR SURVIVAL

TERMINAL LEARNING OBJECTIVE Given a survival situation in any type of environmental condition and minimal equipment and resources, apply the requirements for survival to facilitate individual or group survival and recovery per the references. (FMST.07.19)

ENABLING LEARNING OBJECTIVES

1. Without the aid of references and from a given list, choose one of the survival stressors, in accordance with the references. (WMC.A1.a)

2. Without the aid of references and from a given list, choose one of the priorities of work in a survival situation, in accordance with the references. (WMC.A1.b)

1. REQUIREMENTS FOR SURVIVAL

a. This mental "mind-set" is important in many ways. We usually call it the "will to survive" although you might call it "attitude" just as well. This basically means that, if you do not have the right attitude, you may still not survive.

A1

b. A guideline that can assist you is the acronym " SURVIVAL".

(1) <u>S</u>ize up.

(a) Size up the situation.

1. Conceal yourself from the enemy.

2. Use your senses to hear, smell, and see to determine and consider what is developing on the battlefield before you make your survival plan.

(b) Size up your surroundings.

1. Determine the rhythm or pattern of the area.

2. Note animal and bird noises and their movement.

3. Note enemy traffic and civilian movement.

(c) Size up your physical condition.

1. Check your wounds and give yourself first aid.

2. Take care to prevent further bodily harm.

3. Evaluate condition of self and unit prior to developing survival plan.

(d) Size up your equipment.

1. Consider how available equipment may affect survival senses; tailor accordingly.

(2) <u>U</u>ndue haste makes waste.

(a) Plan your moves so that you can move out quickly without endangering yourself if the enemy is near.

(3) <u>**R**</u>emember where you are.

(a) If you have a map, spot your location and relate it to the surrounding terrain.

(b) Pay close attention to where you are and where you are going. Constantly orient yourself.

(c) Try to determine, at a minimum, how your location relates to the following:

1. The location of enemy units and controlled areas.

2. The location of friendly units and controlled areas.

3. The location of local water sources.

4. Areas that will provide good cover and concealment.

(4) **V**anquish fear and panic.

(a) Realistic and challenging training builds self-confidence and confidence for a unit's leadership.

(b) The feeling of fear and panic will be present. The survivor must control these feelings.

(5) **I**mprovise and improve.

(a) Use tools designed for one purpose for other applications.

(b) Use objects around you for different needs. (i.e. use a rock for a hammer)

(6) **V**alue living.

(a) Place a high value on living.

(b) Refuse to give into the problem and obstacles that face you.

(c) Draw strength from individuals that rise to the occasion.

(7) **A**ct like the natives.

(a) Observe the people in the area to determine their daily eating, sleeping, and drinking routines.

(b) Observe animal life in the area to help you find sources of food and water.

NOTE: Remember that animal reactions can reveal your presence to the enemy. Animals cannot serve as an absolute guide to what you can eat and drink.

(8) **L**ive by your wits, **but for now**, learn basic skills.

(a) Practice basic survival skills during all training programs and exercises.

2. **STRESS**. Stress has many positive benefits. Stress provides us with challenges: it gives us chances to learn about our values and strengths. Too much stress leads to distress. While many of these signs may not be self-identified, it remains critical that all survivors remain attentive to each other's signs of distress. Listed are a few common signs of distress found when faced with too much stress:

   a. Difficulty in making decisions (do not confuse this sign for a symptom of hypothermia).

   b. Angry outbursts.

   c. Forgetfulness.

   d. Low energy level.

   e. Constant worrying.

   f. Propensity for mistakes.

   g. Thoughts about death or suicide.

   h. Trouble getting along with others.

   i. Withdrawing from others.

   j. Hiding from responsibilities.

   k. Carelessness.

3. **SURVIVAL STRESSORS**. (FMST.07.19a) **Any** event can lead to stress. Often, stressful events occur simultaneously. These events are not stress, but they produce it and are called "stressors". In response to a stressor, the body prepares either to "fight or flee". Stressors add up. Anticipating stressors and developing strategies to cope with them are the two ingredients in the effective management of stress. It is essential that the survivor be aware of the types of stressors they will encounter.

   a. Injury, Illness, or Death: Injury, illness, and death are real possibilities a survivor has to face. Perhaps nothing is more stressful than being alone in an unfamiliar environment where you could die from hostile action, an accident, or from eating something lethal. Illness and injury can also add to stress by limiting your ability to maneuver, get food and drink, find shelter, and defend yourself.

   b. Uncertainty and Lack of Control: Some people have trouble operating in settings where everything is not clear-cut. The only guarantee in a survival situation is that nothing is guaranteed. This uncertainty and lack of control also add to the stress of being ill, injured, or killed.

c. Environment: A survivor will have to contend with the stressors of weather, terrain, and the variety of creatures inhabiting an area. Heat, cold, rain, winds, snow, mountains, insects, and animals are just a few of the challenges awaiting the Marine working to survive. Depending on how a survivor handles the stress of environment, his surroundings can be either a source of food and protection or a cause of extreme discomfort leading to injury, illness, or death.

d. Hunger and Thirst: Without food and water a person will weaken and eventually die. Getting and preserving food and water takes on increasing importance as the length of time in a survival setting increases. With the increased likelihood of diarrhea, replenishing electrolytes becomes critical. For a Marine used to having his provisions issued, foraging can be a significant source of stress.

e. Fatigue: It is essential that survivors employ all available means to preserve mental and physical strength. While food, water, and other energy builders may be in short supply, maximizing sleep to avoid deprivation is a very controllable factor. Training data collected at MCCDC demonstrates that individuals who lack sleep for 24 hours or more suffer a 25% decrease in effective performance. Further, sleep deprivation directly correlates with increased fear. Effective survival is highly unlikely when fatigue builds to a level where staying awake becomes a stressful evolution.

f. Isolation: Although Marines complain about higher headquarters, we become used to the information and guidance it provides, especially during times of confusion. Being in contact with others provides a greater sense of security and a feeling someone is available to help if problems occur. A significant stress in survival situations is that often a person or team has to rely solely on its own resources.

4. **NATURAL REACTIONS**. Man has been able to survive many shifts in his environment throughout the centuries. His ability to adapt physically and mentally to a changing world kept him alive. The average person will have some psychological reactions in a survival situation. These are some of the major internal reactions you might experience with the survival stresses.

a. Fear: Fear is our emotional response to dangerous circumstances that we believe have the potential to cause death, injury, or illness. Fear can have a positive function if it encourages us to be cautious in situations where recklessness could result in injury. Unfortunately, fear can also immobilize a person. It can cause us to become so frightened that we fail to perform activities essential for survival.

b. Anxiety: Anxiety can be an uneasy, apprehensive feeling we get when faced with dangerous situations. When used in a healthy way, anxiety urges us to act to end, or at least master, the dangers that threaten our existence. A survivor reduces his anxiety by performing those tasks that will ensure his coming through the ordeal alive. Anxiety can overwhelm a Marine to the point where he becomes easily confused and has difficulty thinking.

c. Anger and Frustration: Frustration arises when a person is continually thwarted in his attempts to reach a goal. The goal of survival is to stay alive until you can reach help or until help can reach you. To achieve this goal, Marines must complete some tasks with minimal resources. One outgrowth of frustration is anger. Getting lost, damaged or forgotten equipment, the weather, inhospitable terrain, enemy patrols, and physical limitations are just a few sources of frustration and anger. Frustration and anger encourage impulsive reactions, irrational behavior, poorly thought-out decisions, and, in some instances, an "I quit" attitude. If the Marine does not properly focus his angry feelings, he can waste much energy in activities that do little to further either his chances of survival or the chances of those around him.

d. Depression: Depression is closely linked with frustration and anger when faced with the privations of survival. A destructive cycle between anger and frustration continues until the person becomes worn down-physically, emotionally, and mentally. When a person reaches this point, he starts to give up, and his focus shifts from "What can I do" to "There is nothing I can do." If you allow yourself to sink into a depressed state, then it can sap all your energy and, more important, your will to survive.

e. Loneliness and Boredom: Man is a social animal and enjoys the company of others. Loneliness and boredom can be another source of depression. Marines must find ways to keep their minds productively occupied.

f. Guilt: The circumstances leading to your survival setting are sometimes dramatic and tragic. It may be the result of an accident or military mission where there was a loss of life. Perhaps you were the only, or one of a few, survivors. While naturally relieved to be alive, you simultaneously may be mourning the deaths of others that were less fortunate. Do not let guilt feelings prevent you from living.

5. **PRIORITIES OF WORK IN A SURVIVAL SITUATION.** (FMST.07.19b) Each survival situation will have unique aspects that alter the order in which tasks need to be accomplished. A general guideline is to think in blocks of time.

a. First 24 hours: The first 24 hours are critical in a survival situation. You must make an initial estimate of the situation. Enemy, weather, terrain, time of day and available resources will determine which tasks need to be accomplished first. They should be the following:

(1) Shelter.

(2) Fire.

(3) Water.

(4) Signaling.

b. Second 24 hours: After the first 24 hours have passed, you will now know you can survive. This time period needs to be spent on expanding your knowledge of the area. By completing the following tasks, you will be able to gain valuable knowledge.

(1) Tools and weapons: By traveling a short distance from your shelter to locate the necessary resources, you will notice edible food sources and game trails.

(2) Traps and snares: Moving further away from your shelter to employ traps and snares, you will be able to locate your shelter area from various vantagepoints. This will enable you to identify likely avenues of approach into your shelter area.

(3) Pathguards: Knowing the likely avenues of approaches, you can effectively place noise and casualty producing pathguards to ensure the security of your shelter area.

c. Remainder of your survival situation. This time is spent on continuously improving your survival situation until your rescue.

6. **GROUP SURVIVAL.** In-group survival, the group's survival depends largely on its ability to organize activity. An emergency situation does not bring people together for a common goal; rather, the more difficult and disordered the situation, the greater are the disorganized group's problems.

a. Groups Morale: High morale must come from internal cohesiveness and not merely through external pressure. The moods and attitudes can become wildly contagious. Conscious, well-planned organization and leadership on the basis of delegated or shared responsibility often can prevent panic. High group morale has many advantages.

(1) Individual feels strengthened and protected since he realizes that his survival depends on others whom he trusts.

(2) The group can meet failure with greater persistency.

(3) The group can formulate goals to help each other face the future.

b. Factors that Influence Group Survival: There are numerous factors that will influence whether a group can successfully survive.

(1) Organization of Manpower - Organized action is important to keep all members of the group informed; this way the members of the group will know what to do and when to do it, both under ordinary circumstances and in emergencies.

(2) Selective Use of Personnel - In well-organized groups, the person often does the job that most closely fits his personal qualifications.

(3) Acceptance of Suggestion and Criticisms - The senior man must accept responsibility for the final decision, but must be able to take suggestion and criticisms from others.

(4) Consideration of Time - On-the-spot decisions that must be acted upon immediately usually determines survival success.

(5) Check Equipment - Failure to check equipment can result in failure to survive.

(6) Acquiring survival knowledge and skills increases survival Knowledge and Skills - Confidence in one's ability to survive.

UNITED STATES MARINE CORPS
Mountain Warfare Training Center
Bridgeport, California 93517-5001

FMST.
07.20
04/03/02

STUDENT OUTLINE

SURVIVAL KIT

TERMINAL LEARNING OBJECTIVE. In a summer mountainous environment, construct a survival kit, in accordance with the references. (FMST.07.20)

ENABLING LEARNING OBJECTIVES

(1) Without the aid of references, choose from a given list in the criteria for a survival kit, in accordance with the references. (FMST.07.20a)

(2) Without the aid of references, choose from a given list one example of each criteria for a survival kit, in accordance with the references. (FMST.07.20b)

1. COMPONENTS FOR A SURVIVAL KIT

a. The environment is the key to the types of items you will need in your survival kit. How much equipment you put in your kit depends on how you will carry the kit. A kit on your body will have to be much smaller than one carried in a vehicle.

b. Always layer your survival kit, keeping the most important items on your body.

c. In preparing your survival kit, select items that can be used for more than one purpose.

A2

d. Your survival kit does not need to be elaborate. You only need functional items that will meet your needs and a case to hold them. The case might be a first aid case, an ammunition pouch, or another suitable case. This case should be:

(1) Water repellent or waterproof.

(2) Easy to carry or attach to your body.

(3) Suitable to accept various sized items.

(4) Durable.

e. When constructing a survival kit, you should have the following components: (FMST.07.20a)

(1) Fire starting items.

(2) Water procurement items.

(3) Food procurement items.

(4) Signaling items.

(5) First aid items.

(6) Shelter items.

2. **ITEMS CONTAINED WITHIN EACH COMPONENT** (FMST.07.20b)

a. Fire Starting Items.

(1) Matches.

(2) Magnifying glass.

(3) Flint and Steel.

(4) Lighter.

(5) Potassium Permanganate, with a container of sugar or anti-freeze.

(6) Prepackaged Tinder.

(a) Commercially Manufactured

(b) Cotton Balls and Petroleum Jelly

b. Water Procurement Items

(1) Water Disinfecting Chemicals.

(a) Iodine Tablets

(b) Betadine Solution

(c) Iodine Solution

(2) Metal Container. (Serves for boiling water)

(a) Canteen Cup

(b) Survival Kit Container

(c) Any suitable can that contained no petroleum products.

(3) Water Carrying Items.

(a) Canteen

(b) Plastic Bag

(c) Plastic/Metal/Glass Container which contained no petroleum products

c. Food Procurement Items

(1) Fish.

(a) Various sized hooks

(b) Various sized sinkers/weights

(c) Metal leaders and swivels

(d) Small weighted jigs

(e) Fishing line

i. Think about the size of fish for that environment when selecting weights and sizes.

(2) Game.

(a) Snares

i. Commercially Manufactured

ii. Aircraft Cable

iii. Tie Wire

(b) Bait

i. MRE Cheese Spread or Peanut Butter Package

(c) 550 Cord for Gill Net and Trap Construction

i. Engineer/Marking Tape

ii. Sling shot rubber and pouch

d. Signaling Items

(1) Day.

(a) Mirror

(b) Whistle

(c) Pyrotechnics (Smoke, Pen Flares)

(d) Air Panels

(2) Night.

(a) Pyrotechnics (Pen Flares, Star Clusters)

(b) Lights (Flashlight, Strobe, Chemlight)

(c) Whistle

e. Shelter Items

(1) Cordage.

(a) 550 Cord

(b) Wire

(c) Communication wire

(d) Tie wire

(2) Finger Saw.

(3) Sewing Kit with Needles for construction/repair of clothing.

(4) Tentage.

(a) Poncho

(b) Tarp

(c) Space blanket

(d) Plastic trash bags

f. First Aid Items

(1) Band-Aids.

(a) Steristrips

(b) Adhesive Tape

(c) Non-stick pads, 4x4's, Gauze, Battle Dressings

(d) Muslin Bandage

(2) Ointments.

(a) Burn

(b) Anti-septic

(3) Miscellaneous.

(a) Salt

(b) Sugar

(c) Eye Wash

(d) Alcohol prep pads

(e) Suture Kit

(f) Scalpel

(g) Vile of Yarrow

g. Miscellaneous items

(1) Fingernail clippers

(2) Compass

(3) Notebook with pen or pencil

(4) Wood eye screws and nails

(5) Surgical tubing

NOTE: It is assumed that the Marine is always carrying a high quality fixed bladed knife, a multi-tool knife, and a sharpening stone.

UNITED STATES MARINE CORPS
Mountain Warfare Training Center
Bridgeport, California 93517-5001

FMST.07.23
04/05/02

STUDENT OUTLINE

SIGNALLING AND RECOVERY

TERMINAL LEARNING OBJECTIVE Given a survival situation in any type of environmental condition and minimal equipment and resources, signal for aid to facilitate individual or group recovery per the references. (FMST 07.23)

ENABLING LEARNING OBJECTIVES

(1) Without the aid of references and from a given list, choose the audio international distress Signal, in accordance with the references. (FMST.07.23a)

(2) Without the aid of references and from a given list, choose the visual international distress signal, in accordance with the references. (FMST.07.23b)

(3) Without the aid of references and given a list of messages, match in writing the international code symbols to each corresponding message, in accordance with the references. (FMST.07.23c)

OUTLINE

1. **SIGNALING DEVICES**. The equipment listed below are items that may be on your body or items inside an aircraft. Generally, these items are used as signaling devices while on the move. They must be accessible for use at any moment's notice. Additionally, in a summer mountainous environment, Marines may experience areas that are snow covered and must be familiar with the effects that snow will have on specific signaling devices.

   a. Pyrotechnics. Pyrotechnics include star clusters and smoke grenades. When using smoke grenades in snow pack, a platform must be built. Without a platform, the smoke grenade will sink into the snow pack and the snow will absorb all smoke. A rocket parachute flare or a hand flare has been sighted as far away as 35 miles, with an average of 10 miles. Pyrotechnic flares are effective at night, but during daylight their detectability ranges are reduced by 90 percent.

A3

b. M-186 Pen Flare: The M-186 Pen Flare is a signaling device carried in the vest of all crew chiefs and pilots. Remember to cock the gun prior to screwing in the flare.

c. Strobe Light: A strobe light is generally carried in the flight vests of all crew chiefs and pilots. It can be used at night for signaling. Care must be taken because a pilot using goggles may not be able to distinguish a flashing strobe from hostile fire.

d. Flashlight: By using flashlights, a Morse code message can be sent. A SOS distress call consists of sending three dots, three dashes, and three dots. Keep repeating this signal.

e. Whistle: The whistle is used in conjunction with the audio international distress signal. It is used to communicate with forces on the ground.

f. AN/PRC-90 & AN/PRC-112: The AN/ PRC 90 survival radio is a part of the aviator's survival vest. The AN/PRC-112 will eventually replace the AN/PRC-90. Both radios can transmit either tone (beacon) or voice. Frequency for both is **282.8 for voice**, and **243.0 for beacon**. Both of these frequencies are on the UHF Band.

g. Day/Night Flare: The day/night flare is a good peacetime survival signal. The flare is for night signaling while the smoke is for day. A red cap identifies the older version flare with three nubbins while the new generation has three rings around the body for identification during darkness. The flare burns for approximately 20 second while the smoke burns for approximately 60 seconds.

NOTE: Once one end is used up, douse in water to cool and save the other end for future use.

h. Signal Mirror: A mirror or any shiny object can be used as a signaling device. It can be used as many times as needed. Mirror signals have been detected as far away as 45 miles and from as high as 16,000', although the average detection distance is 5 miles. It can be concentrated in one area, making it secure from enemy observation. A mirror is the best signaling device for a survivor; however, it is only as effective as its user. Learn how to use one now, before you find yourself in a survival situation.

(1) Military signal mirrors have instructions on the back showing how to use it. It should be kept covered to prevent accidental flashing that may be seen by the enemy.

(2) Any shiny metallic object can be substituted for a signal mirror.

(3) Haze, ground fog, or a mirage may make it hard for a pilot to spot signals from a flashing object. So, if possible, get to the highest point in your area when flashing. If you can't determine the aircraft's location, flash your signal in the direction of the aircraft noise.

AIMING THE SIGNAL MIRROR

2. METHODS OF COMMUNICATION

a. Audio: Signaling by means of sound may be good, but it does have some disadvantages.

(1) It has limited range unless you use a device that will significantly project the sound.

(2) It may be hard to pinpoint one's location due to echoes or wind.

(3) International Distress Signal. (FMST.07.23a) The survivor will make six blasts in one minute, returned by three blasts in one minute by the rescuer.

b. Visual: Visual signals are generally better than audio signals. They will pinpoint your location and can been seen at a greater distances under good weather conditions.

(1) The visual international distress symbol is recognized by a series of three evenly spaced improvised signaling devices. (FMST.07.23b)

3. **IMPROVISED SIGNALING DEVICES**. Improvised signaling devices are generally static in nature. They must be placed in a position to be seen by rescuers. They are made from any resources available, whether natural or man-made.

a. Smoke Generator: The smoke generator is an excellent improvised visual signaling device. It gives the survivor the flexibility to signal in either day or night conditions. This type of signal has been sighted as far away as 12 miles, with an average distance of 8 miles. Smoke signals are most effective in calm wind conditions or open terrain, but effectiveness is reduced with wind speeds above 10 knots. Build them as soon as time and the situation permits, and protect them until needed.

(1) Construct your fire in a natural clearing or along the edge of streams (or make a clearing). Signal fires under dense foliage will not be seen from the air.

(2) Find two logs, 6 - 10 inches in diameter, and approximately five feet long. Place the two logs parallel to each other with 3 - 4 feet spacing.

(3) Gather enough sticks, approximately two inches in diameter and four feet long, to lie across the first two logs. This serves as a platform for the fire.

(4) Gather enough completely dry branches to build a pyramid fire. The pyramid fire should be 4 feet by 4 feet by 2 feet high.

(5) Place your tinder under the platform.

(6) Gather enough pine boughs to lay on top of the pyramid fire. This serves to protect the fire and the tinder.

(7) To light, remove the pine bough and ignite the tinder. If available, construct a torch to speed up the lighting process, especially for multiple fires.

SMOKE GENERATOR

(8) To create a smoke effect during the day light hours, place the pine bough on the ignited fire.

(9) Placing a smoke grenade or colored flare under the platform will change the color of the smoke generated. Remember that you want the fire to draw in the colored smoke, which will create a smoke color that contrasts with the background will increase the chances of success.

b. Arrangement or alteration of natural materials: Such things as twigs or branches, can be tramped into letters or symbols in the snow and filled in with contrasting materials. To attract more attention, ground signals should be arranged in big geometric patterns.

(1) International symbols. The following symbols are internationally known. (FMST.07.23c)

Number	Message	Code symbol
1	REQUIRE ASSISTANCE	V
2	REQUIRE MEDICAL ASSISTANCE	X
3	NO OR NEGATIVE	N
4	YES OR AFFIRMATIVE	Y
5	PROCEED IN THIS DIRECTION	↑

INTERNATIONAL SYMBOLS

(2) Shadows: If no other means are available, you may have to construct mounds that will use the sun to cast shadows. These mounds should be constructed in one of the International Distress Patterns. To be effective, these shadow signals must be oriented to the sun to produce the best shadows. In areas close to the equator, a North—South line gives a shadow anytime except noon. Areas further north or south of the equator require the use of East—West line or some point of the compass in between to give the best result.

(3) Size: The letters should be large as possible for a pilot or crew to spot. Use the diagram below to incorporate the size to ratio for all letter symbols.

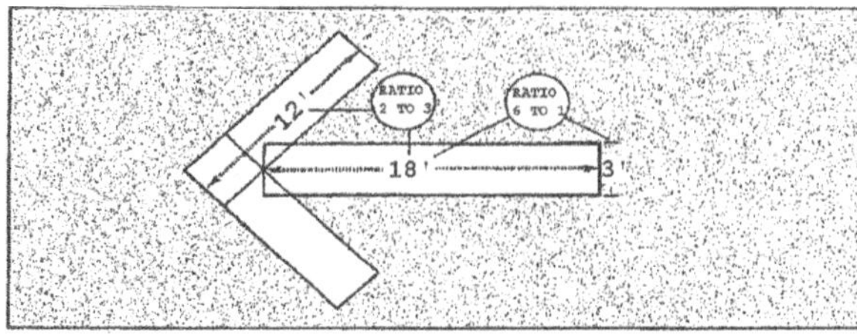

SIZE AND RATIO

(4) Contrast: When constructing letter symbols, contrast the letter from the surrounding vegetation and terrain. Ideally, bring material from another location to build the letter. This could be clothing, air panels, space blanket, etc.

(a) On snow, pile bough or use sea dye from a LPP (Life preserver, personal). Fluorescent sea dye markers have been sighted as far away as 10 miles, although the average detection distance is 3 miles.

4. **AIR TO GROUND COMMUNICATIONS**. Air to ground communications can be accomplished by standard aircraft acknowledgments.

a. Aircraft will indicate that ground signals have been seen and understood by:

(1) Rocking wings from side to side: This can be done during the day or in bright moonlight.

b. Aircraft will indicate that ground signals have been seen but not understood by:

(1) Making a complete, clockwise circle during the day or in bright moonlight.

5. **RECOVERY**. Marines trapped behind enemy lines in future conflicts may not experience quick recovery. Marines may have to move to a place that minimizes risk to the recovery force. No matter what signaling device a Marine uses, he must take responsibility for minimizing the recovery force's safety.

a. Placement Considerations: Improvised signaling devices, in a hostile situation, should not be placed near the following areas due to the possibility of compromise:

(1) Obstacles and barriers.

(2) Roads and trails.

(3) Inhabited areas.

(4) Waterways and bridges.

(5) Natural lines of drift.

(6) Man-made structures.

(7) All civilian and military personnel.

b. Tactical Consideration: The following tactical considerations should be adhered to prior to employing an improvised signaling device.

(1) Use the signals in a manner that will not jeopardize the safety of the recovery force or you.

(2) Locate a position, which affords observation of the signaling device and facilitates concealed avenues of escape (if detected by enemy forces). Position should be located relatively close to extract site in order to minimize "time spent on ground" by the recovery force.

(3) Maintain continuous security through visual scanning and listening while signaling devices is employed. If weapon systems are fire and/or observation should cover available, signaling devices.

(4) If enemy movement is detected in the area, attempt to recover the signaling device, if possible.

(5) Employ improvised signaling devices only during the prescribed times, if briefed in the mission order.

c. Recovery Devices: In mountainous terrain, a helicopter landing may be impossible due to ground slope, snow pack, or vegetation. The survivor must be familiar with recovery devices that may be aboard the aircraft.

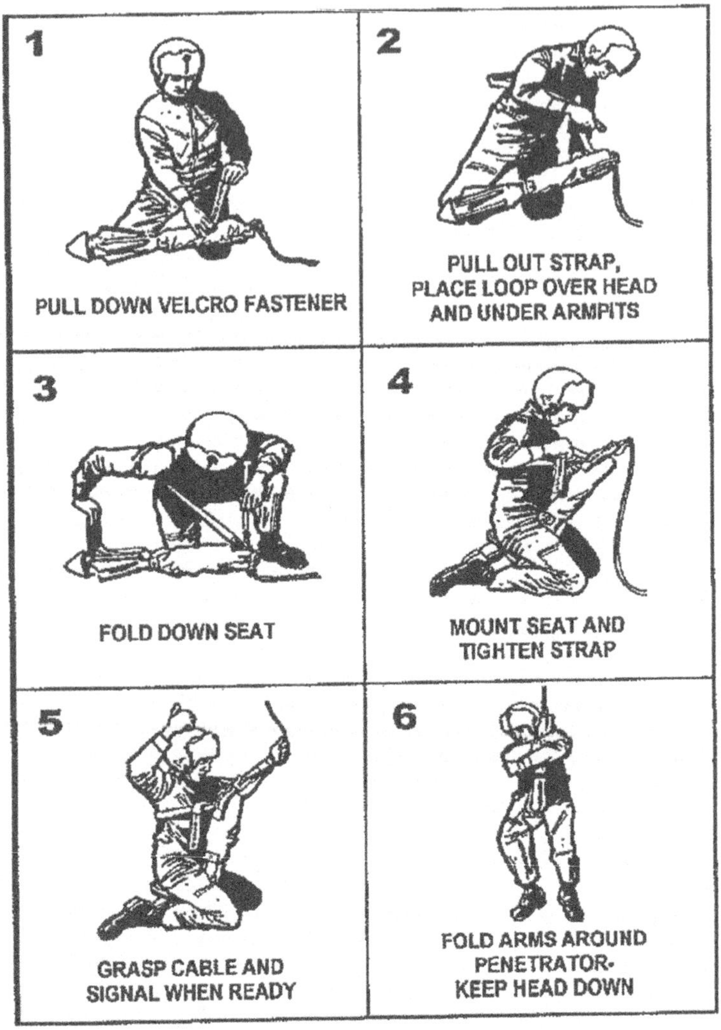

JUNGLE PENETRATOR

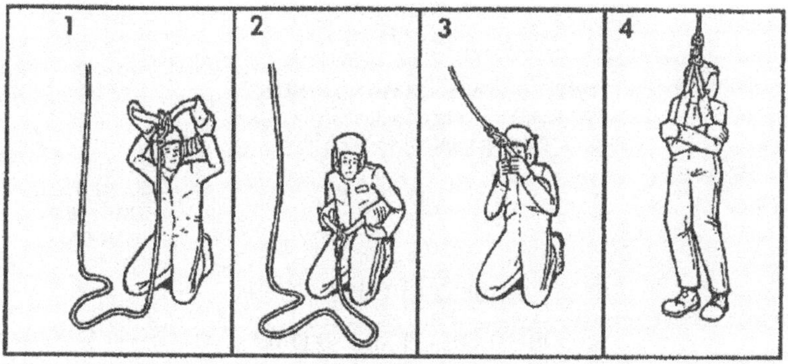

SLING HOIST

d. Recovery by other than aircraft. Recovery by means other than aircraft may occur. Unit Sop's should include signaling and link-up with forces at the following locations:

(1) Border Crossings. The evader who crosses into a neutral country is subject to detention by that country for the duration of the war.

(2) FEBA/FLOT.

(a) Static: Recovery along a static FEBA is always difficult. Under these conditions, enemy and friendly forces can be expected to be densely arrayed and well camouflaged, with good fields of fire. Attempts to penetrate the FEBA should be avoided.

(b) Advancing: Individuals isolated in front of advancing friendly units should immediately take cover and wait for the friendly units to overrun their position.

(c) Retreating: Individuals between opposing forces should immediately take cover and wait for enemy units to pass over their position. After most enemy units have moved on, evaders should try to link up with other isolated friendly elements and return to friendly forces.

(3) Link-up with friendly patrols. Unit authentication numbers and/or locally developed codes may assist the evader to safely make contact in or around the FEBA and when approached by friendly forces.

UNITED STATES MARINE CORPS
Mountain Warfare Training Center
Bridgeport, California 93517-5001

FMST.07.21
FMST.07.04
04/07/02

STUDENT OUTLINE

EXPEDIENT SHELTERS AND FIRES

TERMINAL LEARNING OBJECTIVES

(1) In a summer mountainous environment, construct expedient shelters, in accordance with the references. (FMST.07.21)

(2) In a summer mountainous environment, construct fires, in accordance with the references. (FMST.07.04)

b. **ENABLING LEARNING OBJECTIVES**

(1) Without the aid of references and from a given list, choose the four characteristics of a safe expedient shelter, in accordance with the references. (FMST.07.21a)

(2) Without the aid of references and from a given list, choose the four hazards avoid when using natural shelters, in accordance with the references. (FMST.07.21b)

(3) Without the aid of references and from a given list, choose one of man-made expedient shelters, in accordance with the references. (FMST.07.21c)

(4) Without the aid of references, start a fire using a primitive method, in accordance with the references. (FMST.07.04a)

A4

OUTLINE

1. **BASIC CHARACTERISTICS FOR SHELTER**. (FMST.07.21a) Any type of shelter, whether it is a permanent building, tentage, or an expedient shelter must meet six basic characteristics to be safe and effective. The characteristics are:

    a. Protection from the Elements: The shelter must provide protection from rain, snow, wind, sun, etc.

    b. Heat Retention: It must have some type of insulation to retain heat; thus preventing the waste of fuel.

    c. Ventilation: Ventilation must be constructed, especially if burning fuel for heat. This prevents the accumulation of carbon monoxide. Ventilation is also needed for carbon dioxide given off when breathing.

    d. Drying Facility: A drying facility must be constructed to dry wet clothes.

    e. Free from Natural Hazards: Shelters should not be built in areas of avalanche hazards, under rock fall or "standing dead" trees have the potential to fall on your shelter.

    f. Stable: Shelters must be constructed to withstand the pressures exerted by severe weather.

2. **NATURAL SHELTERS**. Natural shelters are usually the preferred types because they take less time and materials construct. The following may be made into natural shelters with some modification.

    a. Caves or Rock Overhangs: Can be modified by laying walls of rocks, logs or branches across the open sides.

    b. Hollow Logs: Can be cleaned or dug out, then enhanced with ponchos, tarps or parachutes hung across the openings.

    c. Hazards of Natural Shelters: (FMST.04.21b)

        (1) Animals: Natural shelters may already be inhabited (i.e. bears, coyotes, lions, rats, snakes, etc.). Other concerns from animals may be disease from scat or decaying carcasses.

        (2) Lack of Ventilation: Natural shelters may not have adequate ventilation. Fires may be built inside for heating or cooking but may be uncomfortable or even dangerous because of the smoke build up.

        (3) Gas Pockets: Many caves in a mountainous region may have natural gas pockets in them.

(4) Instability: Natural shelters may appear stable, but in reality may be a trap waiting to collapse.

3. **MAN-MADE SHELTERS**. (FMST.04.21c) Many configurations of man-made shelters may be used. Over-looked man-made structures found in urban or rural environments may also provide shelter (i.e. houses, sheds, or barns). Limited by imagination and materials available, the following man-made shelters can be used in any situation.

a. Poncho Shelter.

b. Sapling Shelter.

c. Lean-to.

d. Double lean-to.

e. A-frame Shelter.

f. Fallen Tree Bivouac.

4. **CONSTRUCTION OF MAN-MADE SHELTERS**. To maximize the shelter's effectiveness, Marines should take into consideration the following prior to construction.

a. Considerations.

(1) Group size.

(2) Low silhouette and reduced living area dimensions for improved heat conservation.

(3) Avoid exposed hill tops, valley floors, moist ground, and avalanche paths.

(4) Create a thermal shelter by applying snow, if available, to roof and sides of shelter.

(5) Location of site to fire wood, water, and signaling, if necessary.

(6) How much time and effort needed to build the shelter.

(7) Can the shelter adequately protect you from the elements (sun, wind, rain, and snow). Plan on worst case scenario.

(8) Are the tools available to build it. If not, can you make improvised tools?

(9) Type and amount of materials available to build it.

(10) When in a tactical environment, you must consider the following:

(a) Provide concealment from enemy observation.

(b) Maintain camouflaged escape routes.

(c) Use the acronym BLISS as a guide.

B – Blend in with the surroundings.
L –Low silhouette.
I – Irregular shape.
S – Small.
S – Secluded located.

b. Poncho Shelter: This is one of the easiest shelters to construct. Materials needed for construction are cord and any water-repellent material (i.e. poncho, parachute, and tarp). It should be one of the first types of shelter considered if planning a short stay in any one place.

(1) Find the center of the water-repellent material by folding it in half along its long axis.

(2) Suspend the center points of the two ends using cordage.

(3) Stake the Four Corners down, with sticks or rocks.

c. Sapling Shelter: This type of shelter is constructed in an area where an abundance of saplings are growing. It is an excellent evasion shelter.

(1) Find or clear an area so that you have two parallel rows of saplings at least 4' long and approximately 1 ½' to 2' apart.

(2) Bend the saplings together and tie them to form several hoops, which will form the framework of the shelter.

(3) Cover the hoop with a water-repellent covering.

(4) The shelter then may be insulated with leaves, brush, snow, or boughs.

(5) Close one end permanently. Hang material over the other end to form a door.

d. Lean-to: A lean-to is built in heavily forested areas. It requires a limited amount of cordage to construct. The lean-to is an effective shelter but does not offer a great degree of protection from the elements.

(1) Select a site with two trees (4-12" in diameter) spaced far enough apart that a man can lie down between them. Two sturdy poles can be substituted by inserting them into the ground the proper distance apart.

(2) Cut a pole to support the roof. It should be at least 3-4" in diameter and long enough to extend 4-6" past both trees. Tie the pole horizontally between the two trees, approximately 1 meter off the deck.

(3) Cut several long poles to be used as stringers. They are placed along the horizontal support bar approximately every 1 ½' and lay on the ground. All stringers may be tied to or laid on the horizontal support bar. A short wall or rocks or logs may be constructed on the ground to lift the stringers off the ground, creating additional height and living room dimensions.

(4) Cut several saplings and weave them horizontally between the stringers. Cover the roof with water-repellent and insulating material.

LEAN-TO

c. Double lean-to: The double lean-to shelter is constructed for 2-5 individuals. It is constructed by making two lean-tos and placing them together.

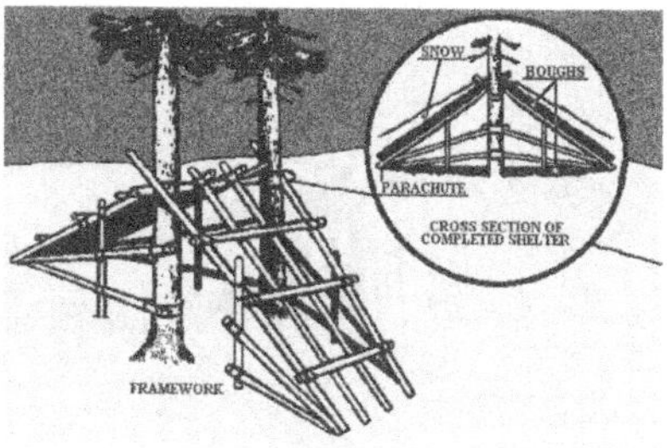

DOUBLE LEAN-TO

f. A-Frame Shelter: An A-Frame shelter is constructed for 1-3 individuals. After the framework is constructed, bough/tentage is interwoven onto the frame and snow, if available, is packed onto the outside for insulation.

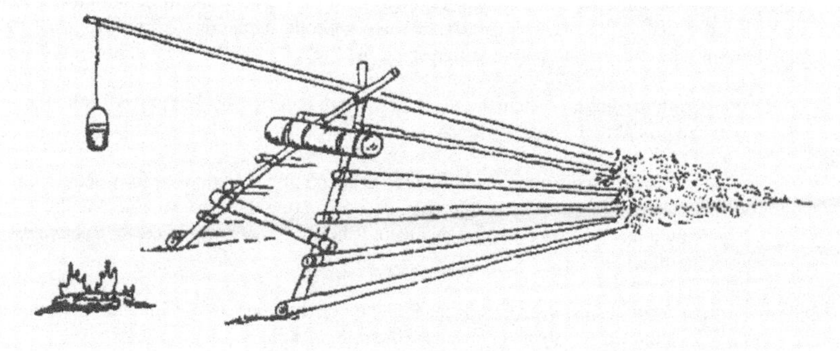

A-FRAME

g. Fallen Tree Bivouac: The fallen tree bivouac is an excellent shelter because most of the work has already been done.

(1) Ensure the tree is stable prior to constructing.

(2) Branches on the underside are cut away making a hollow underneath.

(3) Place additional insulating material to the top and sides of the tree.

(4) A small fire is built outside of the shelter.

FALLEN TREE BIVOUAC

5. **REFLECTOR WALLS.** Heating a shelter requires a slow fire that produces lots of steady heat over a long period of time. A reflector wall should be constructed for all open-ended shelters. A reflector wall is constructed with a flat rock or a stack of green logs propped behind the fire. A surprising amount of heat will bounce back from the fire into the shelter.

6. **FIRES**. Fires fall into two main categories: those built for cooking and those built for warmth and signaling. The basic steps are the same for both: preparing the fire lay, gathering fuel, building the fire, and properly extinguishing the fire.

   a. Preparing the fire lay: There are two types of fire lays: fire pit and Dakota hole. Fire pits are probably the most common.

      (1) Create a windbreak to confine the heat and prevent the wind from scattering sparks. Place rocks or logs used in constructing the fire lay parallel to the wind. The prevailing downwind end should be narrower to create a chimney effect.

      (2) Avoid using wet rocks. Heat acting on the dampness in sandstone, shale, and stones from streams may cause them to explode.

      (3) Dakota Hole: The Dakota Hole is a tactical fire lay. Although no fire is 100% tactical, this fire lay will accomplish certain things:

         (a) Reduces the signature of the fire by placing it below ground.

         (b) Provides more of a concentrated heat source to boil and cook, thus preserving fuel and lessening the amount of burning time.

         (c) By creating a large air draft, the fire will burn with less smoke than the fire pit.

         (d) It is easier to light in high winds.

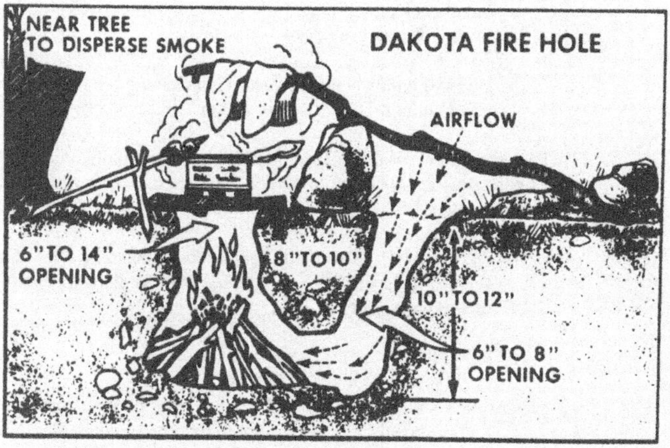

DAKOTA HOLE

b. Gather Fuel: Many Marines take shortcuts when gathering firewood. Taking a few extra minutes can mean the difference between ease and frustration when building a fire.

(1) Tinder: Tinder is the initial fuel. It should be fine and dry. Gather a double handful of tinder for the fire to be built and an extra double handful to be stored in a dry place for the following morning. Dew can moisten tinder enough to make lighting the fire difficult. Some examples are:

(a) Shredded cedar/juniper bark, pine needles.

(b) Dry grass.

(c) Slivers shaved from a dry stick.

(d) Hornet's nest.

(e) Natural fibers from equipment supplemented with pine pitch (i.e., cotton battle dressing).

(f) Cotton balls and petroleum jelly or Char-cloth.

NOTE: Sticks used for tinder should be dry and not larger than the diameter of a toothpick.

(2) Kindling: This is the material that is ignited by the tinder that will burn long enough to ignite the fuel.

(a) Small sticks/twigs pencil-thick up to the thickness of the thumb. Ensure that they are dry.

(b) Due to a typically large resin content, evergreen limbs often make the best kindling. They burn hot and fast, but typically do not last long.

(3) Fuel Wood: Fuel Wood is used to keep the blaze going long enough to fulfill its purpose. Ideally, it should burn slow enough to conserve the woodpile, make plenty of heat, and leave an ample supply of long-lasting coals.

(a) Firewood broken from the dead limbs of standing trees or windfalls held off the ground will have absorbed less moisture and therefore should burn easily.

(b) Refrain from cutting down live, green trees.

(c) Softwoods (evergreens and conifers) will burn hot and fast with lots of smoke and spark, leaving little in the way of coals. Hardwoods (broad leaf trees) will burn slower with less smoke and leave a good bed of coals.

(d) Learn the woods indigenous to the area. Birch, dogwood, and maple are excellent fuels. Osage orange, ironwood, and manzanita, though difficult to break up, make terrific coals. Aspen and cottonwood burn clean but leave little coals.

(e) Stack your wood supply close enough to be handy, but far enough from the flames to be safe. Protect your supply from additional precipitation.

(f) If you happen to go down in an aircraft that has not burned a mixture of gas and oil may be used. Use caution when igniting this mixture.

c. Building the Fire. The type of fire built will be dependent upon its intended use, either cooking or heating and signaling.

(1) Cooking Fires. The following listed fires are best used for cooking:

(a) Teepee Fire: The teepee fire is used to produce a concentrated heat source, primarily for cooking. Once a good supply of coals can be seen, collapse the teepee and push embers into a compact bed.

TEEPEE FIRE

(2) Heating and Signaling Fires.

(a) Pyramid Fire: Pyramid fires are used to produce large amounts of light and heat. They will dry out wet wood or clothing.

PYRAMID FIRE

d. Starting Fires. Starting a fire is done by a source of ignition and falls into two categories, modern igniters and primitive methods.

(1) Modern Methods: Modern igniters use modern devices we normally think of to start a fire. Reliance upon these methods may result in failure during a survival situation. These items may fail when required to serve their purpose.

(a) Matches and Lighters. Ensure you waterproof these items.

(b) Convex Lens. Binocular, camera, telescopic sights, or magnifying lens are used on bright, sunny days to ignite tender.

(c) Flint and Steel: Sometimes known as metal matches or "Mag Block". Scrape your knife or carbon steel against the flint to produce a spark onto the tinder. Some types of flint & steel designs will have a block of magnesium attached to the device, which can be shaved onto the tinder prior to igniting. Other designs may have magnesium mixed into the flint to produce a higher quality of spark.

(2) Primitive Methods. Primitive fire methods are those developed by early man. There are numerous techniques that fall into this category. The only method that will be taught at MCMWTC is the Bow & Drill.

(3) Bow & Drill. The technique of starting a fire with a bow & drill is a true field expedient fire starting method, which requires a piece of cord and knife from your survival kit to construct. The components of the bow & drill are bow, drill, socket, and fireboard, ember patch, and bird's nest.

(a) Bow: The bow is a resilient, green stick about 3/4 of an inch in diameter and 30-36 inches in length. The bowstring can be any type of cord, however, 550 cord works best. Tie the string from one end of the bow to the other, without any slack.

(b) Drill: The drill should be a straight, seasoned hardwood stick about 1/2 to 3/4 of an inch in diameter and 8 to 12 inches in length. The top end is tapered to a blunt point to reduce friction generated in the socket. The bottom end is slightly rounded to fit snugly into the depression on the fireboard.

(c) Socket: The socket is an easily grasped stone or piece of hardwood or bone with a slight depression on one side. Use it to hold the drill in place and to apply downward pressure.

(d) Fire board: The fire board is a seasoned softwood board which should ideally be 3/4 of an inch thick, 2-4 inches wide, and 8-10 inches long. Cut a depression 3/4 of an inch from the edge on one side of the fireboard. Cut a U-shape notch from the edge of the fireboard into the depression. This notch is designed to collect and form an ember, which will be used to ignite the tinder.

(e) Ember Patch: The ember patch is made from any type of suitable material (i.e., leather, aluminum foil, and bark). It is used to catch and transfer the ember from the fireboard to the bird's nest. Ideally, it should be 4 inches by 4 inches in size.

(f) Bird's Nest: The bird's nest is a double handful of tinder, which will be made into the shape of a nest. Tinder must be dry and finely shredded material (i.e., outer bark from juniper/cedar/sage brush or inner bark from cottonwood/aspen or dry grass/moss). Lay your tinder out in two equal rows about 4 inches wide and 8-12 inches long. Loosely roll the first row into a ball and knead the tinder to further break down the fibers. Place this ball perpendicular onto the second row of tinder and wrap. Knead the tinder until all fibers of the ball are interwoven. Insert the drill half way into the ball to form a partial cylinder. This is where the ember will be placed.

(4) Producing a fire using the bow & drill.

(a) Place the ember patch under the U-shaped notch.

(b) Assume the kneeling position, with the left foot on the fireboard near the depression.

(c) Load the bow with the drill. Ensure the drill is between the wood of the bow and bow string. Place the drill into the depression on the fireboard. Place the socket on the tapered end of the drill.

(d) Use the left hand to hold the socket while applying downward pressure.

(e) Use the right hand to grasp the bow. With a smooth sawing motion, move the bow back and forth to twirl the drill.

(f) Once you have established a smooth motion, smoke will appear. Once smoke appears, apply more downward pressure and saw the bow faster.

(g) When a thick layer of smoke has accumulated around the depression, stop all movement. Remove the bow, drill, and socket from the fire board, without moving the fireboard. Carefully remove your left foot off the fireboard.

(h) Gently tap the fireboard to ensure all of the ember has fallen out of the U-shaped notch and is lying on the ember patch. Remove the fireboard.

(i) Slowly fan the black powder to solidify it into a glowing ember. Grasping the ember patch, carefully drop the ember into the cylinder of the bird's nest.

(j) Grasp the bird's nest with the cylinder facing towards you and parallel to the ground. Gently blow air into the cylinder. As smoke from the nest becomes thicker, continue to blow air into the cylinder until fire appears.

(5) Trouble Shooting the Bow & Drill

(a) Drill will not stay in depression- Apply more downward pressure and/or increase width/depth of depression.

(b) Drill will not twirl- lessens the amount of downward pressure and/or tighten bowstring.

(c) Socket smoking- Lessen the amount of downward pressure. Wood too soft when compared to hardness of drill. Add some lubrication: animal fat, oil, or grease.

(d) No smoke- Drill and fireboard are the same wood. Wood may not be seasoned. Check drill to ensure that it is straight. Keep left hand locked against left shin while sawing.

(e) Smoke but no ember- U-shaped notch not cut into center of the depression.

(f) Bowstring runs up and down drill- use a locked right arm when sawing. Check drill to ensure that it is straight. Ensure bowstring runs over the top of the left boot.

(g) Birds nest will not ignite- Tinder not dry. Nest woven too tight. Tinder not kneaded enough. Blowing too hard (ember will fracture).

e. Extinguishing the Fire. The fire must be properly extinguished. This is accomplished by using the drown, stir, and feel method.

(1) Drown the fire by pouring at water in the fire lay.

(2) Stir the ember bed to ensure that the fire is completely out.

(3) Check the bed of your fire by feeling for any hot spots.

(4) If any hot spots are found, start the process all over again.

UNITED STATES MARINE CORPS
Mountain Warfare Training Center
Bridgeport, California 93517-5001

FMST.07.22
04/07/02

STUDENT OUTLINE

SURVIVAL NAVIGATION

LEARNING OBJECTIVES

TERMINAL LEARNING OBJECTIVE. In a summer mountainous environment, navigate in a survival situation, in accordance with the references. (FMST.07.22)

ENABLING LEARNING OBJECTIVES

(1)Without the aid of references and from a given list, choose one of the considerations for travel, in accordance with the references. (FMST.07.22a)

(2)Without the aid of references and from a given list, choose two methods for locating the North Star, in accordance with the references. (FMST.07.22b)

OUTLINE

1. **CONSIDERATIONS FOR STAYING OR TRAVELLING**. (FMST.07.22a)

   a. Stay with the aircraft or vehicle if possible. More than likely somebody knows where it was going. It is also a ready-made shelter.

   b. Leave only when:

      (1) Certain of present location; have known destination and the ability to get there.

      (2) Water, food, shelter, and/or help can be reached.

      (3) Convinced that rescue is not coming.

A5

c. If the decision is to travel, the following must also be considered:

(1) Which direction to travel and why.

(2) What plan is to be followed.

(3) What equipment should be taken

(4) How to mark the trail.

(5) Predicted weather.

d. If the tactical situation permits leave the following information at the departure point:

(1) Departure time.

(2) Destination.

(3) Route of travel/direction.

(4) Personal condition.

(5) Available supplies.

2. DAYTIME SURVIVAL NAVIGATION

a. Sun Movement. It is generally taken for granted that the sun rises in the east and sets in the west. This rule of thumb, however, is quite misleading. In fact, depending on an observer's latitude and the season, the sun could rise and set up to 50 degrees off of true east and west.

b. The following diagram and terms are essential to understanding how the sun and stars can help to determine direction:

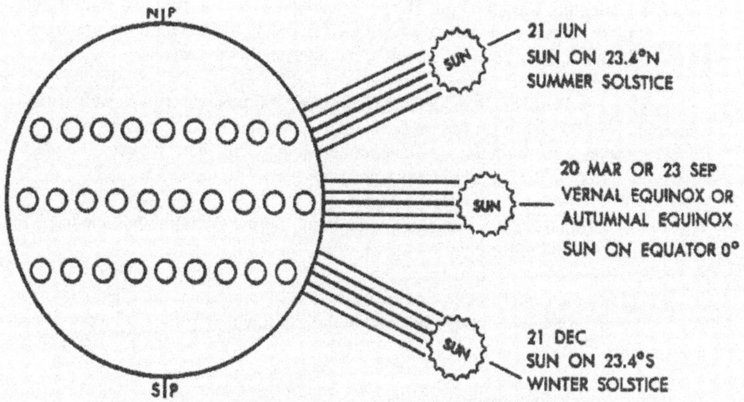

Position of the Sun at Equinox and Solstice

(1) Summer/Winter Solstice: (21 June/21 December) Two times during the year when the sun has no apparent northward or southward motion.

(2) Vernal/Autumnal Equinox: (20 March/23 September) Two times during the year when the sun crosses the celestial equator and the length of day and night are approximately equal.

c. Sun's Movement. As reflected in the diagram above, the sun continuously moves in a cycle from solstice to equinox, throughout each day, however, the sun travels a uniform arc in the sky from sunrise to sunset. Exactly halfway along its daily journey, the sun will be directly south of an observer (or north if the observer is in the Southern Hemisphere). This rule may not apply to observers in the tropics (between 23.5 degrees north and south latitude) or in the Polar Regions (60 degrees latitude). It is at this point that shadows will appear their shortest. The time at which this occurs is referred to as "local apparent noon."

d. Local Apparent Noon. Whenever using any type of shadow casting device to determine direction, "local apparent noon" (or the sun's highest point during the day) must be known. Local apparent noon can be determined by the following methods.

(1) Knowing sunrise and sunset from mission orders, i.e., sunrise 0630 and sunset 1930. Take the total amount of daylight (13 hours), divide by 2 (6 hours 30 minutes), and add to sunrise (0630 plus 6 hours 30 minutes). Based on this example, local apparent noon would be 1300.

(2) Using the string method. The string method is used to find two equidistant marks before and after estimated local apparent noon. The center point between these two marks represents local apparent noon.

e. Sun's Bearing. With an understanding of the sun's daily movement, as well as its seasonal paths, a technique is derived that will determine the true bearing of the sun at sunrise and sunset. With the aid of a circular navigational chart, we can accurately navigate based on the sun's true bearing:

(1) Determine the sun's maximum amplitude at your operating latitude using the top portion of the chart.

(2) Scale the center baseline of the chart where 0 appears as the middle number; write in the maximum amplitude at the extreme north and south ends of the baseline.

(3) Continue to scale the baseline; you should divide the baseline into 6 to 10 tick marks that represent equal divisions of the maximum amplitude.

(4) From today's date along the circumference, draw a straight line down until it intersects the baseline.

(5) The number this line intersects is today's solar amplitude. If the number is left of 0, it is"north" amplitude; if the number is right of 0, it is a "south" amplitude. Use the formula at the bottom of the chart to determine the sun's bearing at sunrise or sunset.

LATITUDE (N OR S)	5	10	15	20	25	30	35	40	45	50	55	60
MAXIMUM AMPLITUDE	24°	24°	24°	25°	26°	27°	29°	31°	34°	38°	44°	53°

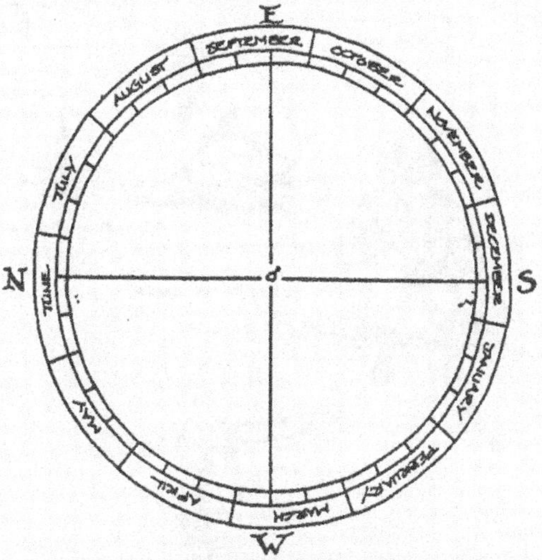

NORTH AMPLITUDES	SOUTH AMPLITUDES
SUNRISE BEARING = 90° - AMPLITUDE	SUNRISE BEARING = 90° + AMPLITUDE
SUNSET BEARING = 270° + AMPLITUDE	SUNSET BEARING = 270° - AMPLITUDE

Circular Navigational Chart

f. Shadow Stick Construction. This technique will achieve a cardinal direction within 10 degrees of accuracy if done within two hours of local apparent noon. Once again, this technique may be impractical near the Polar Regions, as shadows tend to be very long; similarly, in the tropics shadows are generally very small.

(1) Get a straight, 3-6 foot stick free of branches and pointed at the ends and 3-5 small markers: i.e., sticks, rocks, or nails.

(2) Place the stick upright in the ground and mark the shadow tip with a marker.

(3) Wait 10-15 minutes and mark shadow tip again with a marker.

(4) Repeat this until all of the markers are used.

SHADOW STICK METHOD

(5) The markers will form a West – East line.

(6) Put your left foot on the first marker and your right foot on the last marker, you will then be facing north.

3. **POCKET NAVIGATOR** The only material required is a small piece of paper or other flat-surface material upon which to draw the trace of shadow tips and a 1 to 2 inch pin, nail, twig, wooden matchstick, or other such device to serve as a shadow-casting rod.

a. Set this tiny rod upright on your flat piece of material so that the sun will cause it to cast a shadow. Mark the position where the base of the rod sits so it can be returned to the same spot for later readings. Secure the material so that it will not move and mark the position of the material with string, pebbles, or twigs, so that if you have to move the paper you can put it back exactly as it was. Now, mark the tip of the rod shadow.

b. As the sun moves, the shadow-tip moves. Make repeated shadow-tip markings every 15 minutes. As you make the marks of the shadow tip, ensure that you write down the times of the points.

c. At the end of the day, connect the shadow-tip markings. The result will normally be a curved line. The closer to the vernal or autumnal equinoxes (March 21 and September 23) the less pronounced the curvature would be. If it is not convenient or the tactical situation does not permit to take a full day's shadow-tip markings, your observation can be continued on subsequent days by orienting the pocket navigator on the ground so that the shadow-tip is aligned with a previously plotted point.

d. The markings made at the sun's highest point during the day, or solar noon, is the north—south line. The direction of north should be indicated with an arrow on the navigator as soon as it is determined. This north-south line is drawn from the base of the rod to the mark made at solar noon. This line is the shortest line that can be drawn from the base of the pin to the shadow-tip curve.

e. To use your pocket navigator, hold it so that with the shadow-tip is aligned with a plotted point at the specified point. I.e.; if it is now 0900 the shadow-tip must be aligned with that point. This will ensure that your pocket navigator is level. The drawn arrow is now oriented to true north, from which you can orient yourself to any desired direction of travel.

f. The pocket navigator will work all day and will not be out of date for approximately one week.

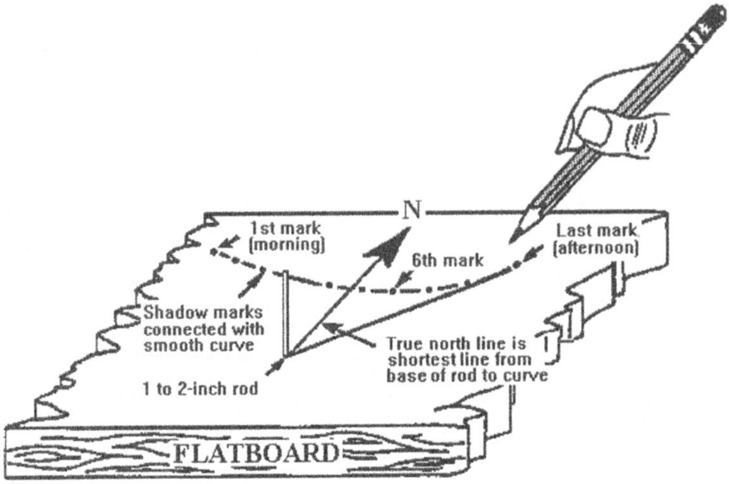

POCKET NAVIGATOR

4. NIGHTTIME SURVIVAL NAVIGATION

a. Mark North. To aid you in navigating at night, it is beneficial to watch where the sun goes down. If you're going to start moving after dark mark the northerly direction.

b. Locating the North Star. There are two methods used in locating the North Star. (FMST.07.22b)

(1) Using the Big Dipper (*Ursa Major*). The best indictors are the two "dippers ". The North Star is the last star in the handle of the Little Dipper, which is not the easiest constellation to find. However, the Big Dipper is one of the most prominent constellations in the Northern Hemisphere. The two lowest stars of the Big Dipper's cup act as pointers to the North Star. If you line up these two stars, they make a straight line that runs directly to the North Star. The distance to the North Star along this line is 5 times that between the two pointer stars.

(2) Using Cassiopeia (*Big M or W*). Draw a line straight out from the center star, approximately half the distance to the Big Dipper. The North Star will be located there.

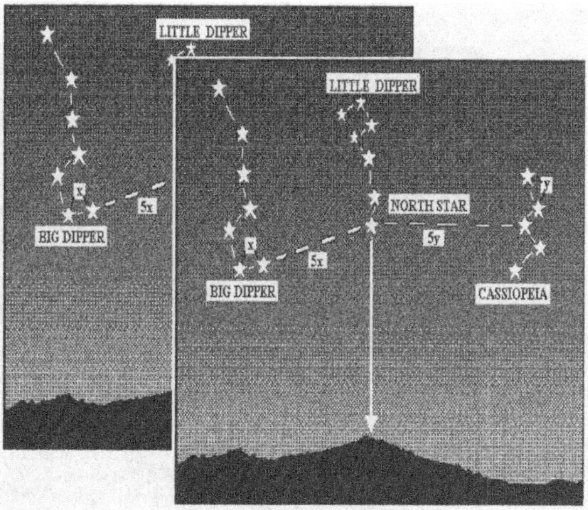

LOCATING THE NORTH STAR

NOTE: Because the Big Dipper and Cassiopeia rotate around the North Star, they will not always appear in the same position in the sky. In the higher latitudes, the North Star is less effective for the purpose of orienting because it appears higher in the sky. At the center of the Arctic Circle, it would be directly overhead and all directions lead south.

c. Southern Cross. In the Southern Hemisphere, Polaris is not visible. There, the Southern Cross is the most distinctive constellation. An imaginary line through the long axis of the Southern Cross, or True Cross, points towards a dark spot devoid of stars approximately three degrees offset from the South Pole. The True Cross should not be confused with the larger cross nearby know as the False Cross, which is less bright, more widely spaced, and has five stars. Two closely spaced, very bright stars that trail behind the crosspiece can confirm the True Cross. These two stars are often easier to pick out than the cross itself. Look for them. Two of the stars in the True Cross are among the brightest stars in the heavens; they are the stars on the southern and eastern arms. The stars on the northern and western arms are not as conspicuous, but are bright.

NOTE: The imaginary point depicted in the picture is the dark spot devoid of stars.

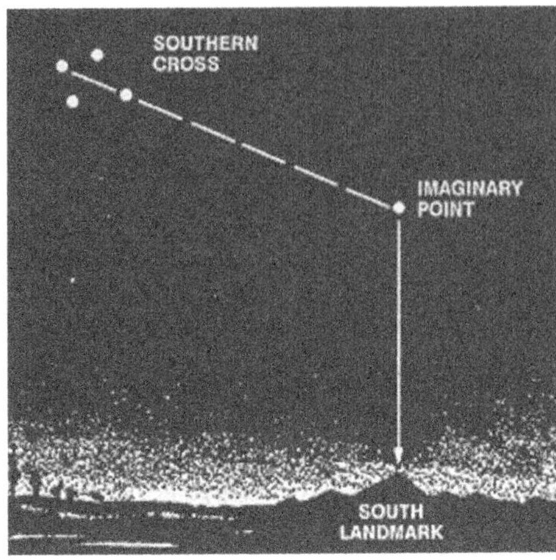

SOUTHERN CROSS

d. Moon Navigator. Like the sun, the moon rises in the east and sets in the west. Use the same method of the shadow stick as you did during the day.

5. **IMPROVISED COMPASSES**. There are three improvised techniques to construct a compass.

a. Synthetic technique. The required items are a piece of synthetic material, (i.e., parachute cloth), and a small piece of iron or steel that is long, thin, and light. Aluminum or yellow metals won't work (only things that rust will do). A pin or needle is perfect, but a straightened paper clip, piece of steel baling wire or barbed wire could also work.

(1) Stroke the needle repeatedly in one direction against the synthetic material. Ensure that you lift the material a few inches up into the air at the end of each stroke, returning to the beginning of the needle before descending for another stroke in the same direction. Do this approximately 30 strokes. This will magnetize the needle.

(2) Float the metal on still water using balled up paper, wood chip, or leaf. Gather some water in a non-magnetic container or a scooped out recess in the

ground, such as a puddle. Do not use a "tin can" which is made of steel. (Aluminum can would be fine.) Place the float on the water, then the metal on it. It will slowly turn to orient itself.

b. Magnet technique. You will achieve the same results by using a magnet. Follow the same steps as you did with the synthetic material. The magnets you are most likely to have available to you are those in a speaker or headphones of a radio.

c. Magnetization through a battery. A power source of 2 volts or more from a battery can be used with a short length of insulated wire to magnetize metal. Coil the wire around a needle. If the wire is non-insulated, wrap the needle with paper or cardboard. Attach the ends to the battery terminals for 5 minutes.

d. Associated problems with improvised compasses. The following are common problems with all improvised compasses.

(1) Soft steel tends to lose its magnetism fairly quickly, so you will have to demagnetize your needle occasionally, though you should not have to do this more than two or three times a day.

(2) Test your compass by disturbing it after it settles. Do this several times. If it returns to the same alignment, you're OK. It will be lined up north and south, though you will have to determine by other means which end is north. Use the sun, stars, or any other natural signs in the area.

(3) Remember that this will give magnetic north. In extreme northern latitudes, the declination angle can be extreme.

6. SURVIVAL NAVIGATION TECHNIQUES

a Navigator

(1) Employ a navigation method.

(2) Find the cardinal direction.

(3) Pick a steering mark in the desired direction of travel.

b. Maintain a Log: The possibility may arise when you will not have a map of the area. A log will decrease the chance of walking in circles.

(1) Construction.

(a) Use any material available to you i.e., paper, clothing, MRE box, etc.

(b) Draw a field sketch annotating North, prominent terrain features, and Friendly/enemy position.

(2) Maintenance.

(a) Annotate distance traveled, elevation gained and lost, and cardinal directions.

(b) Maintain and update field sketch as movement progresses.

(c) Ensure readability of your field sketch. (I.e., don't clutter the sketch so much that it can't be read.)

c. <u>During Movement Constantly Refer To</u>.

(1) Log.

(2) Steering marks.

d. <u>Actions If You Become Lost</u>.

(1) Immediate action

(a) Orient your sketch. This will probably make your mistake obvious.

(2) Corrective action

(a) Backtrack using steering marks until you have determined the location of your error.

(b) Re-orient your sketch.

(c) Select direction of travel and continue to march.

UNITED STATES MARINE CORPS
Mountain Warfare Training Center
Bridgeport, California 93517-5001

FMST.07.08
04/07/02

STUDENT OUTLINE

WATER PROCUREMENT

LEARNING OBJECTIVES

TERMINAL LEARNING OBJECTIVE Given a unit in a cold weather environment and the necessary equipment and supplies, perform cold weather preventive medicine, in accordance with the references. (FMST.07.08)

ENABLING LEARNING OBJECTIVES

(1) Without the aid of references from a given list, choose one of the techniques for finding incidental water, in accordance with the references. (FMST.07.08i)

(2) Without the aid of references from a given list, choose one of the four hazardous fluids to avoid substituting for potable water, in accordance with the references. (FMST.07.08j)

(3) Without the aid of references, construct a water-producing device, in accordance with the references. (FMST.07.08k)

OUTLINE

1. WATER INTAKE

a. Thirst is not a strong enough sensation to determine how much water you need.

b. The best plan is to drink, utilizing the OVER DRINK method. Drink plenty of water anytime it is available and particularly when eating.

A6

c. Dehydration is a major threat. A loss of only 5 % of your body fluids causes thirst, irritability, nausea, and weakness; a 10% loss causes dizziness, headache, inability to walk, and a tingling sensation in limbs; a 15% loss causes dim vision, painful urination, swollen tongue, deafness, and a feeling of numbness in the skin; also a loss of more than 15% body fluids could result in death.

d. Your water requirements will be increased if:

(1) You have a fever.

(2) You are experiencing fear or anxiety.

(3) You evaporate more body fluid than necessary. (i.e., not using the proper shelter to your advantage)

(4) You have improper clothing.

(5) You ration water.

(6) You overwork.

2. **INCIDENTAL WATER**. (FMST.07.08.i)

a. During movement, you may have to ration water until you reach a reliable water source. Incidental water may sometimes provide opportunities to acquire water. Although not a reliable or replenished source, it may serve to stretch your water supply or keep you going in an emergency. The following are sources for incidental water:

(1) Dew: In areas with moderate to heavy dew, tying rags or tuffs of fine grass around your ankles can collect dew. While walking through dewy grass before sunrise, the rags or grass will saturate and can be rung out into a container. The rags or grass can be replaced and the process is repeated.

(2) Rainfall: Rainwater collected directly in clean container or in plants that contain no harmful toxins is generally safe to drink without disinfecting. The survivor should always be prepared to collect rainfall at a moment's notice. An inverted poncho works well to collect rainfall.

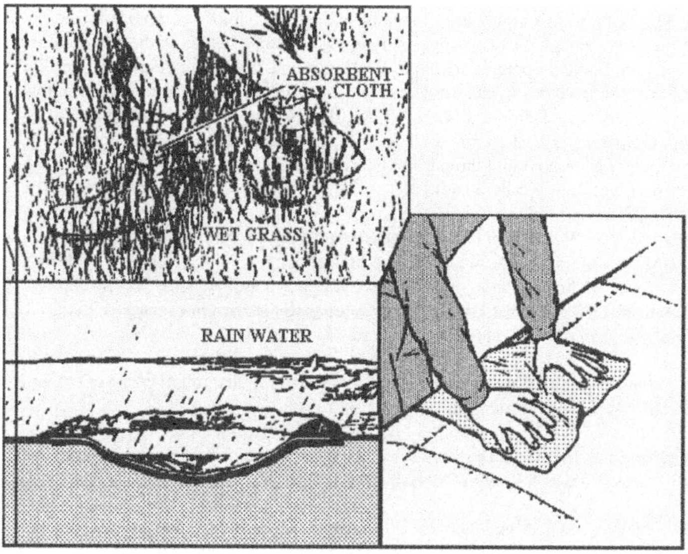

3. **HAZARDOUS FLUIDS** (FMST.07.08j)

a. Survivors have occasionally attempted to augment their water supply with other fluids, such as alcoholic beverages, urine, blood, or seawater. While it is true that each of these fluids has a high water content, the impurities they contain may require the body to expend more fluid to purify them. Some hazardous fluids are:

(1) Seawater: Seawater in more than minimal quantities are actually toxic. The concentration of sodium and magnesium salts is so high that fluid must be drawn from the body to eliminate the salts and eventually the kidneys cease to function.

(2) Alcohol: Alcohol dehydrates the body and clouds judgment. Super-cooled liquid, if ingested, can cause immediate frostbite of the throat, and potential death.

(3) Blood: Blood, besides being salty, is a food. Drinking it will require the body to expend additional fluid to digest it.

(4) Urine: Drinking urine is not only foolish, but also dangerous. Urine is nothing more than the body's waste. Drinking it only places this waste back into the body, which requires more fluid to process it again.

4. **WATER QUALITY**. Water contains minerals, toxins, and pathogens. Some of these, consumed in large enough quantities may be harmful to human health. Pathogens are our primary concern. Pathogens are divided into Virus, Cysts, Bacteria, and Parasites. Certain pathogens are more resistant to chemicals and small enough to move through microscopic holes in equipment (i.e., T-shirt, parachute). Certain pathogens also have the ability to survive in extremely cold water temperatures. Pathogens generally do not live in snow and ice. Water quality is divided into three levels of safety with disinfecting as the most desired level, then purified, followed by potable.

a. Disinfecting. Water disinfecting removes or destroys harmful microorganisms. Giardia cysts are an ever-present danger in clear appearing mountain water throughout the world. By drinking non-potable water you may contract diseases or swallow organisms that could harm you. Examples of such diseases or organisms are Dysentery, Cholera, Typhoid, Flukes, and Leeches.

b. Remember impure water, no matter how overpowering the thirst, is one of the worst hazards in a survival situation.

c. The first step in disinfecting is to select a treatment method. The two methods we will discuss are as follows: (FMST.07.08j)

(1) Heat. The Manual of Naval Preventive Medicine (P-5010) states that you must bring the water to a rolling boil before it is considered safe for human consumption. This is the most preferred method.

(a) Bringing water to the boiling point will kill 99.9% of all Giardia cysts. The Giardia cyst dies at 60^{0}C and Cryptosporidium dies at 65C. Water will boil at 14,000' at 86^{0}C and at 10,000' at 90C. With this in mind you should note that altitude does not make a difference unless you are extremely high.

(2) Chemicals. There are numerous types of chemicals that can disinfect water. Below are a few of the most common. In a survival situation, you will use whatever you have available.

(a) Iodine Tablets.

(b) Chlorine Bleach.

(c) Iodine Solution.

(d) Betadine Solution.

(e) Military water purification tablets. These tablets are standard issue for all DOD agencies. These tablets have a shelf life of four years from the date of manufacture, unless opened. Once the seal is broken, they have a shelf life of one year, not to exceed the initial expiration date of four years.

49703 Month / Year / Batch Number

(3) Water Disinfecting Techniques and Halogen Doses.

Iodination techniques Added to 1 liter or quart of water	Amount for 4 PPM	Amount for 8 PPM
Iodine tablets Tetraglcine hydroperiodide EDWGT Potable Aqua Globaline	½ tablet	1 tablet
2% iodine solution (tincture)	0.2 ml 5 gtts	0.4 ml 10 gtts
10% povidone-iodine solution*	0.35 ml 8 gtts	0.70 ml 16 gtts
Chlorination techniques	Amount for 5 PPM	Amount for 10 PPM
Household bleach 5% Sodium hypochloric	0.1 Ml 2 gtts	0.2 ml 4 gtts
AquaClear Sodium dichloroisocyanurate		1 tablet
AquaCure, AquaPure, Chlor-floc Chlorine plus flocculating agent		8 PPM 1 tablet

*Providone-iodine solutions release free iodine in levels adequate for disinfecting, but scant data is available.

Measure with dropper (1 drop=0.05 ml) or tuberculin syringe

PPM-part per million gtts-drops ml-milliliter

Concentration of **Contact time in minutes at various water temperatures**

Halogen

	5 C / 40 F	15 C / 60 F	30 C / 85 F
2 PPM	240	180	60
4 PPM	180	50	45
8 PPM	60	30	15

NOTE: These contact times have been extended from the usual recommendations to account for recent data that prolonged contact time is needed in very cold water to kill *Giardia* cysts.

NOTE: Chemicals may not destroy Cryptosporidium.

d. Purification. Water purification is the removal of organic and inorganic chemicals and particulate matter, including radioactive particles. While purification can eliminate offensive color, taste, and odor, it may not remove or kill microorganisms.

(1) Filtration. Filtration purifying is a process by which commercial manufacturers build water filters. The water filter is a three-tier system. The first layer, or grass layer, removes the larger impurities. The second layer, or sand layer, removes the smaller impurities. The final layer, or charcoal layer (**not the ash but charcoal from a fire**), bonds and holds the toxins. All layers are placed on some type of straining device and the charcoal layer should be at least 5-6 inches thick. Layers should be changed frequently and straining material should be boiled. Remember that this is not a disinfecting method, cysts can possibly move through this system.

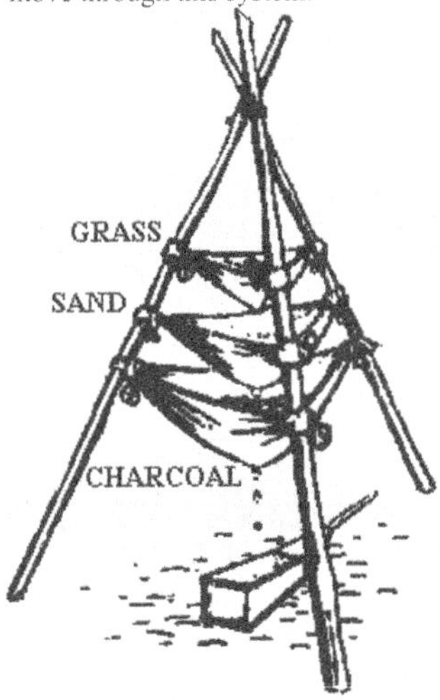

WATER FILTER

(2) Commercial Water Filters. Commercial water filters are generally available in most retail stores and may be with you. Understanding what the filter can do is the first step in safeguarding against future illnesses.

(a) A filter that has a .3 micron opening or larger will not stop Cryptosporidium.

(b) A filter system that does not release a chemical (i.e., iodine) may not kill all pathogens.

(c) A filter that has been overused may be clogged. Usage may result in excessive pumping pressure that can move harmful pathogens through the opening.

c. Potable. Potable indicates only that a water source, on average over a period of time, contains a "minimal microbial hazard," so the statistical likelihood of illness is acceptable.

(1) Sedimentation. Sedimentation is the separation of suspended particles large enough to settle rapidly by gravity. The time required depends on the size of the particle. Generally, 1 hour is adequate if the water is allowed to sit without agitation. After sediment has formed on the bottom of the container, the clear water is decanted or filtered from the top. Microorganisms, especially cysts, eventually settle, but this takes longer and the organisms are easily disturbed during pouring or filtering. Sedimentation should not be considered a means of disinfecting and should be used only as a last resort or in an extreme tactical situation.

5. **SOLAR STILLS.** (FMST.07.08k)

a. Solar stills are designed to supplement water reserves. Contrary to belief, they will not provide enough water to meet the daily requirement for water.

b. AboveGround Solar Still. This device allows the survivor to make water from vegetation. To make the aboveground solar still, locate a sunny slope on which to place the still, a clear plastic bag, green leafy vegetation, and a small rock.

(1) Construction.

(a) Fill the bag with air by turning the opening into the breeze or by "scooping" air into the bag.

(b) Fill the bag half to three-quarters full of green leafy vegetation. Be sure to remove all hard sticks or sharp spines that might puncture the bag.

CAUTION Do not use poisonous vegetation. It will provide poisonous liquid.

(c) Place a small rock or similar item in the bag.

(d) Close the bag and tie the mouth securely as close to the end of the bag as possible to keep the maximum amount of air space. If you have a small piece of tubing, small straw, or hollow reed, inserts one end in the mouth of the bag before tying it securely. Tie off or plug the tubing so that air will not escape. This tubing will allow you to drain out condensed water without untying the bag.

(e) Place the bag, mouth downhill, on a slope in full sunlight. Position the mouth of the bag slightly higher than the low point in the bag.

(f) Settle the bag in place so that the rock works itself into the low point in the bag.

(g) To get the condensed water from the still, loosen the tie and tip the bag so that the collected water will drain out. Retie the mouth and reposition the still to allow further condensation.

(h) Change vegetation in the bag after extracting most of the water from it.

(i) Using 1-gallon zip-loc. bag instead of trash bags is a more efficient means of construction.

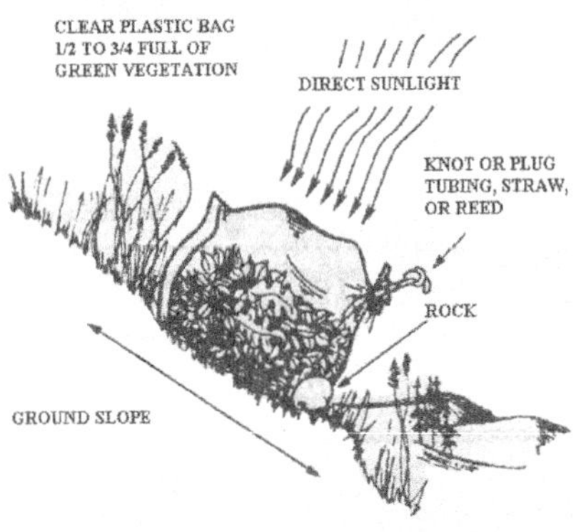

<u>ABOVE GROUND SOLAR STILL</u>

c. <u>BelowGround Solar Still</u>. Materials consist of a digging stick, clear plastic sheet, container, rock, and a drinking tube.

(1) Construction.

(a) Select a site where you believe the soil will contain moisture (such as a dry streambed or a low spot where rainwater has collected). The soil should be easy to dig, and will be exposed to sunlight.

(b) Dig a bowl-shaped hole about 1 meter across and 24 inches deep.

(c) Dig a sump in the center of the hole. The sump depth and perimeter will depend on the size of the container you have to place in it. The bottom of the sump should allow the container to stand upright.

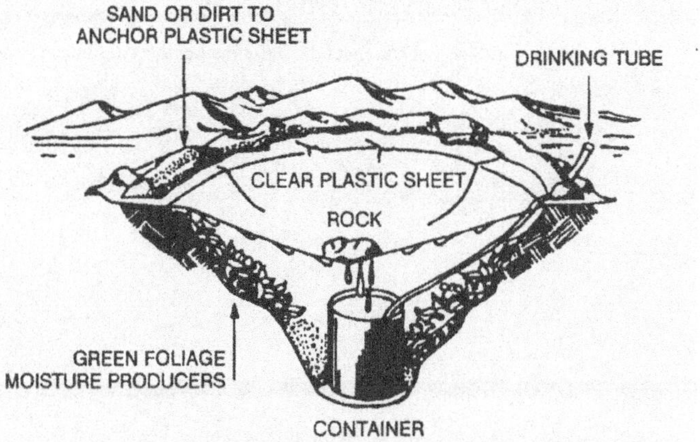

<u>**BELOW GROUND SOLAR STILL**</u>

(d) Anchor the tubing to the container's bottom by forming a loose overhand knot in the tubing. Extend the unanchored end of the tubing up, over, and beyond the lip of the hole.

(e) Place the plastic sheet over the hole, covering its edges with soil to hold in place. Place a rock in the center of the plastic sheet.

(f) Lower the plastic sheet into the hole until it is about 18 inches below ground level. Make sure the cone's apex is directly over the container. Ensure the plastic does not touch the sides of the hole because the earth will absorb the moisture.

(g) Put more soil on the edges of the plastic to hold it securely and prevent the loss of moisture.

(h) Plug the tube when not in use so that moisture will not evaporate.

(i) Plants can be placed in the hole as a moisture source. If so, dig out additional soil from the sides.

(j) If polluted water is the only moisture source, dig a small trough outside the hole about 10 inches from the still's lip. Dig the trough about 10 inches deep and 3 inches wide. Pour the polluted water in the trough. Ensure you do not spill any polluted water around the rim of the hole where the plastic touches the soil. The trough holds the polluted water and the soil filters it as the still draws it. This process works well when the only water source is salt water.

(k) Three stills will be needed to meet the individual daily water intake needs.

UNITED STATES MARINE CORPS
Mountain Warfare Training Center
Bridgeport, California 93517-5001

WMC
04/07/02

STUDENT HANDOUT

FORAGING PLANTS AND INSECTS FOR SURVIVAL USES

PURPOSE The purpose of this period of instruction is to familiarize the Wilderness Medicine Student with the potential resources for survival sustenance.

OUTLINE

1. GENERAL CONSIDERATIONS. There are few places without some type of edible vegetation. Plants contain vitamins, minerals, protein, carbohydrates, and dietary fiber. Some plants also contain fats. The following are general considerations:

a. Do not assume that because birds or animals have eaten a plant, it is edible by humans.

b. Poor plant recognition skills will seriously limit your ability to survive.

c. Plant dormancy and snowfall make foraging plants difficult during the winter months.

d. Plants generally poison by:

(1) Ingestion: When a person eats a part of a poisonous plant.

(2) Contact: When a person makes contact with a poisonous plant that cause any type of skin irritation or dermatitis.

(3) Absorption: When a person absorbs the poison through the skin, which can interrupt a bodily function.

(4) Inhalation: Poisoning can occur through the inhalation of smoke that contains poisonous plant residue.

A7

e. Plant properties can change throughout the growing season. Plants can be edible during certain periods while poisonous in others.

2. PLANT RECOGNITION. Plant recognition can take years of study to learn. The following is used as a general guideline.

a. Types of plants that should be avoided are:

(1) All mushrooms.

(2) Plants with a milky sap.

(3) White and yellow berries should be avoided, as they are almost always poisonous.

(4) Plants with shiny leaves should be considered poisonous.

(5) Plants that are irritants to the skin should not be eaten.

(6) All beans and peas should be avoided.

b. Edible Plants: There are several edible plants found at the Marine Corps Mountain Warfare Training Center. All are easy to recognize and many of the same plants are located throughout the United States and abroad.

(1) Wild Onions - Complete plant.

(2) Dandelion - Roots and leaves.

(3) Mallow - Leaves, stem and flowers.

(4) Watercress -Complete plant.

(5) Cattail - Root, stalk and stem.

(6) Bull Thistle - The flower.

(7) Juniper Tree - Berries and cambium layer.

(8) All Conifers - Cambium layer, needles, and nuts within the cones.

(9) Currants - Berries.

c. Preparing an edible plant. All edible plants should be boiled twice with a change of water to remove any bitterness, acids, or bacteria that the plant may contain.

3. PLANT TESTING PROCEDURE. Always adopt the following procedure when trying unknown plants as food.

a. Wait 8 hours without eating.

b. Select a plant that grows in sufficient quantity in the local area. Separate the part of the plant you wish to test root, stem, leaf, or flower. Certain parts of plants are poisonous while the, other parts may be edible.

c. Rub a portion of the plant you have selected on your inner forearm. Wait 15 minutes and look for any swelling, rash, or irritation.

d. Boil the plant or plant part in changes of water. The toxic properties of many plants are water-soluble or can be destroyed by heat. Cooking and discarding two changes of water can lessen the amount of poisonous material or remove it completely. These boiling periods should last at least 5 minutes each.

e. Place 1 teaspoon of the prepared plant food in the mouth and chew for 5 minutes, but do not swallow. If unpleasant effects occur (burning, bitter, or nauseating taste), remove the plant from the mouth at once and discard it as a food source. If no unpleasant effects occur, swallow the plant material and wait 8 hours.

f. If after 8 hours no unpleasant effects have occurred (nausea, cramps, and diarrhea), eat two tablespoonfuls and wait 8 hours.

g. In no unpleasant effects have occurred at the end of this 8-hour period, the plant may be considered edible.

h. Completely document and sketch the plant in a logbook to refer to for future use. This will aid in future procurement of this plant. If plant properties have changed, you will have to repeat the plant testing procedure.

NOTE: Have charcoal ready to eat. If any side effects occur, attempt to induce vomiting and immediately eat charcoal. Charcoal will absorb toxins and reduce the chance of Death.

4. MEDICINAL PLANTS One of the many things that hinder a survivor's ability to survive is medical problems. Injuries incurred will reduce survival expectancy and the ability to evade.

a. Coyote Willow: Coyote Willow is a thick forming shrub with clustered stems and very narrow leaves. It habits wet soils, especially riverbanks, sandbars, and silt flats.

(1) Drinking tea made from the inner bark can reduce fever.

(2) The dried and powdered inner bark can be used to stop bleeding.

(3) Administering a half-cup of willow charcoal dissolved in water can treat diarrhea.

(4) The Coyote Willow roots can be mashed and applied to tooth aches.

b. Yarrow: Yarrow is a plant (tall, usually not branched, with many white, slat topped group flowers). Do not confuse Yarrow with Water Hemlock.

(1) Rubbing a handful of crushed Yarrow flowers and leaves on any exposed can make insect repellent skin.

(2) Placing a Yarrow leaf poultice on the wound can stop bleeding.

(3) Relief from many rashes, including poison oak and ivy, can be achieved by applying a Yarrow leaf compress to the effected area.

(4) The Yarrow root can be chewed to relieve the pain of a toothache or break a fever.

(5) A potent anesthetic can be made by scrubbing fresh Yarrow roots in water to clean them. Once the roots are clean, crush them into a spongy mass and apply gently to the wounded area.

c. Tea To make a tea, place the plants in a container and pour in two cups of boiling water. Let the mixture steep for 20 minutes and drink.

d. Poultice: Place plant (Coyote Willow, Yarrow, etc.) in gauze or other similar material and fold it so the gauze will hold the plant in place. Put gauze in a larger cloth, about 6"x8", and roll the sides inward. Fold the cloth over without losing the plant. While boiling water, dip the bottom portion of the cloth containing the plant into the hot water by holding the edges. Keep the plant submerged in the boiling water until it becomes saturated. Bring the cloth straight up and with a twisting motion, wring the excess water back into the pot. Apply the poultice to the affected area as soon as it has cooled enough to place on the wound. To be effective, the poultice should be as hot as you can tolerate it.

e. Compress: A compress is made just like a poultice; except it is cold when applied to the wound. NOTE: Poultice or Compress should be applied for 1 to 24 hours, as needed. When applying a poultice, you may experience a throbbing pain as it draws out the infection and neutralizes toxins. When the pain subsides, the poultice has accomplished its task and should be removed. Apply a fresh poultice as needed until the desired level of healing has been reached.

5. POISONOUS PLANTS. Rarely will a survivor have an ideal means of killing large game, however there are certain plants that can aid the survivor. The two plants that we will talk about are the Water Hemlock and Monkshood.

a. Preparing Poisonous Plants: Once a poisonous plant has been located, dig up the root of the plant. The roots of the Water Hemlock and Monkshood generally grow 8 to 10 inches deep. Extreme caution must be used when handling the root. Do not handle the root without a barrier between your hands and the root. A barrier can be gloves, socks, T-shirt, or even moss. Split the root lengthwise to expose the inside of the root where the toxin is located. With the root split, rub the tip of your spear/arrow inside of the root opening. In a slow controlled manner, work from the bottom of the tip to the top. Once the tip is thoroughly coated, allow the toxin to dry and apply another coat to the tip. Continue to apply coat after coat of toxin, until the root is completely drained of its' toxin. You are now finished and the tip is ready for use.

NOTE: Do not eat the meat in the vicinity of the poisoned wound.

CAUTION: Do not apply poison to the actual point and edges of tip. This avoids accidental puncture to you. Wash and boil all gloves and clothing used as a barrier before wearing

6. INSECTS. Insects are the most abundant life form on earth and are an excellent survival food. They are easy to catch and provide 65-80% protein; compared to 20% for beef. They aren't too appetizing, but personal bias has no place in a survival situation. The focus must remain on maintaining your health.

a. Insects to avoid.

(1) All adults that sting or bite.

(2) Hairy or brightly colored insects.

(3) All Caterpillars.

(4) Insects that have a pungent odor.

(5) All spiders.

(6) Disease carriers like ticks, flies, or mosquitoes.

b. Edible Insects.

(1) Insect Larvae.

(2) Grasshoppers.

(3) Beetles.

(4) Grubs.

(5) Ants.

(6) Termites.

(7) Worms.

c. Foraging for Insects. One must be careful not to expend more energy harvesting food than can be replaced. For example, catching insects such as grasshoppers can become exasperating and tiring.

(1) At night grasshoppers climb tall plants and clings to the stalks near the top. They can be picked from the plants in the early morning while they are chilled and dormant.

(2) Dig for worms in damp humus soil, under rocks/logs or look for them on the ground after it has rained.

(3) Carpenter ants are found in dead trees and stumps, which can be gathered by hand.

(4) Most other insects can be found in rotten logs, under rocks, and in open grassy areas.

d. Preparing Edible Insects.

(1) Insects with a hard outer shell have parasites. Remove the wings and barbed legs before cooking.

(2) Drop worms into potable water for at least a half-hour. They will naturally purge themselves. You can either cook or eat them raw.

(3) Most other insects can be eaten raw. Cooking insects will improve their taste. If the thought of eating insects is unbearable, grind them into a paste and mix with other foods.

UNITED STATES MARINE CORPS
Mountain Warfare Training Center
Bridgeport, California 93517-5001

WMC.A8.0
05/02/01

STUDENT HANDOUT

SURVIVAL TRAPS AND SNARES

PURPOSE: The purpose of this class is to familiarize the student with some of the techniques used to take game. Game can be useful for both food and clothing.

OUTLINE

1. GENERAL TIPS AND TECHNIQUES FOR TRAPPING

a. General Tips. Knowing a few general hints and tips will make the trapping of animals much easier and considerably more effective. The eight general tips for trapping are:

(1) Know your game: Knowing the habits of the animal you want to trap will help lure it into your trap. Such things as when and where they move, feed, and water will help you determine where the trap can be most effectively placed.

A8

(2) Keep things simple: You don't have time in a survival situation to construct elaborate traps and they do not necessarily do a better job.

(3) Set traps in the right place: Animals will travel and stop in certain locations. That is where to best employ traps.

(4) Cover up your scent: Animals will avoid an area, which smells strange to them. Smoke from your fire is the best cover to use.

(5) Use the right type of trap: Some traps work better than others do (i.e. a deadfall works better on squirrels than a spring pole).

(6) Use the right size trap: Adjust the size of your traps to the size of the animal.

(7) Check traps: Check your traps twice daily: morning and evening. Checking your traps less than twice a day can allow your game to escape, rot, or be taken by other predators.

(8) Bait your traps: Bait of any type will add to your chances of success.

b. General Techniques. A general technique is the method in which the trap is intended to kill or hold the animal.

(1) Strangulation: This method strangles the animal, such as a snare.

(2) Mangling: This method crushes the animal, such as a deadfall.

(3) Entanglement: This method entangles the animal, such as a net.

(4) Live: This method holds the animal, such as a box trap.

c. Types of Triggers. There are three basic triggers used for all traps and path guards. Depending on the situation, variations of these triggers can be used.

(1) Puite Figure 4.

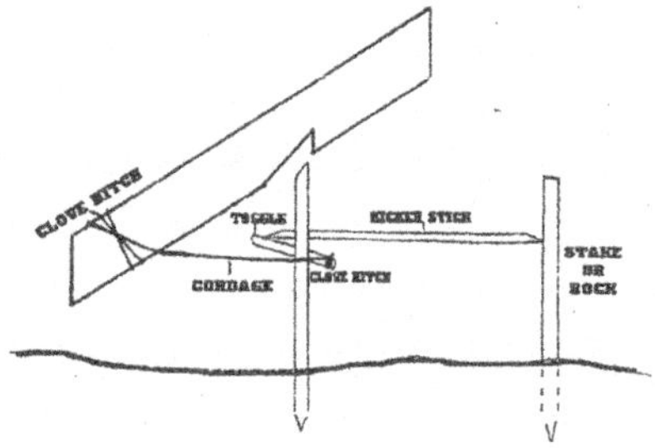

(2) Toggle.

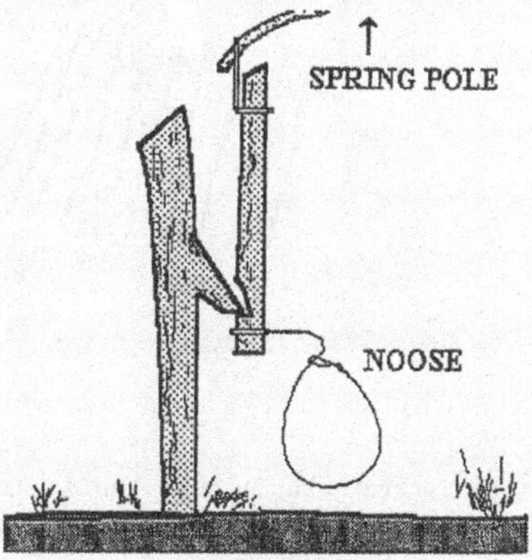

TOGGLE

(c) Universal.

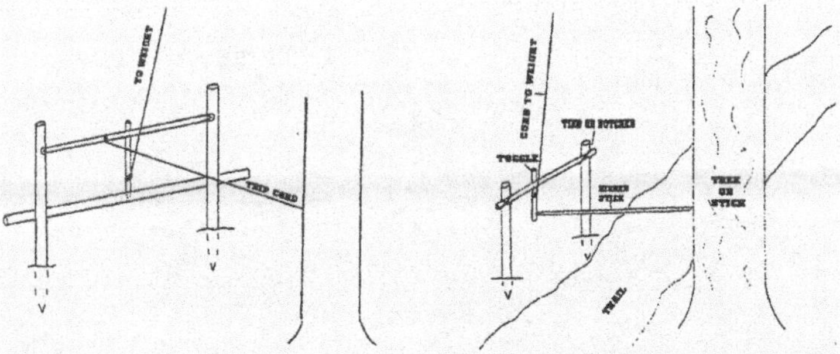

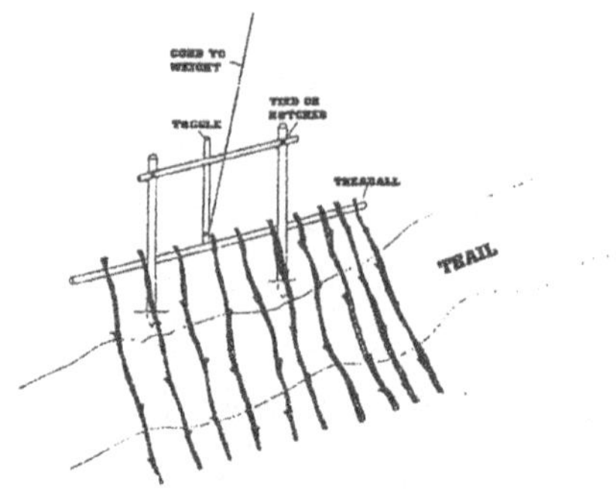

2. TRAPS. Traps are designed to hold or kill animals by use of some type of action. This action is generally caused by either a weight or spring loaded device.

a. Puite Deadfall: The Puite deadfall requires a knife and piece of cord to construct. It is designed to mangle small rodents. The trigger for the deadfall is the Puite figure 4.

b. Spring Pole: The spring pole requires a small sapling and cordage to construct. The trigger for the spring pole is the toggle. It is designed to lift the animal off the ground; not allowing predatory animals to take your game. Remember, the trigger can not be so tight that the intended game can not set it off.

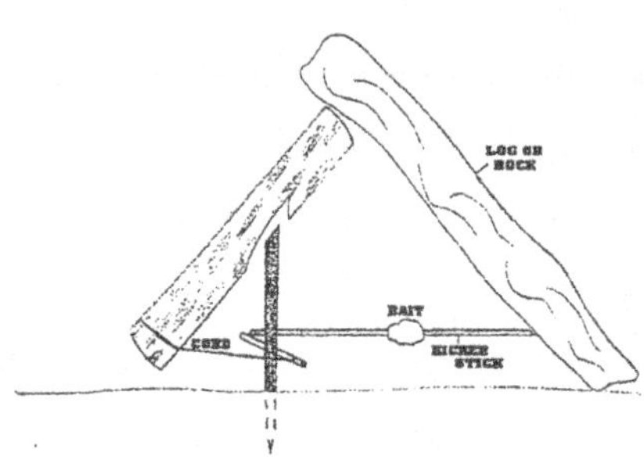

SPRING POLE

c. Box Trap: The box trap requires limited cordage to construct. It is designed to hold live small rodents and birds. The box trap trigger is the Puite figure 4.

3. PATH GUARDS Path guards is designed to protect and provide security for your shelter area against the enemy and predatory animals. They are classified into noise and casualty producing path guards.

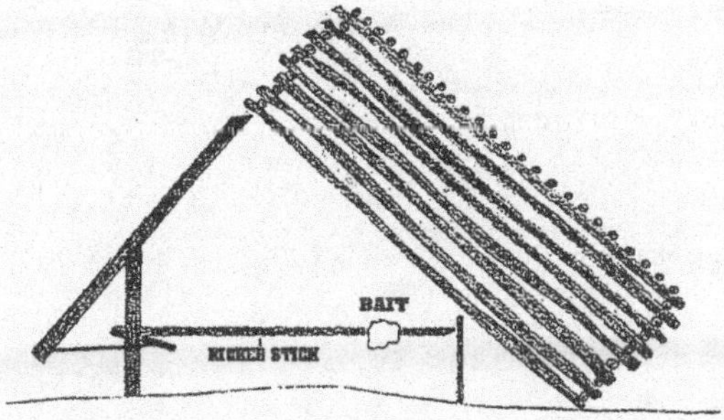

BOX TRAP

a. Noise producing path wards. Noise producing pathguards serve as an alarm for your shelter area. When triggered, it should produce some type of loud noise. Although construction can vary, depending on materials available, one example is as follows:

(1) Secure a young sapling to a universal trigger.

(2) At the end of the sapling, tie several pieces of metal to the sapling. Use what ever is available for metal.

(3) Camouflage the metal on the ground.

(4) When triggered, the sapling should swing back and forth, causing the metal rattle.

b. Casualty producing path wards. Casualty producing path guards, when triggered, should cause death or injury to the enemy or predatory animal. Tips should be poisoned as discussed in Survival Plant Uses class. Triggers for this type of path guard should be the universal triggers.

(1) Log Jerk

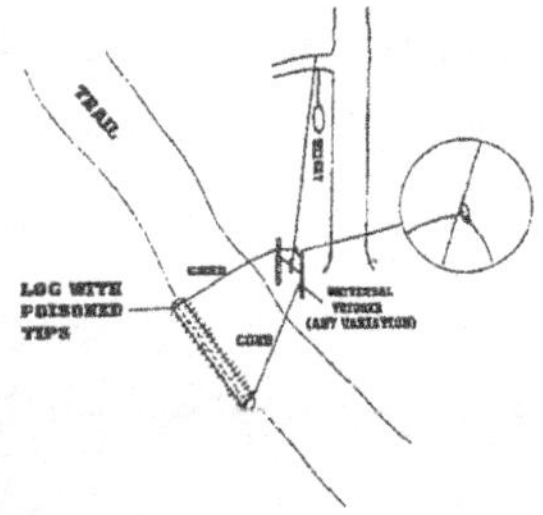

(2) Fish Hook Nightmare

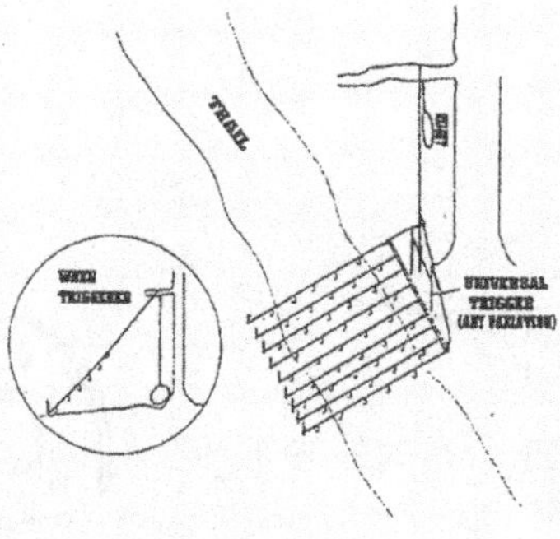

4. SNARING. A snare in nothing more than a piece of wire, rope, or cord with a loop at one end, which tightens, down around an animals neck. Snares are much easier and less time consuming to construct than traps; while producing more results.

a. Requirements for snaring there are five requirements too effectively employ snares. They are as follows:

(1) Wire: Although snares can be used with rope or cord, they are less effective than wire. Wire should have memory and resist kinking. Aircraft cable type in size 1/32 to 3/16 inch should be used. This type of wire prevents animal chew out and resists breakage. Remember, you want the smallest diameter cable capable of holding the animal.

(2) Locking device: A locking device is imperative for a snare to work properly. Locking devices secure the snare around the animal's neck. There are several methods available for a locking device.

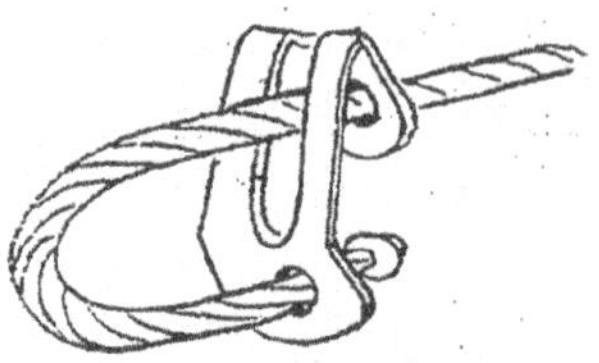

COMMERCIAL **BENT WASHER**

(3) Channelization: The objective of channelization is to force the animal to move through the snare. Remember, animals will take the path of least resistance.

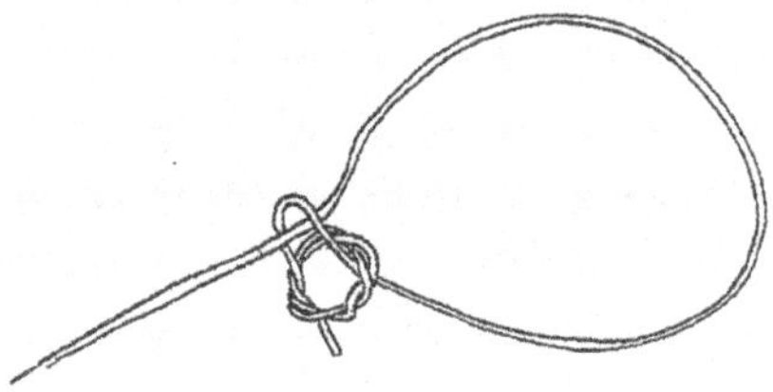

KNOT

(4) Loop size: Probably the single most important requirement in snaring. A correctly employed snare will have the snare holding the animal around the neck. Loop size is placed on the snare according to the intended animal. Too large will result in a body or leg catch, resulting in possible chew out or breakage. Too small will enable the animal to force the snare to the side, resulting in a miss. Additionally, the loop must be placed with specific ground clearance. The following chart will give a general guideline.

ANIMAL	NOOSE SIZE	GROUND CLEARANCE
SQUIRREL	2 1/2 TO 3 INCHES	1/2 TO 1 1/2 INCHES
RABBIT	4 TO 5 1/2 INCHES	1 1/2 TO 3 INCHES
RACCOON	6 INCHES	3 TO 4 INCHES
FOX	7 TO 10 INCHES	8 TO 10 INCHES
COYOTES	12 TO 14 INCHES	12 INCHES
BOBCAT	9 INCHES	8 INCHES

Note: (1) Noose size is the diameter of the snare loop.

(2) Ground clearance is measured from the bottom of the loop to the ground.

(5) Bait. In a survival situation, you will not be able to employ numerous snares. Baiting all snares increases your chances of success.

b. Snares although there are numerous ideas to employ snares, here are a few ideas.

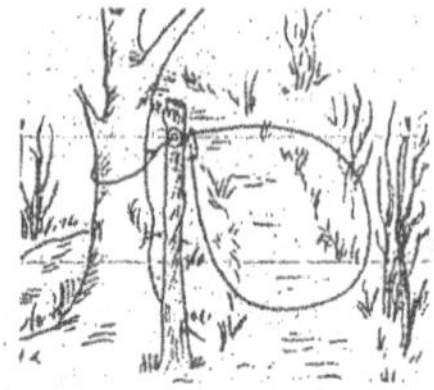

TRAIL SET

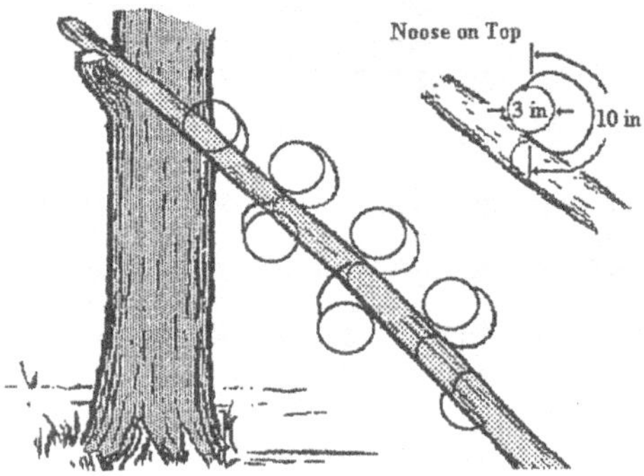

SQUIRREL POLE

DEN SET

CUBBIE SET

5. FISH. In a mountainous region, fish are normally an abundant resource. Not only are they a food source, the "left-overs" provide an excellent bait for traps and snares.

a. Fishing locations: Fishing in mountain streams is generally best done with a hand line. When fishing these streams, always look for these places to fish.

FISHING LOCATIONS

b. Expedient hooks: Although hooks should be carried in a survival kit, the survivor should be able to construct additional hooks if the situation arises. Expedient hooks are made to become lodged in the throat of the fish. Below are a few examples.

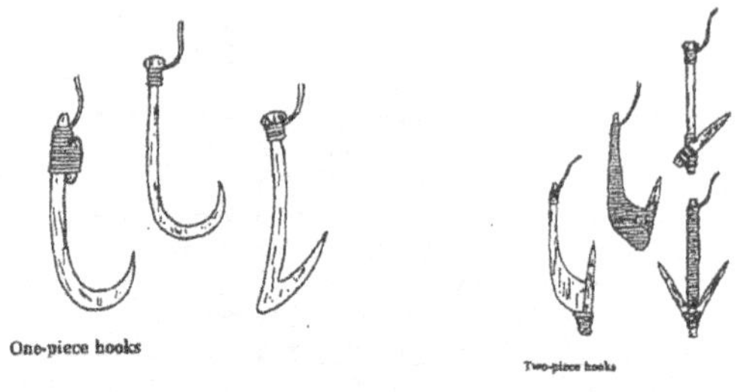

EXPEDIENT HOOKS

c. Fish traps: A fish trap can be effective if you have a shallow stream and time to construct it. A basic fish trap in nothing more than a barricade of rocks or sticks across a stream with another barricade using a funnel-type entrance which fish can be driven into but have a difficult time finding their way out. Once fish are trapped between these two barricades they may be speared, clubbed, or grabbed. This can be very effective when certain types of fish are moving in large groups to spawn. This type of trap is very effective in catching fish. A door can be constructed at the mouth of the trap so that excess fish can be kept live until needed.

FISH TRAP

d. Set Lines: Set lines are an effective method of fishing which conserves energy. Put them out over night with several boated hooks attached. Place them with the hooks either on the bottom or suspended off the bottom, until you have determined where the fish are feeding.

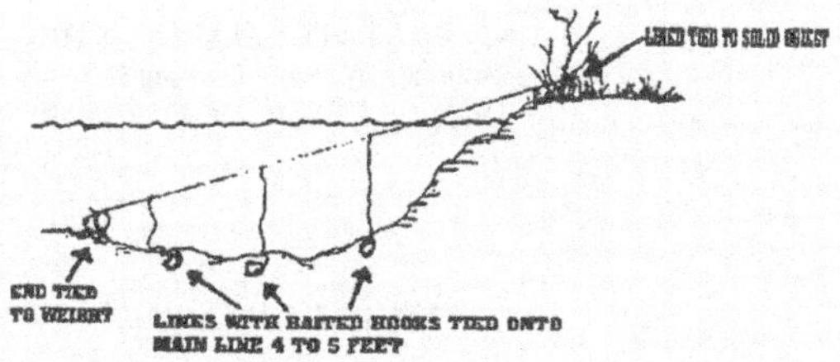

SET LINES

c. Netting: A gill net is most effective in still water, e.g., a lake (near the inlet and outlet are good locations) or back water in a large stream (for survival doesn't hesitate to block the stream). Nets can be constructed using the inner cords of parachute shroud lines. Place floats on top and weights on the bottom to keep the net vertical in the water. When the ice is on a lake, the fish are inclined to stay deeper. The smaller the mesh, the smaller the fish you can catch, but a small mest will still entangle a large fish. A mesh of 3 inches is a good standard. A gill net is very time consuming and requires allot of material to construct.

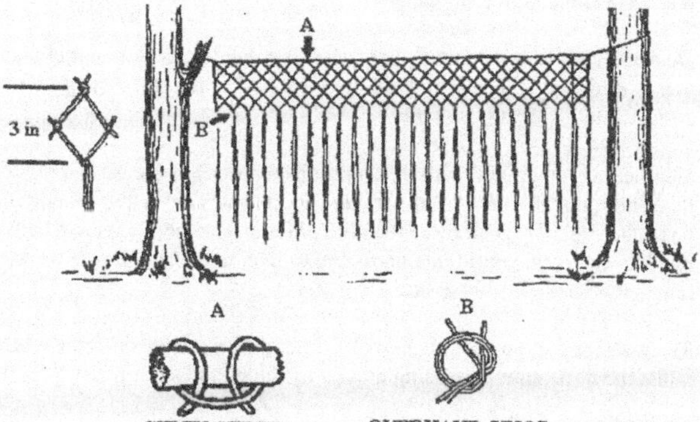

GILL NET

UNITED STATES MARINE CORPS
Mountain Warfare Training Center
Bridgeport, California 93517-5001

WMC
04/07/02

STUDENT HANDOUT

MOUNTAIN WEATHER

LESSON PURPOSE: The purpose of this period of instruction is to introduce you to weather patterns, the invisible aspects of weather, and some visible clues to help you forecast the weather. This lesson relates to all operations in the mountains at MWTC.

OUTLINE

1. GENERAL

a. The earth is surrounded by the atmosphere, which is divided, into several layers. The world's weather systems are in the troposphere, the lower of these layers. This layer reaches as high as 40,000 feet.

b. Dust and clouds in the atmosphere absorb or bounce back much of the energy that the sun beams down upon the earth. Less than one half of the sun's energy actually warms the earth's surface and lower atmosphere.

c. Warmed air, combined with the spinning (rotation) of the earth, produces winds that spread heat and moisture more evenly around the world. This is very important because the sun heats the Equator much more than the poles and without winds to help restore the balance, much of the earth would be impossible to live on. Where the air-cools, you can get clouds, rain, snow, hail, fog, frost, etc.

d. The weather that you find in any place depends on many things, i.e. how hot the air is, how moist the air is, how it is being moved by the wind, and especially, is it being lifted or not?

A10

2. PRESSURE

a. All of these factors are related to air pressure, which is the weight of the atmosphere at any given place. The lower the pressure, the more likely are rain and strong winds.

b. In order to understand this we can say that the air in our atmosphere acts very much like a liquid.

c. Areas with a high level of this liquid would exert more pressure on the Earth and be called a "high pressure area".

d. Areas with a lower level would be called a "low pressure area".

e. In order to equalize the areas of high pressure it would have to push out to the areas of low pressure.

f. The characteristics of these two pressure areas are as follows:

(1) High-pressure area. Flows out to equalize pressure.

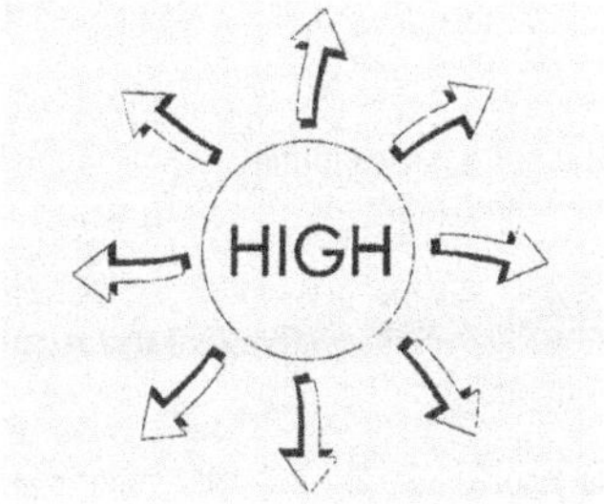

(2) Low pressure area. Flows in to equalize pressure.

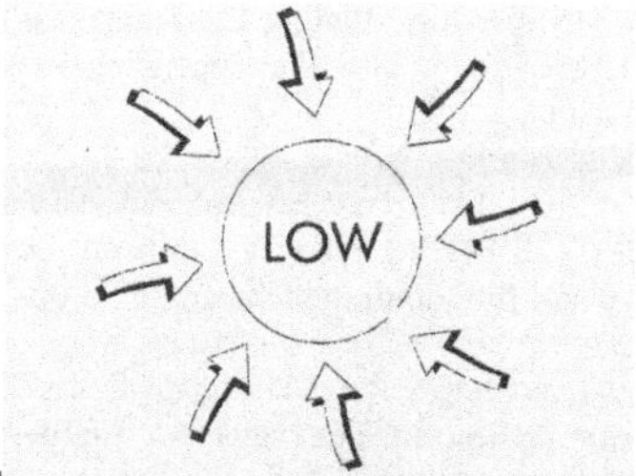

g. The air from the high-pressure area is basically just trying to gradually flow out to equalize its pressure with the surrounding air, while the low pressure is beginning to build vertically. Once the low has achieved equal pressure, it can't stop and continues to build vertically; causing turbulence, which results in bad weather.

NOTE: When looking on the weather map, you will notice that these differences in pressure resemble contour lines. These contour lines are called isobars and are translated to mean equal pressure area.

h. Isobars: Pressure is measured in millibars or another more common measurement is inch mercury.

i. Fitting enough, areas of high pressure are called ridges and areas of low pressure are called troughs.

NOTE: The average air pressure at sea level is:

29.92 inches mercury.
1,013 millibars.

j. As we go up in altitude, the pressure (or weight) of the atmosphere decreases.

EXAMPLE: At 18,000 feet in elevation it would be 500 millibars vice 1,013 millibars at sea level.

3. HUMIDITY Humidity is the amount of moisture in the air. All air holds water vapor, although it is quite invisible.

a. Air can hold only so much water vapors, but the warmer the air, the more moisture it can hold. When the air has all the water vapor that it can hold, the air is said to be saturated (100% relative humidity).

b. If the air is then cooled, any excess water vapor condenses; that is, it's molecules join to build the water droplets we can see.

c. The temperature at which this happens is called the "condensation point". The condensation point varies depending on the amount of water vapor and the temperature of the air.

d. If the air contains a great deal of water vapor, condensation will form at a temperature of 20°C (68°F). But if the air is rather dry and does not hold much moisture, condensation may not form until the temperature drops to 0°C (32°F) or even below freezing.

e. Adiabatic Lapse Rate: The adiabatic lapse rate is the rate that air will cool on ascent and warm on descent. The rate also varies depending on the moisture content of the air.

(1) Saturated Air = 2.2°F per 1,000 feet.

(2) Dry Air = 5.5°F per 1,000 feet.

4. WINDS. As we stated earlier, the uneven heating of the air by the sun and rotation of the earth causes winds. Much of the world's weather depends on a system of winds that blow in a set direction. This pattern depends on the different amounts of sun (heat) that the different regions get and also on the rotation of the earth.

a. Above hot surfaces, air expends (air molecules spread out), and move to colder areas where it cools and becomes denser, and sinks to the earth's surface. This forms a circulation of air from the poles along the surface or the earth to the Equator, where it rises and moves towards the poles again.

b. Once the rotation of the earth is added to this, the pattern of the circulation becomes confusing.

c. Because of the heating and cooling, along with the rotation of the earth, we have these surface winds. All winds are named from the direction they originated from:

(1) Polar Easterlies: These are winds from the polar region moving from the east. This is air that has cooled and settled at the poles.

(2) Prevailing Westerlies: These winds originate from approximately 30 degrees North Latitude from the west. This is an area where prematurely cooled air, due to the earth's rotation, has settled back to the surface.

(3) Northeast Tradewinds: These are winds that originate from approximately 30 degrees North from the Northeast. Also prematurely cooled air.

d. Jet Stream: A jet stream can be defined as a long, meandering current of high speed winds near the tropopause (transition zone between the troposphere and the stratosphere) blowing from generally a westerly direction and often exceeding 250 miles per hour. The jet stream results from:

    (1) Circulation of air around the poles and Equator.

    (2) The direction of airflow above the mid latitudes.

(3) The actual path of the jet stream comes from the west, dipping down and picking up air masses from the tropical regions and going north and bringing down air masses from the Polar Regions.

NOTE: The average number of long waves in the jet stream is between three and five depending on the season. Temperature differences between polar and tropical regions influence this. The long waves influence day to week changes in the weather; there are also short waves that influence hourly changes in the weather.

5. AIR MASSES. As we know, all -of these patterns move air. This air comes in parcels known as "air masses". These air masses can vary in size from as small as a town too as large as a country. These air masses are named for where they originate:

a. Maritime: Over water.

b. Continental: Over land.

c. Polar: Above 60 degrees North.

d. Tropical: Below 60 degrees North.

e. Combining these give us the names and description of the four types of air masses:

(1) Continental Polar: Cold, dry air mass.

(2) Maritime Polar: Cold wet air mass.

(3) Continental Tropical: Dry, warm air mass.

(4) Maritime Tropical: Wet, warm air mass.

ABBREVIATED MEANINGS
CP = Continental Polar.
MP = Maritime Polar.
CT = Continental
Tropical. MT =
Maritime Tropical.

f. The thing to understand about air masses; they will not mix with another air mass of a different temperature and moisture content. When two different air masses collide, we have a front, which will be covered in more detail later in this period of instruction.

g. Here are some other types of winds that are peculiar to mountain environments but don't necessarily affect the weather:

(1) Anabolic wind: These are winds that blow up mountain valleys to replace warm rising air and are usually light winds.

(2) Catabolic wind: These are winds that blow down mountain valley slopes caused by the cooling of air and are occasionally strong winds.

6. LIFTING/COOLING. As we know, air can only hold so much moisture depending on its temperature. If we cool this air beyond its saturation point, it must release this moisture in one form or another, i.e. rain, snow, fog, dews, etc. There are three ways that air can be lifted and cooled beyond its saturation point.

a. Orographic uplift: This happens when an air mass is pushed up and over a mass of higher ground such as a mountain. Due to the adiabatic lapse rate, the air is cooled with altitude and if it reaches its saturation point we will receive precipitation.

OROGRAPHIC UPLIFT

b. Convection effects: This is normally a summer effect due to the sun's heat radiating off of the surface and causing the air currents to push straight up and lift air to a point of saturation.

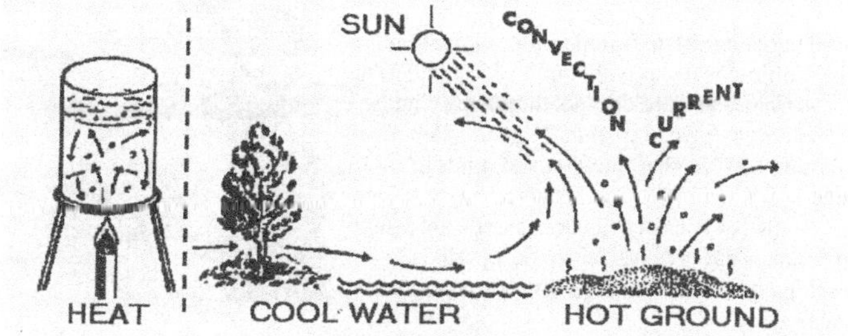

CONVECTION EFFECTS

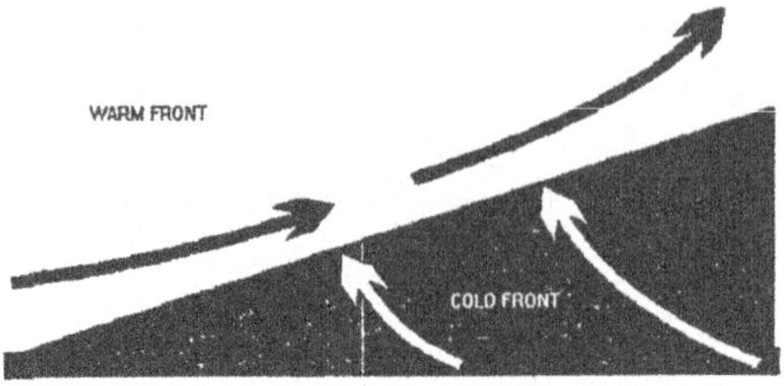

FRONTAL LIFTING

c. Frontal lifting: As we know when two air masses of different moisture and temperature content collide, we have a front. Since the air masses will not mix, the warmer air is forced aloft; from there it is cooled and then reaches its saturation point. Frontal lifting is where we receive the majority of our precipitation. A combination of the different types of lifting is not uncommon.

7. CLOUDS. Anytime air is lifted or cooled beyond its saturation point (100% relative humidity), clouds are formed. Clouds are one of our signposts to what is happening. Clouds can be described in many different ways; they can also be classified by height or appearance, or even by the amount of area covered, vertically or horizontally.

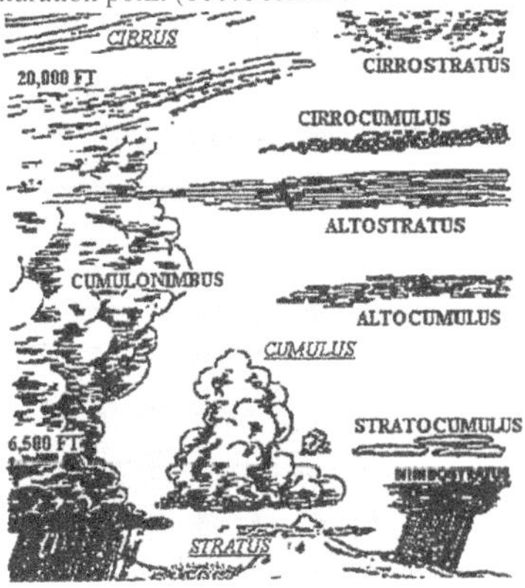

a. Cirrus: These clouds are formed of ice crystals at very high altitudes (usually 20,000 to 35,000 feet) in the mid-latitudes and are thin, feathery type clouds. These clouds can give you up to 24 hours warning of approaching bad weather, hundreds of miles in advance of a 'warm front. Frail, scattered types, such as "mare-tails" or dense cirrus layers, tufts are a sign of fair weather but predictive may be a prelude to approaching lower clouds, the arrival of precipitation and the front.

b. Cumulus: These clouds are formed due to rising air currents and are prevalent in unstable air that favors vertical development. These currents of air create cumiliform clouds that give them a piled or bunched up appearance, looking similar to cotton balls. Within the cumulus family there are three different types to help us to forecast the weather:

(1) Cotton puffs of cumulus are Fair Weather Clouds but should be observed for possible growth into towering cumulus and cumulonimbus.

(2) Towering cumulus are characterized by vertical development. Their vertical lifting is caused by some type of lifting action, such as connective currents found on hot summer afternoons or when wind is forced to rise up the slope of a mountain or possibly the lifting action that

may be present in a frontal system. The towering cumulus has a puffy and "cauliflower-shaped" appearance.

(3) Cumulonimbus clouds are characterized in the same manner as the towering cumulus, form the familiar "thunderhead" and produce thunderstorm activity. These clouds are characterized by violent updrafts, which carry the tops of the clouds to extreme elevations. Tornadoes, hail and severe rainstorms are all products of this type of cloud. At the top of the cloud, a flat anvil shaped form appears as the thunderstorm begins to dissipate.

c. Stratus: Stratus clouds are formed when a layer of moist air is cooled below its saturation point. Stratiform, clouds lie mostly in horizontal layers or sheets, resisting vertical development. The word "stratus" is derived from the Latin word "layer". The stratus cloud is quite uniform and resembles fog. It has a fairly uniform base and a dull, gray appearance. Stratus clouds make the sky appear heavy and will occasionally produce fine drizzle or Very light snow with fog. However, because there is little or no vertical movement in the stratus clouds, they usually do not produce precipitation in the form of heavy rain or snow.

d. As previously stated, clouds are formed when air is lifted to a point where it cools to its saturation point. We also know that frontal lifting affects our fronts, which produce the largest portion of our precipitation.

8. FRONTS. As we know, fronts often happen when two air masses of different moisture and temperature content interact. One of the ways we can identify that this is happening is by the progression of the clouds.

a. Warm Front: A warm front occurs when warm air moves into and over a slower (or stationary) cold air mass. Since warm air is less dense, it will rise naturally so that it will push the cooler air down and rise above it. The cloud you will see at this stage is cirrus. From the point where it actually starts rising, you will see stratus. As it continues to rise, this warm air-cools by the cold air and, this, receiving moisture at the same time. As it builds in moisture, it darkens becoming "nimbus-stratus", which means rain of thunderclouds. At that point some type of moisture will generally fall.

b. Cold Front. A cold front occurs when a cold air mass (colder than the ground that it is traveling over) overtakes a warm air mass that is stationary or moving slowly. This cold air, being denser, will go underneath the warm air, pushing it higher. Of course, no one can see this "air", but they can see clouds and the clouds themselves can tell us what is happening. The cloud progression to look for is cirrus to cirrocumulus to cumulus and, finally, to cumulonimbus.

c. Occluded Front. Cold fronts move faster than warm ones so that eventually a cold front overtakes a warm one and the warm air becomes progressively lifted from the surface. The zone of division between cold air ahead and cold air behind is called a "cold occlusion". If the air behind the front is warmer than ahead, it is a warm occlusion. Most land areas experience more occlusions than other types of fronts. In the progression of clouds leading to fronts, orographic uplift can play part in deceiving you of the actual type of front, i.e. progression of clouds leading to a warm front with orographic cumulus clouds added to these. The progression of clouds in an occlusion is a combination of both progressions from a warm and cold front.

9. USING PRESSURE AS AN INDICATOR. A very important factor of telling us what might happen is the pressure. As we know, low pressure or dropping pressure normally indicates deteriorating weather whereas high pressure usually gives us more good weather or clearing of bad weather. There are a couple of ways to monitor our pressure and are as follows:

a. Barometer: A barometer could be described as a pan of mercury with a tube leading out of the pan. Pressure from the atmosphere causes the mercury to rise in the tube.

(1) The tube is marked in millibars and the station that's reading these millibars will know how much it should rise for that location. Once again, if it rises more than normal, it would be considered a high-pressure reading.

b. Altimeter: Another means that is used to measure pressure is an altimeter, which is commonly used by mountaineers. It works like this:

(1) As you rise in elevation the pressure becomes less, thus allowing the needle in the altimeter to rise. If the needle rises without you rising with it, there is less pressure in the atmosphere than before and thus, a low-pressure area.

c. Contrail Lines: A basic way of identifying a low-pressure area is to note the contrail lines from jet aircraft. If they don't dissipate within two hours, that indicates a low-pressure area in your area. This usually occurs about 24 hours prior to an oncoming front.

d. Lenticulars: These are optical, lens-shaped cumulus clouds that have been sculpted by the winds. This indicates moisture in the air and high winds aloft. When preceding a cold front, winds and clouds will begin to lower.

10. USING SIGNS FROM NATURE. These signs will give you a general prediction of the incoming weather conditions. Try to utilize as many signs together as possible, which will improve your prediction. All of these signs have been tested with relative accuracy, but shouldn't be depended on 100%. But in any case you will be right more times than wrong in predicting the weather. From this we can gather as much information as needed and compile it along with our own experience of the area we are working in to help us form a prediction of incoming weather. The signs are as follows:

a. A spider's habits are very good indicators of what weather conditions will be within the next few hours. When the day is to be fair and relatively windless, they will spin long filaments over which they scout persistently. When precipitation is imminent, they shorten and tighten their snares and drowse dully in their centers.

b. Insects are especially annoying two to four hours before a storm.

c. If bees are swarming, fair weather will continue for at least the next half day.

d. Large game such as deer, elk, etc., will be feeding unusually heavy four to six hours before a storm.

e. When the smoke from a campfire, after lifting a short distance with the heated air, beats downward, a storm is approaching. Steadily rising smoke indicates fair weather.

f. A red sun or sky in the morning indicates nearing rain due to the excessive amount of moisture in the air.

g. A red evening sky indicates that the air contains so little moisture that rain within the next 24 hours is highly improbable.

h. A gray, overcast evening sky indicates that moisture carrying dust particles in the atmosphere have become overloaded with water; this condition favors rain.

i. A gray morning sky indicates dry air above the haze caused by the collecting of moisture on the dust in the lower atmosphere; you can reasonably a fair day.

j. When the setting sun shows a green tint at the top as it sinks behind clear horizon, fair weather is probable for most of the next 24 hours.

k. A rainbow in the late afternoon indicates fair weather ahead. However, a rainbow in the morning is a sign of prolonged bad weather.

l. A corona is the circle that appears around the sun or the moon. When this circle grows larger and larger, it indicates that the drops of water in the atmosphere are evaporating and that the weather will probably be clear. When this circle shrinks by the hour, it indicates that the water drops in the atmosphere are becoming larger, forming into clouds, rain is almost sure to fall.

m. In fair weather, air currents flow down streams and hillsides in the early morning and start drifting back up towards sunset. Any reversal of these directions warns of a nearing storm.

n. When the breeze is such that the leaves show their undersides, a storm is likely on the way.

o. It is so quiet before a storm, that distant noises can be heard more clearly. This is due to the inactivity of wildlife a couple of hours before a storm.

p. When in the mountains, the sight of morning mist rising from ravines is a good sign of clear weather the rest of the day.

q. A heavy dew or frost in the morning is a sign of fair weather for the rest of the day. This is due to the moisture in the atmosphere settling on the ground vice in the form of precipitation such as rain, snow, etc.

UNITED STATES MARINE CORPS
Mountain Warfare Training Center
Bridgeport, California 93517-5001

FMST.07.42
03/14/02

STUDENT HANDOUT

RESCUE PHILOSOPHY FOR SWIFTWATER RESCUE TEAMS

TERMIANL LEARNING OBJECTIVE In a swift water rescue scenario, organize a swift water rescue, in accordance with the reference. (FMST.07.42)

ENABLING LEARNING OBJECTIVES.

(1) With out the aid of reference, from a given list select the correct components of rescue preparation, in accordance with the reference. (FMST.07.42a)

(2) With out the aid of reference, from a given list select the correct role description of the rescue team members, in accordance with the reference. (FMST.07.42b)

(3) With out the aid of reference, from a given list select the correct methods of rescue, in accordance with the reference. (FMST.07.42c)

OUTLINE.

1. **COMPONENTS OF RESCUE PREPARATION**. (FMST.07.42a) These components comprise the backbone of both technical skills and the proper mental attitude required of a rescue professional. Remember that this course only provides you with the first component.

a. Training. Provides the students with the basic technical skills required to perform a safe rescue, these can be gained by attending an approved SWIFTWATER RESCUE COURSE.

b. Practice. Use of the skills learned in training.

c. Experience. Gained through advanced training, practical exercises and real-world situations.

B1

d. Judgment. The combination and application of skills for a given situation. Remember that every rescue situation is different. Don't get into a SOP mindset.

2. **PRIORITIES OF SAFETY**. Most of you have heard the saying that "It is better to be judged by twelve then carried by six". With that in mind you must realize that, at times, there is no way to affect a safe rescue. If you place yourself or your team in a hostile environment, from which you cannot escape, you are adding to the problem, not assisting in the solution. For this reason we set priorities of safety.

a. Self-aid. Be able to rescue yourself.

b. Buddy-aid. Ensure all safety measures are in force and that team members can rescue each other in an emergency.

c. Victim-aid. Only after self and buddy-aid are ensured should you attempt to rescue the victim. This will enable you to concentrate and focus your entire effort on the rescue.

3. **ROLES OF THE TEAM MEMBERS**. (FMST.07.24b) Often when dealing with rescues, you will find yourself working with outside agencies and organizations. In order to successfully coordinate the entire effort the Incident Command System (ICS) was developed. In swiftwater rescue situations there are four specific roles that must be filled.

a. Incident Commander. Is in overall command of the rescue efforts. Does not necessarily have the technical knowledge to conduct the rescue, but must be able to coordinate the rescue effort.

b. Rigger. Personnel responsible for equipment and rope systems, i.e. establishing the tension diagonal or the tyrolean high line.

c. Gofer. This group of individuals fills all other supporting roles, i.e. Communications, logistics, medevac, accounting, equipment accountability etc.

d. Rescuer. The individuals who are actually conducting the rescue.

4. **COMPONENTS OF SEARCH AND RESCUE**. In order to successfully complete a rescue there is four objectives that must be met. In order to remember them we use the acronym LAST.

a. Locate. This is the search in the search and rescue business.

b. Access. The determination of the assets, equipment, technical skills, and personnel required safely reaching and extracting the victim.

c. Stabilization. Providing the on scene medical attention required in order to extract the victim without causing further injury or aggravating an existing condition.

d. Transport. Removing the victim from the hostile environment to a safe one where further treatment can be provided.

5. **METHODS OF RESCUE**. (FMST.07.42c) In any rescue, we must access the situation to determine the best techniques to apply. Bearing in mind the self-aid, buddy-aid, victim-aid concept as well as time, weather, victim's condition, etc. In all cases we will attempt the lowest risk methods first, while preparing for a higher risk option. In all cases remember the KISS principle.

a. Reach. Use a ladder, fire hose, stick, paddle, rifle or any other sturdy object to extend your reach in order to reach the victim.

b. Throw. Get some type of floatation device to the victim. Or throw a line to the victim in order to pendulum or pull him/her to the shore.

c. Row. Putting a boat in the water to reach the victim.

d. Go and Tow. These are the contact rescues that, while look good on film, put you at the extreme risk of becoming a second victim.

e. Helo. Most civilian helicopter crews are not trained in SAR evolutions. Military aircraft and crews are a better bet. Several uses for helos are in the support role. Locating the victim, inserting and extracting teams and moving equipment are some of these roles. If using the helo in a direct role it is better to use a qualified SAR swimmer/diver to hook-up or pick-up victims.

UNITED STATES MARINE CORPS
Mountain Warfare Training Center
Bridgeport, California 93517-5001

FMST.07.25
03/14/02

STUDENT HANDOUT

PERSONAL AND TEAM EQUIPMENT

TERMINAL LEARNING OBJECTIVE As a member of a swiftwater rescue team, maintain swiftwater rescue equipment, in accordance with the reference. (FMST.07.25)

ENABLING LEARNING OBJECTIVES.

(1) Without the aid of references, from a given identify the correct items of personal equipment, in accordance with the reference. (FMST.07.25a)

(2) Without the aid of references, from a given identify the correct items of team equipment, in accordance with the reference. (FMST.07.25b)

OUTLINE.

1. **PERSONAL EQUIPMENT**. (FMST.07.25a) The following items of personal equipment should be in each man's possession for a swiftwater rescue.

    a. Dry Suit. Warmest, most versatile, full body protection available. Extra thermal layers can be added for varying water temperatures. Provides wind protection out of the water and provides protection against contaminated water.

    b. Wet Suit. Excellent protection from flotsam. Variety of thick nesses for thermal protection. Insulating quality is degraded in moving water due to the flushing effect.

    c. Footwear. Primary is full sole scuba booties. You can use tennis shoes or jungle boots with heavy wool socks.

    d. Gloves. Neoprene is recommended.

    e. Hoods. Wetsuit hood or pile hood for insulation. Remember 30-50% of your body heat losses can occur through the head and neck region of your body if left unprotected.

B2

f. Helmet. Protects you from flotsam and impact injuries. Should be padded and ventilated to allow water to flow through.

g. Knife. Fixed or locking blade attached to the vest, handle down.

h. Whistle. Required for communication. Best without "pea".

i. Eye Protection. Protects subject from debris.

j. Personal Floatation Device. This is the single most important piece of equipment. Jackets should provide at least 15 lbs. of floatation. Proper fit is a priority. Rescue vests with a "Blow-Out" belt should only be used by trained professionals.

2. **TEAM EQUIPMENT**. (FMST.07.25b) The following is a list of some of the team equipment for a typical four-man team. Of course the situation will dictate which equipment will be required for the mission on hand, but all hands should be familiar with this equipment. This is only a list. The uses and application of this gear, along with practical application exercises later in this course will provide you with the hands-on training and working knowledge to properly and safely use this equipment.

QTY	ITEM
1	300 ft, l Static Rope
4	150 ft, 11 mm Static Rope
25	85 Locking Stubai Carabiners
4	2" Pulleys
2	Rescue Figure 8
6	3 ft, 7mm Prussic
20	15 ft, 1" Tubular Nylon Runners
2	Ascenders
3	Throw Bags
2 pr	Fins
1	Rescue PFD
1	Self Bailing Raft (13 or 14 foot)
1	Oar Frame w/ Oars, or Seven Paddles
3	Radios (waterproofed)
1	Medical Kit
1	Litter

NOTE: This is not a complete list of rescue equipment. However, it is a good basic load.

UNITED STATES MARINE CORPS
Mountain Warfare Training Center
Bridgeport, California 93517-5001

FMST.07.41
03/14/02

STUDENT HANDOUT

SWIFTWATER TERMINOLOGY AND DYNAMICS

TERMINAL LEARNING OBJECTIVES In a swift water rescue scenario, evaluate swift water river conditions in accordance with the reference. (FMST.07.41)

ENABLING LEARNING OBJECTIVES.

1) Without the aid of reference, from a given list correctly identify the swiftwater terminology, in accordance with the reference. (FMST.07.41a)

2) Without the aid of reference, from a given list correctly identify the classifications of rivers, in accordance with the reference. (FMST.07.41b)

OUTLINE.

1. **SWIFTWATER TERMINOLOGY.** (FMST.07.41a)

a. Hydraulic or Hydraulic effect: Movement of water caused by pressure.

b. Eddy: Horizontal reversal. Current is blocked by an obstacle creating a vacuum which, when filled, causes water to flow upstream.

c. Eddy Fence: Reversal line between downstream current and the eddy. This fence can be very strong and the water flow may vary in height by several inches.

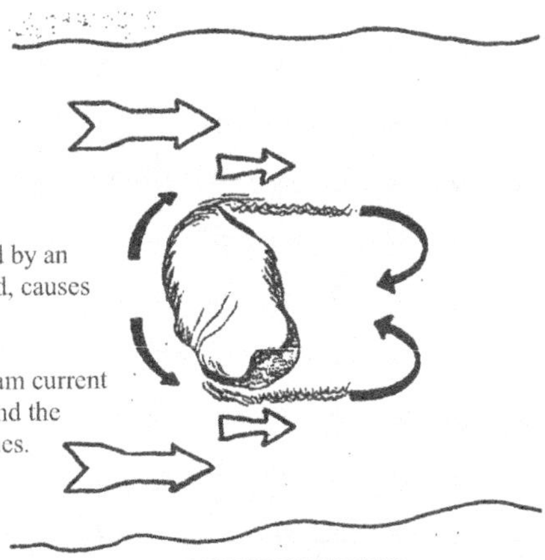

EDDY/EDDY FENCE

B3

d. Hole, Stopper, Keeper: Vertical reversal as water flows over an obstacle the main current follows the base of the river causing a surface vacuum. Downstream water flows back upstream to fill this void. This sets up a recirculating effect.

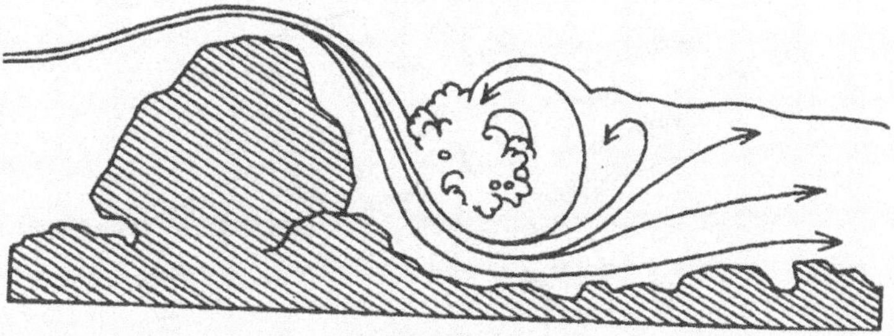

HOLE/STOPPER/KEEPER

e. Standing Wave. Haystack: Series of waves in the main channel caused by obstacles, ledges, or increases in the flow rate, gradient, or volume.

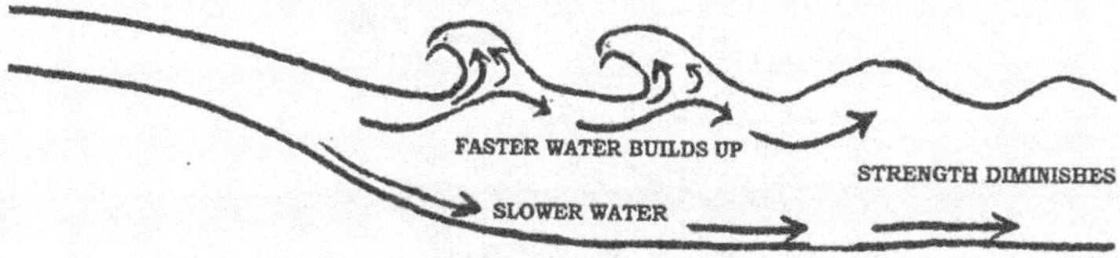

STANDING WAVES/HAYSTACKS

f. Downstream V. Tonal: Represent largest volume of water or main channel.

g. Upstream V: Caused by obstacles submerged just below the surface.

h. Strainer: Debris, logs, small trees on flooded banks, which allow water to flow through but will not allow a body to pass.

STRAINER / GUARD RAIL

STRAINER / TREE

i. Cushion, Pillow: Build-up of water on the upstream side of obstacles or the outside of curves.

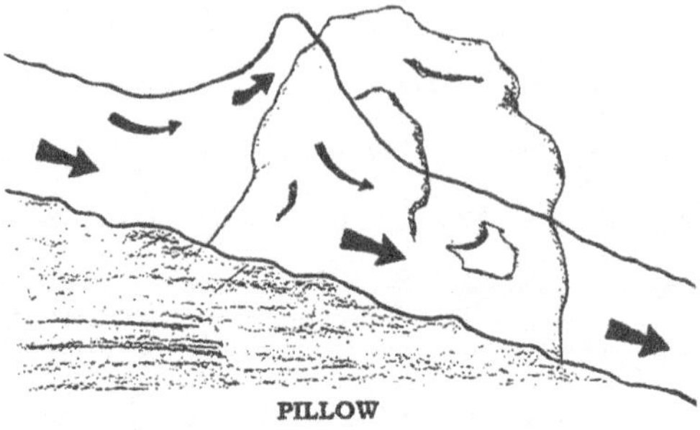

j. Confluence: Merging of streams.

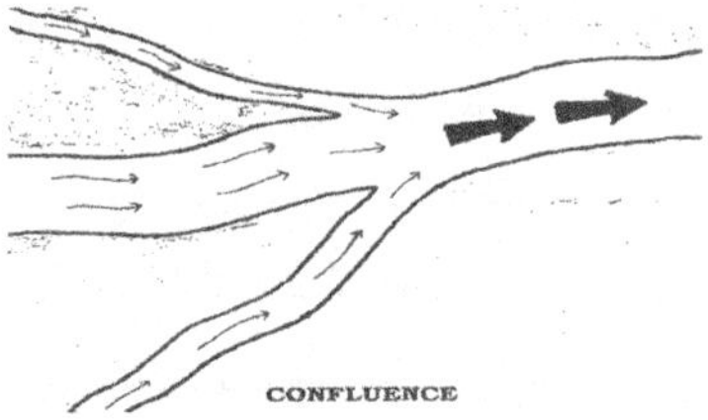

k. River Right. On the right side of the river looking downstream.

l. River Left. On the left side of the river looking downstream.

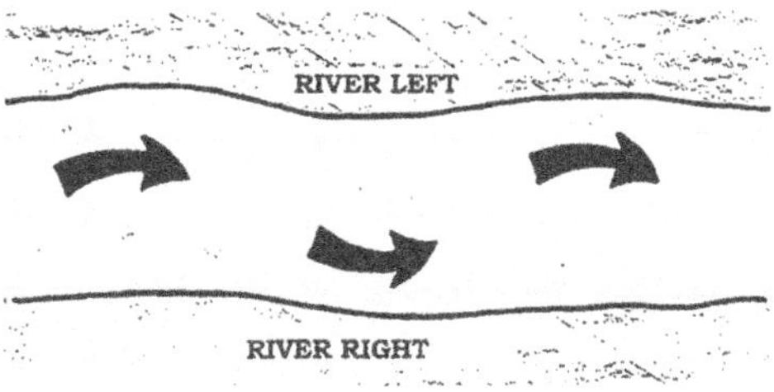

2. **CLASSIFICATIONS OF RIVERS**: (FMST.07.41b)

a. Class 1. Moving water, few obstacles, small ripples or waves.

b. Class 2. Moving water, clear channels, wide waves to 3', no scouting required.

c. Class 3. Rapids, narrow channels that require complex maneuvering, waves capable of flipping rafts or swamping canoes, scouting required.

d. Class 4. Long hazardous rapids, precise boat handling in rough water, scouting required, rescues hazardous.

e. Class 5. Severe, extended, extremely violent rapids, professional boat handlers and scouting required. Rescues extremely hazardous, situations are life threatening.

f. Class 6. Only Expert/Professional River Runners. Rescue may be impossible.

g. CLASSES U DON'T EVEN THINK ABOUT IT! Waters are not navigable.

3. **RIVER STRENGTH**.

a. Volume of Water. The volume of water in a given stretch of the river between two points is based on a simple formula and is measured in Cubic Feet Per Second (CFS). Simply take the river depth x river width x river velocity = volume of water in CFS.

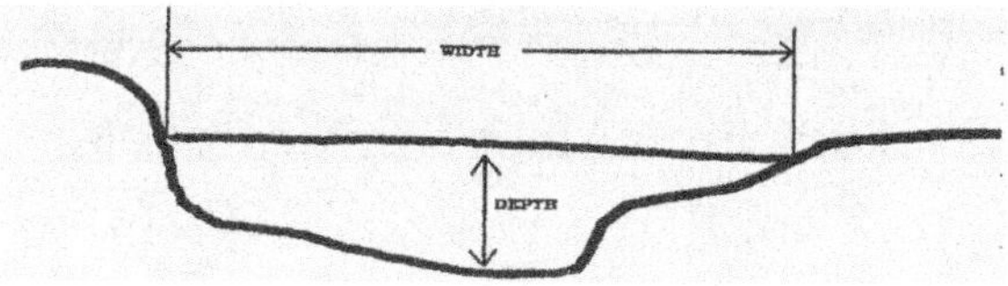

For example: 6 feet deep x 100 feet wide x 2 feet per second = 1,200 CFS

b. River Velocity. In order to determine the velocity or current speed measure a 100-foot section of the river and drop a float at the upstream end. Then, time its movement to the lower end and use the chart below.

TIME IN SECONDS	VELOCITY FT PER SEC	VELOCITY MPH
5	20	13
10	10	7
15	6.7	5
17	5.9	4
20	5	3.5
25	4	3
29	3.5	2.5
37	2.7	2
50	2	1.5
80	1.3	1
110	0.7	0.5

NOTE: Should the shape of the river change by becoming wider or narrower, once you have the flow in CFS, you can determine the new velocity by using the following formula:

$$\text{Velocity} = \frac{\text{Volume in CFS}}{\text{Width x Depth}}$$

c. Force Exerted. The velocity of the water will determine the force exerted X's the amount of exposed area under pressure. For simplicity we will use the following chart:

CURRENT VELOCITY	FORCE EXERTED ON OBJECT		
MPH	LEGS RAFT	BODY	WRAPPED
3	17	34	168
6	67	134	672
9	151	302	1,512
12	269	538	2,688

4. **LOW HEAD DAMS**. Low head dams are used throughout the world for irrigation, flood control, at hydroelectric plants and for many other uses. Basically they are pouring over obstacles that create hydraulics, keeper holes, and recirculating currents at their bases. Unlike natural obstacles, that rarely span the entire river and that allow you to be "flushed" out of the bottom or sides, low head dams create a hydraulic that extends from shore to shore.

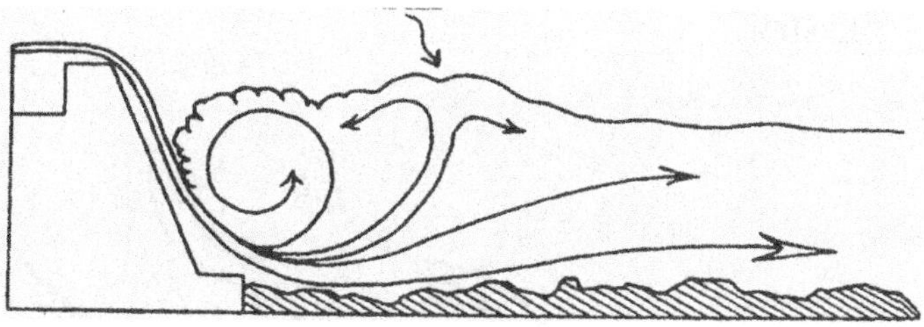

a. The size of the hydraulic created is based on several denominators:

(1) Height of drop.

(2) Angle of drop.

(3) Volume of water CFS.

(4) Velocity of water MPH.

(5) Bottom composition of runout.

b. The degree of difficulty for the rescuer is also compounded by:

(1) Water depth below dam.

(2) Debris build-up.

(3) Dam construction/conformation.

(4) Lack of access.

(5) Lack of natural anchors.

(6) Distance between riverbanks.

c. The basic hydrology is the same as any other pour over obstacle. As the river flows over the dam, the water follows the face of the dam to the river bottom. When sufficient speed diminishes the water rebounds to the surface and continues downstream. This flow has caused a vacuum or pocket directly below the dam that must be filled. Therefore, when the water rebounds to the surface, some of it flows back upstream to form a vertical reversal or pour-over eddy. The line of demarcation for the upstream/downstream flow is called the boil line. The shape and angle of the dam face and the volume and velocity of the water flowing over the dam create the distance and width of the boil line.

5. **REVIEW**.

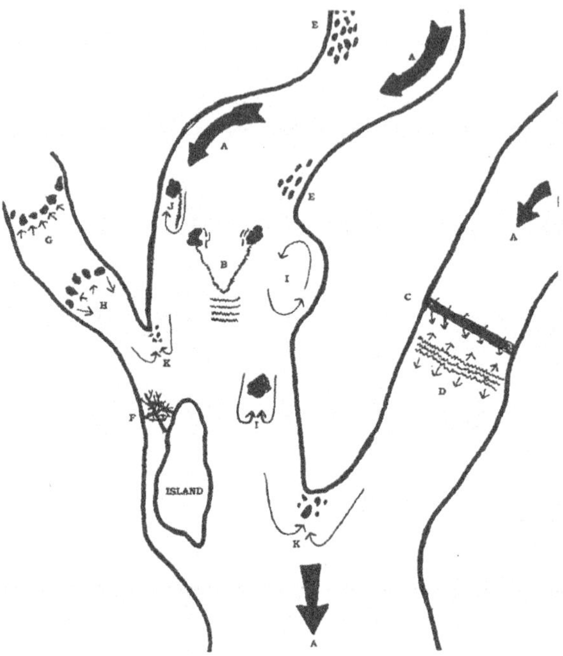

A. Main Current
B. Downstream V
C. Low head Dam
D. Boil Line
E. Rock Garden
F. Debris Pile/Strainer
G. Frowning Hole
H. Smiling Hole
I. Eddy
J. Eddy Fence
K. Confluence

UNITED STATES MARINE CORPS
Mountain Warfare Training Center
Bridgeport, California 93517-5001

FMST.07.24
04/14/02

STUDENT HANDOUT

COMMUNICATIONS

TERMINAL LEARNING OBJECTIVES Given a casualty in a swift water environment, assignment as a member of a rescue team, and a whistle, employ swift water rescue communication signals to support mission requirements per the reference. (FMST.07.24)

ENABLING LEARNING OBJECTIVES.

(1) In a summer mountainous environment, prior to swift water rescue operations, when the instructor verbally states a hand and arm signal, the student will demonstrate that signal, in accordance with the references. (FMST.07.24a)

(2) In a summer mountainous environment, standing at the edge of a swift water stream, when the instructor blasts a whistle a varied amount of times, the student will focus their attention in the proper direction, in accordance with the references. (FMST.07.24b)

OUTLINE.

1. **RELIABLE COMMUNICATION**. In many cases time is critical in a SWR scenario. Therefore immediate, understandable communication is a must. Radios, bull horns, and voice work well in some situations, but wet equipment, dead batteries, and ambient noise levels can render these methods ineffective. Therefore, rescue professionals have developed a series of simple hand and arm, and whistle signals. Any rescue team can develop their own signals. The only requirement is that all hands understand the signals and that they are briefed to any other agency you are working with.

B4

2. **HAND AND ARM SIGNALS**. (FMST.07.24a)

ACTION	MEANING
1 hand extended over head	Require Assistance
Both hands extended forming an "O" with one hand touching The head	Okay
Extend both hands above head, then point left/right	Move in direction indicated
Extend both hands above head, wave, then point left/right	Eddy out in direction indicated
Cross both arms over chest	Require medical assistance or Corpsman

3. **WHISTLE BLASTS**. (FMST.07.24b)

ACTION	MEANING
One Blast	Stop/Attention
Two Blasts	Upstream
Three Blasts	Downstream
Three Blasts Repeated	Emergency

UNITED STATES MARINE CORPS
Mountain Warfare Training Center
Bridgeport, California 93517-5001

FMST.0
7.27
04/14/02

STUDENT HANDOUT

THROW BAGS AND FOOT ENTANGLEMENTS

TERMINAL LEARNING OBJECTIVES In a swiftwater rescue scenario employ a throw bag, in accordance with the reference. (FMST.07.27)

ENABLING LEARNING OBJECTIVES.

1) Without the aid of reference, provided with a throw bag and swimming victim, rescue the victim, in accordance with the reference. (FMST.07.27a)

OUTLINE.

1. **THE THROW BAG**: The throw bag is very simply a cloth bag with a small float insert, packed with 50 to 70 feet of 1/4" to 3/8" floating polypropylene or spectra/poly line. The concept is to hold the exposed end of the line and throw the packed bag to a swimmer. As the bag flies, the rope deploys enabling you to reach a victim 50 to 70 feet from shore. It is used either from a standing position, or a moving position on open shores.

    a. Standing Procedures. (FMST.07.27a)

1. Assume a stable belay position and ensure your footing is solid. Avoid loose, sloped or wet rock.

2. Attract victim by blowing your whistle (1 blast). Show him/her the throw bag and yell, "grab the rope". NOTE: If the victim grabs the bag rather than the rope, the rope will continue to deploy to its full length.

3. Grasping the exposed end of the rope, throw the bag underhand in a smooth motion directly at the victim. You may need to lead them in order to score a direct hit.

4. Tell the victim to place the rope over his/her upstream shoulder. This will establish a good ferry angle.

B5

5. Do not wrap the rope around yourself or tie it off at anytime. You may have to release the victim to avoid an obstacle or strainer. Use a dynamic belay and pull the victim to shore or pendulum the victim to other rescuers downstream.
6. If available have other rescuers downstream to recover the victim or repeat the procedure if you miss!

b. Moving Procedures. The moving procedures are the same as the standing procedures except you are on a smooth, open shoreline, so you move slightly downstream of the victim as you pull them gently ashore.

2. <u>**FOOT ENTRAPMENTS AND VERTICAL PINNINGS**</u>. These scenarios are caused by a variety of actions. Swimmers put their feet down or fishermen, in waders, get their feet stuck between rocks and are pushed over by the current. Swimmers are stuck in a strainer. Rafters get wrapped on a rock with victims pinned between the raft and rock or kayakers drop down a vertical shoot and get lodged in the rocks at the bottom. Whatever way it occurs you must respond quickly in order to save the victim. Initially take the following precautions:

   1. Position upstream spotters, to provide warning of boats or debris.

   2. Plan alternate methods if initial try doesn't work.

   3. Have back-up rescuers downstream in order to retrieve the victim.

   a. Once these actions are complete conduct the rescue using one of the five following methods.

      1. <u>Bechel-Ray One Line Method</u>. The simplest and safest method. Get a rescuer on both sides of the river. Establish a rope between the two rescuers and downstream of the victim. Walk upstream until the rope hooks the victim. If you can position the line near the victims ankles you may be able to pop him/her loose and the downstream safety can pick him/her up. If not you can haul the victim upstream until he/she is clear of the obstacle, then by one rescuer playing out line and the other retrieving it you can bring the victim ashore. Remember to bring the victim to the shore with the medevac capability.

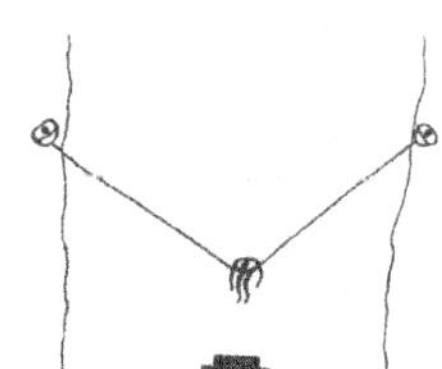

      2. <u>The Strong Swimmer-variation</u>. If the previous method doesn't work, leave the line in place and anchor it to stabilize the victim above the surface. Then have a "strong swimmer" go down one side of the rope, as would a tension diagonal. This rescuer is not tied in. Upon arrival the rescuer can readjust the line on the victim, attempt to pull them free, give them a PFD and helmet, or if a debris pile or rock is near, get on it and assist in further efforts.

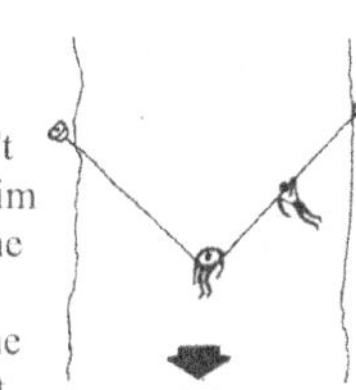

3. Two Line Loop. There are two variations on this method both of which are applied using the initial rope used in the Bechel-Ray attempt.

(1) Deploy a second rope across the river and hook it onto the first line. Allow it to go one-half way down the rope to the victim. Pull at a 90° angle to rope #1. This will establish a mechanical advantage and pull the victim loose.

(2) Once the victim is free walk the line downstream trapping the victim in a loop. Then by releasing slack on line one, you can bring him/her ashore.

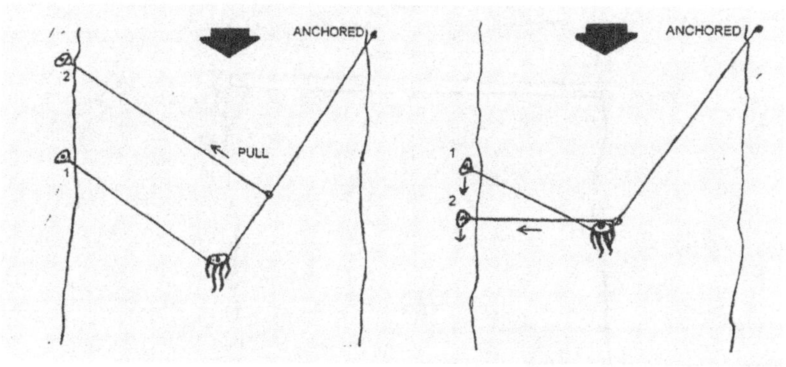

4. European Tethered Rescuer Lower. Using a rescue vest with a "blow-out" belt establish an upstream belay and lower the tethered rescuer to the victim. The rescuer's body will block the water forming an eddy around the victim enabling him to be pulled free. Ensure that the rescuer has a working rescue vest, as this will enable him to cut away if the situation dictates.

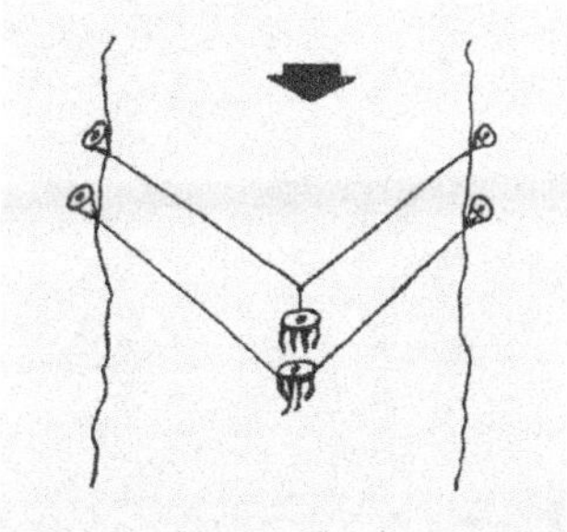

5. Mechanical Advantage Systems. As a last resort you can use a mechanical advantage system. This method is usually used only for body recoveries, as the amount of force exerted on the victim is tremendous and can cause severe injuries.

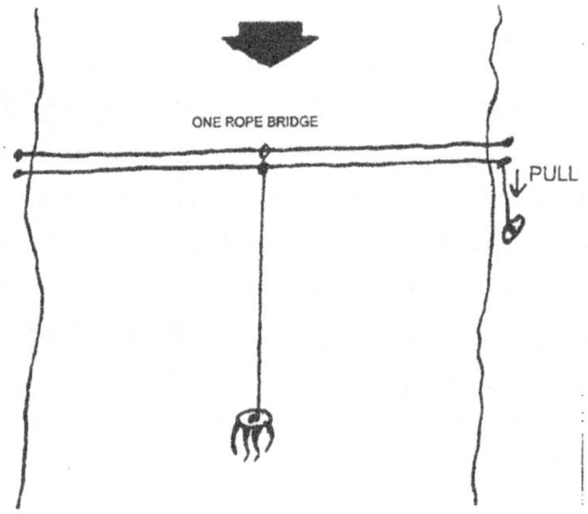
ONE ROPE BRIDGE
PULL

UNITED STATES MARINE CORPS
Mountain Warfare Training Center
Bridgeport, California 93517-5001

FMST.07.28
04/14/02

STUDENT OUTLINE

DEFENSIVE SWIMMING AND FERRY ANGLE

TERMINAL LERNING OBJECTIVES Given a swift water river and throw bags, perform aggressive swift water swimming to meet mission requirements per the reference. (FMST.07.28)

ENABLING LEARNING OBJECTIVES.

(1) Without the aid of reference, demonstrate the defensive swimming position, in accordance with the reference. (FMST.07.28a)

(2) Without the aid of reference, demonstrate use of the ferry angle to cross the river, in accordance with the reference. (FMST.07.28b)

OUTLINE

1. **THE DEFENSIVE SWIMMING POSITION (DSP)**. (FMST.07.28a) The DSP is assumed by lying flat on your back, chin on chest (to read the river & select a route), with your feet up, heels just below the level of your buttocks. Holding your feet in this position will enable you to fend off of shallow rocks and will reduce the likelihood of a foot entrapment or pinning. Your arms and hands should be kept close to the body but may be used like oars on a boat to maneuver. Your head will remain upstream except to cross a strainer or to over take a victim in a swimming rescue.

2. **FERRY ANGLE**. (FMST.07.28b) A ferry angle is used to maneuver diagonally across a river. By placing your body at a 45° angle to the current, the river will push you across the river. This is because the force of the current will travel down the exposed, upstream, side of the body moving it laterally more rapidly then the current pushing straight. This is like traversing a slope with edged skis vice sideslipping down on flat skis. In order to cross faster or to reach an eddy you may roll to your belly and execute the rescue crawl stroke or breaststroke. Remember to maintain your ferry angle with your head upstream this will enable you to travel diagonally across the river.

3. **STRAINER CROSSING**. The first rule is to avoid the strainer or shallow submerged obstacle. Read the river and maneuver to the main channel by using downstream "V's" or tongues. If the obstacle cannot be avoided, roll on your belly, head downstream and use the rescue crawl stroke. You must get your body moving faster than the current. Upon arrival at the obstacle use both hands to "vault" yourself over. A well-timed dolphin kick will assist in keeping your legs clear of entanglements. Once clear of the obstacle, resume the defensive swimming position. Bear in mind that it is usually safer to remain in the main current or to eddy hop downstream than it is to attempt to get out on a flooded riverbank, the outside of a bend or on slippery rocks. Also, use the dolphin kick rather than placing your feet on the bottom and kicking off. This will reduce the likelihood of foot entrapments.

UNITED STATES MARINE CORPS
Mountain Warfare Training Center
Bridgeport, California 93517-5001

FMST.07.25
03/14/02

STUDENT HANDOUT

FEET-WET RESCUES AND C-SPINE ROLLS

TERMINAL LEARNING OBJECTIVES In a swift water rescue scenario, conduct a swimming rescue, in accordance with the reference. (FMST.07.25)

ENABLING LEARNING OBJECTIVES.

(1) Without the aid of reference, while in a river and provided with a victim, perform a contact rescue, in accordance with the reference. (FMST.07.25a)

(2) Without the aid of reference, while in a river and provided with a simulated unconscious victim, perform a C-spine roll, in accordance with the references. (FMST.07.25b)

OUTLINE.

1. **FEET WET RESCUE GENERAL CONSIDERATIONS**. (FMST.07.25a)

   a. Always wear the proper personal equipment, i.e. PFD, helmet, drysuit, etc.

b. Don't make physical contact if it can be avoided. Make him/her "chase" you ashore. Use a life ring, can, or boogie board, and tow it along with the victim. Toss him a throw bag and swim/tow him/her until you reach the shore, then pendulum or haul the victim in. Remember, to a panicked victim YOU are a floatation device.

2. **CONTACT RESCUE**.

a. Approach the victim from upstream. Use the rescue crawl stroke to catch-up to him/her. Talk to the victim and attempt to get him/her to swim to shore or "chase" you to the shore.

B7

b. Explain to the victim the defensive swimming position and attempt to get them in this position. Assume this position yourself, in contact rescues this is called the "Reverse and Ready" position. Tell the victim what you are going to do prior to grabbing them.

c. Gain contact by grabbing the shoulders of the life jacket or utilize a cross-chest carry. If the victim is unconscious and face down use a C-Spine roll down and maintain stability as much as possible.

d. Assume a good ferry angle with the victim downstream of you and use his/her body to protect you from obstacles.

e. If the victim panics and attempts to climb on top of you, break away or lock the victim up and go for an alligator roll. Reestablish control and calm the victim as soon as possible.

f. Once contact is made remember that you may be held accountable under the "abandonment" laws if you quit.

3. **TOWED VICTIM SWIM**. When towing a victim it is essential that you establish a ferry angle. If the victim is on a boogie board or float, explain to him/her what you are doing and have them help by kicking their feet. If you are towing by the PFD grasp the shoulder towards the far bank and remain in the DSP (defensive swimming position) in case your victim panics.

4. **LIVE BAIT RESCUES**. The final go and tow method is the "live bait" rescue. Using a rescue vest with a "blow out" belt and a mule team on the shore to belay you, swim to the victim as explained in contact rescues. Once you establish contact you are, in effect, a throw bag with hands. The belayer will then either pendulum you ashore or hauls you in. In an emergency you can cut away and conduct a normal contact rescue.

5. **C-SPINE ROLLS AND STABILITY**.

a. Basic Rules: (FMST.07.25b)

(1) Suspect C-Spine involvement in any head injury.

(2) Stabilize and immobilize, use C-Collar, Sam Splints, or a field expedient method.

(3) Keep the victim in the water until a spine board or equivalent is available, or there are enough rescuers to move the victim safely.

(4) Observe for other signs, i.e. cranial bleeding, bleeding from ears, eyes, nose (Do not attempt to stop this bleeding) or unequal/unresponsive pupil response.

b. Methods:

(1) First, approach the victim from behind. Reach under the arms and grasp the victim's head with both hands, then roll the victim. This is the least desirable method as you cannot swim properly or evaluate the ABC's of the victim.

(2) Second, approach the victim from the side, extend his/her arms above and alongside his/her head, reach across and roll the victim onto your arm maintaining the victims arms in place to ensure stabilization. Evaluate the ABC's, provide rescue breathing and use your free arm to swim.

(3) Finally, approach a victim wearing a PFD from the head. If face down, cross your arms thumbs inboard and use your forearms to immobilize the neck. Then roll the victim over. If face up, simply approach as if gaining contact on a contact rescue and maintain control of the C-Spine.

UNITED STATES MARINE CORPS
Mountain Warfare Training Center
Bridgeport, California 93517-5001

WMC

04/14/02

STUDENT HANDOUT

BOATS AND BOAT HANDLING

LESSON PURPOSE. The purpose of this period of instruction is to introduce the student to the boats available at MWTC for river rescue evolutions.

OUTLINE

1. **GENERAL**. There are a wide variety of boats and hybrid river craft available to rescue crews. Whitewater rafts, rigid raiders, zodiacs, airboats, and regular pleasure boats can all be used in the proper situations. All have been used both safely and have caused deaths and injuries when used improperly. Wherever you are, learn the equipment available, both its assets and its limitations.

2. **WHITEWATER RAFTS**. The whitewater raft of today is an inflatable, self-bailing, multi-chambered vessel with several tie rings and foot cups. Additionally, the materials used in construction resist rips, tears, and punctures much better than their predecessors. The size most often found on local rivers is 12 to 14 feet long due to the size of the rivers. In the MWTC SAR locker, we have a 13-foot self-bailer. This will be our primary rescue boat.

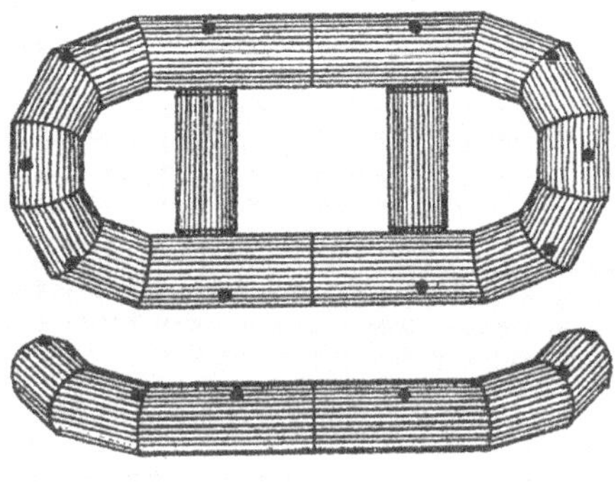

3. **RAFTING COMMANDS.** In order to control the direction of a peddler powered boat the coxswain or guide must use a series of simple, concise, and understandable commands.

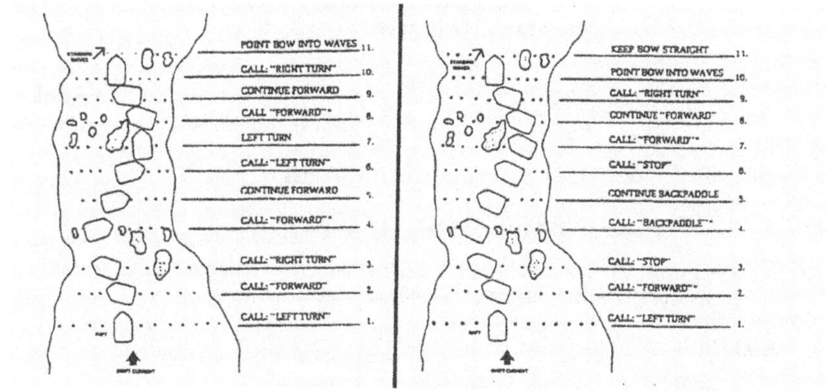

NORMAL TACTICS APPROACH — CONSERVATIVE/FAST CURRENT

a. Forward - All hands paddle forward together.

b. Back Paddle - All hands back paddle together.

c. Stop - Stop paddling - Remove blades from water.

d. Left Turn - Left side back paddles, right side forward.

e. Right Turn - Right side back paddles, left side forward.

f. High Side - Move quickly to the down streamside of the boat.

4. **PADDLE STROKES**. In order to control the raft from the coxswain position, either with a crew or alone on a tether system, you will need to apply the following strokes:

a. Sweep - Stroke in a wide arc around the fantail of the boat.

b. Draw - Reach out 90° away from the boat and "draw" it straight back to you.

c. Pry - Put paddle blade beside raft and "pry" the water away from the boat.

d. "J" - Used by lone guide to keep boat straight. Regular straight stroke finished with an outboard stroke or tail.

UNITED STATES MARINE CORPS
Mountain Warfare Training Center
Bridgeport, California 93517-5001

WMC
04/14/02

STUDENT HANDOUT

ROPE SYSTEMS AND BOAT ANCHORS

LESSON PURPOSE: The purpose of this period of instruction is to introduce the student to the rope systems and boat anchors used in swiftwater rescue situations.

OUTLINE

1. **TWO AND FOUR POINT TETHERING SYSTEMS**.

a. With all rope systems the first order of business is to get a rope across the river. Best case is to throw a rope bag across to someone already on the other side. Next best is to use another boat or raft to ferry the rope across. Last we use a swimmer to drag the rope across. Remember to use a smaller line on the swimmer and not to tie it to him/her. When swimming, start high and pendulum across using a good ferry angle.

b. Clip in on the side "D" rings forward for a two point or on the two forward and two aft "D" rings for a four-point tether system. For a two point you now need to simply load in a rescuer (if used) and move the boat into the river by taking slack on the far side and giving slack from the near side. For a four point man the two nearside lines and load a far side aft lineman and a rescuer (if used) into the boat and ferry the boat across. The far side aft belayer will get out and the boat is on a four point system ready for action.

c. This system is good for Class II Rivers with clear banks, few mid stream obstacles, and little floating debris. Use more belayers as required by river speed and boatload.

d. In all rope systems the whistle and hand and arm signals used to communicate are those taught in the communications class earlier in this course.

e. To break down the system simply reverse the procedures. The only difference is that the last man rides in the boat instead of swimming.

B9

2. **THREE POINT AND BOATMEN'S SELF-EQUALIZING ANCHORS.**

a. These anchors have in common the fact that they are self equalizing. This means that, regardless of the direction of pull, all anchor points ("D" rings, thwarts, etc) will remain under equal tension.

b. Using ropes the knots required are the double figure 8 on a bight, a directional figure 8, and a figure 8 on a bight.

c. Using webbing (one inch tubular nylon) a water tape knot is required.

d. If you have an abundance of carabineers, simply clip into the anchor points and back into the small loop on the double figure 8 on a bight, or into the directional figure 8. If not run the rope through the small loop. Remember that this is not preferred as it causes rope on rope stress and rubbing that can melt the rope.

e. When using webbing the same concept applies. Ensure you put a twist in the webbing at the contact point to ensure that a single blow-out will not collapse the entire system.

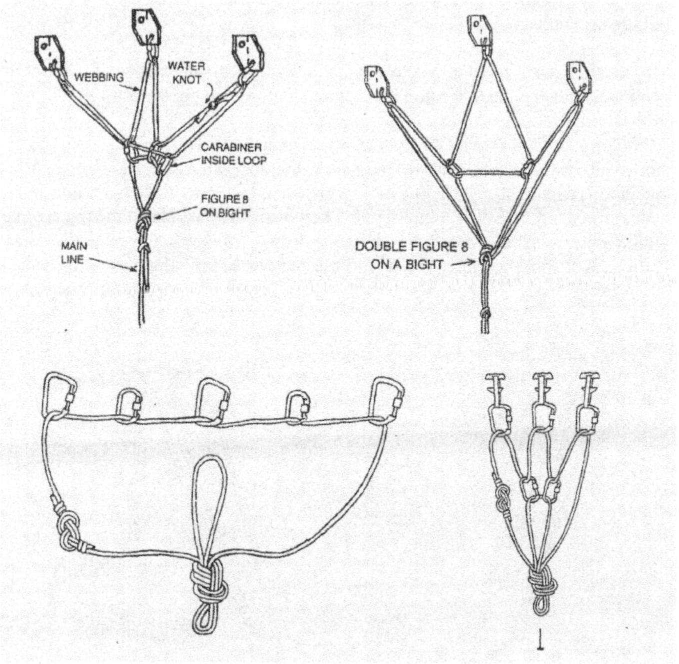

3. **PIG-RIG FOR MECHANICAL ADVANTAGE.** At times you will find yourself with ropes that are long enough to span the river, but not long enough to establish a mechanical advantage system. In these situations, you can use a piggy-back line or "pig-rig" by following the procedures below:

a. Tie off the mainline on either ends, or tie off the far side and set up a belay device on the near side using a munter hitch or rescue 8.

b. Establish an anchor for the pig-rig below the near side anchor.

c. Tie a prussic on the mainline and clip in a carabiner. Tie a figure 8 on a bight into the end of the pig-rig line and clip it into the prussic's carabineer.

d. Run the pig-rig line back to the anchor carabineer and then back through the prussic carabiner.

e. Pull on the running end towards the anchor. This will give you a 4:1 mechanical advantage on the main line.

f. If more tension is required you can either retie the nearside anchor (if tied), or use your belaying device to tension the mainline before moving the prussic.

g. If no prussic cord is available, you can accomplish this with a directional figure 8 on the mainline. However, this will only allow you to tighten the system once. A better method is to use the pig-rig line to tie a prussic directly on to the main line and run your system off of a directional figure 8 tied below the prussic on the pig-rig line. Remember that if the lines are of the same diameter this system will not work as the prussic will slip, therefore, you should use a kragur knot with ropes of the same diameter.

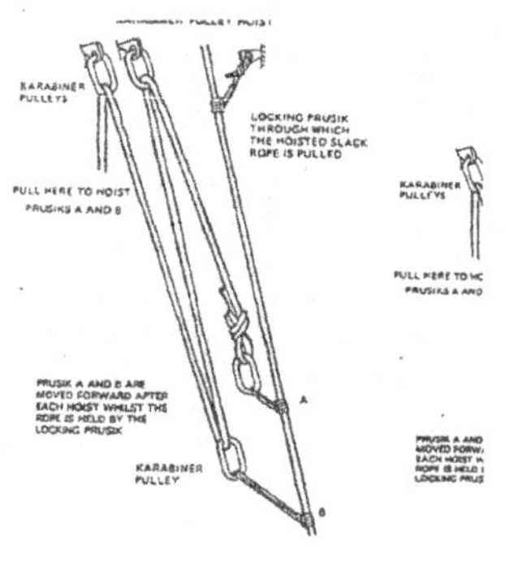

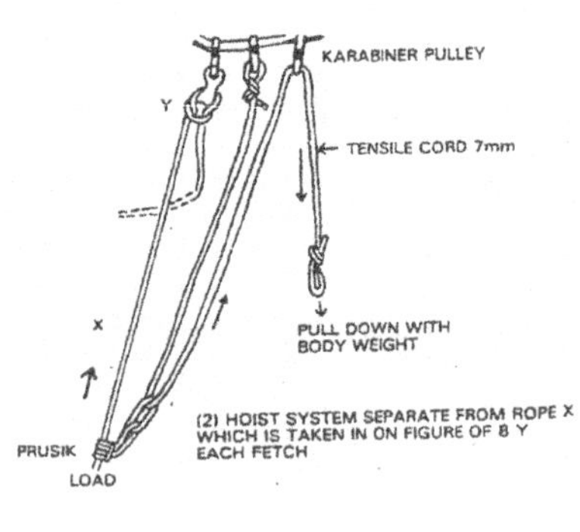

4. **TYROLEAN HIGHLINE AND TELFER LOWERING SYSTEMS** Both of these systems are used to maintain a boat as a stable rescue platform. The lateral movement (side to side/shore to shore) in both systems is controlled by mule teams on the shore. The main difference is that in the tyrolean highline the movement up and down-stream is also controlled from the shore, while in the telfer lowering system, the rescuer in the boat controls that aspect of movement. Both systems have good and bad points and both should be mastered. The situation will dictate which system you will use. Follow the steps below to establish these systems:

a. First you must establish TWO lines across the river. You can use any method taught to accomplish this.

b. Establish one line as the anchor line and tighten it as taught in the "one rope bridge" class or use a pig-rig, if required. If you have uneven mule teams set it at a slight down-stream angle towards the weak team's shore (no more then 10 to 15 degrees) to assist the weak team.

c. Rig the boat with a self-equalizing anchor system.

d. Rig the lowering systems as shown in the diagrams. Either tyrolean or telfer.

e. Load a rescuer or rescue team into the boat and execute. Remember the rescuer can act as coxswain and assist in maneuvering the boat by establishing a proper ferry angle.

f. Communications will be as taught earlier. Ensure you have a shore commander who maintains eye contact with the rescuers and mule teams on both the near and far shores.

g. When using the telfer system, a counter hitch, stitch plate or rescue 8 can be used as a belay device. This enables you to tie-off when you reach the victim so that you can affect the rescue without worrying about controlling the boat.

h. When using either of these systems it is a good idea to set-up a mechanical advantage system between the control (central) pulleys aid the raft/rescue platform.

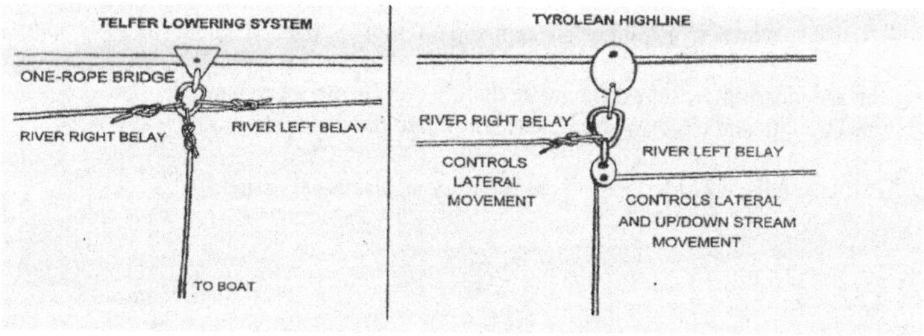

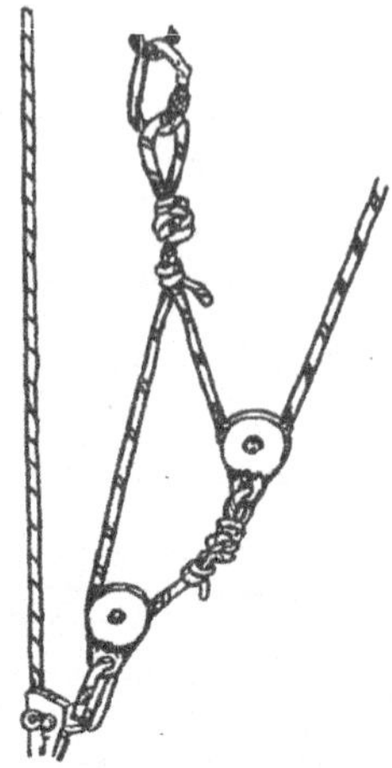

5. **BOAT WRAPS**. As we discussed in the "Swiftwater Terminology and Dynamics" class, the river can put an extreme amount of force on a boat that is wrapped around an obstacle in the river. You first job is to ensure that NO ONE is trapped between the boat and the obstacle. If you do not have full accountability of personnel, don't hesitate! CUT THE BOAT! The easiest method to do this is to cut the floor out of the older "bucket" boats, or to cut the laces holding in the floor of a "self bailing" raft. Depending on how tightly wrapped the boat is you have several options available:

   a. First ensure the crew is accounted for and treated for any injuries.

   b. Establish a tag-line and attempt to bounce and pull the raft free.

   c. Set a Z-drag inside the boat to dump water.

   d. Use an oar or tree to lever the boat off the obstacle.

e. Set up a mechanical advantage system and attempt to "peal" the boat free.

f. Release air from the tubes. Be aware that this can cause more harm than good by reducing tube strength and exposing more surface area to the water pressure.

g. If all else fails, consider cutting out the floor as discussed earlier.

UNITED STATES MARINE CORPS
Mountain Warfare Training Center
Bridgeport, California 93517-5001

WMC
04/14/02

STUDENT HANDOUT

VEHICLE RESCUES

LESSON PURPOSE. The purpose of this period of instruction is to train the students in the procedures used to rescue personnel from vehicles trapped in swift water.

1. **VEHICLE BEHAVIOR**. There are three types of hazards in a river. Floating objects such as trees and branches, suspended objects or tumbling objects such as sawyers, rocks, and other debris, and stationary objects such as live trees on flooded banks and boulders. A vehicle will become all of these "loads". For the first minute, or so, they will float, as they become partially filled with water, they will be suspended. This is normally when they tumble or roll, and finally they will become stationary as they lodge against another stationary object or fill completely. Once they become stationary, depending on the water depth, they may form an eddy, a pour through or pour over, creating a hydraulic or a standing wave if fully covered. Also, the straighter a vehicle is (head-on) to the water flow the more stable it is. Finally, bear in mind that changes in water flow or displacement (weight) of the vehicle (taking a victim off it) may cause the vehicle to once again become a suspended load and tumble.

2. **RESCUE PROCEDURES**

a. The initial action is to secure the vehicle. Ensure that you have adequate anchors. Trees, guardrails, and bridge abutments are good. Avoid using another vehicle if possible.

b. Approach the vehicle on the downstream side by using the eddy created by the vehicle. Use a boat, either tethered or under power, a strong swimmer or boogie board. Another method is to use the line securing the vehicle as a tension diagonal to reach the vehicle. Then establish a second tension diagonal or endless rope to move the victims ashore. Ensure your upstream spotters are in place to warn of other suspended debris coming your way.

B10

c. If the vehicle is sealed tight, crack one window slightly to allow water to get in the vehicle slowly and equalize the pressure. Warning: If you pop or shatter a window you may cause a sympathetic detonation of the remaining windows.

d. Equip the victims with PFD's and remove to the boat or onto the rope system. Consider the following points:

(1) The victims weight might be the only thing stopping it from rolling. You might need to transfer a rescuer in a one for one exchange.

(2) On hard pack or asphalt the vehicle will shift more easily.

(3) If the vehicle is sideways keep the victims on the upstream side. The high side rule is reversed.

(4) Vehicles roll slowly.

(5) Helo's are a last resort.

UNITED STATES MARINE CORPS
Mountain Warfare Training Center
Bridgeport, California 93517-5001

FMST.07.3
0
04/14/02

STUDENT HANDOUT

STREAM CROSSING

TERMINAL LEARNING OBJECTIVE: Given a mountain swift water environment, and a staff, conduct a mountain stream crossing to meet mission requirements per the reference. (FMST.07.30)

ENABLING LEARNING OBJECTIVES:

(1) Without the aid of references, choose from a given list the considerations in site selection for a stream crossing, in accordance with the references. (FMST.07.30a)

(2) Without the aid of the reference, choose from a given list the safety precautions taken while conducting a stream crossing, in accordance with the reference. (FMST.07.30b)

(3) In a summer mountainous environment, execute the individual preparations taken prior to crossing a stream, in accordance with the reference. (FMST.07.30c)

(4) In a summer mountainous environment and given a staff, cross a mountain stream using the staff method, in accordance with the reference. (FMST.07.30d)

(5) In a summer mountainous environment, cross a mountain stream using team methods, in accordance with the reference. (FMST.07.30e)

1. **SITE SELECTION** (FMST.07.30a)

   a. Mountain streams and rivers are military obstacles and therefore are danger areas for units crossing them. In order to reduce the time in the vicinity of the danger area, a recon team should precede the main body and select the best crossing site. The site selection for a stream crossing should include these eight considerations:

B11

(1) Look for logjams, rocks or fallen trees that will provide a dry crossing if possible.

(2) If a dry crossing is not possible, select a crossing point at a wide and shallow point where the current is slower.

(3) Avoid sharp bends. They can be deep with a strong current on the outside of the bend.

(4) Look for a firm, smooth bottom. This is because large rocks and boulders provide poor footing and cause a great deal of turbulence in the water.

(5) It may be easier to cross several small channels of water rather than one large one.

(6) Do not cross just above rapids, falls, or logjams, taking a fall or slipping could have serious consequences.

(7) Cross in the early morning. The water level will be lower since there has been less daylight for the snow to melt. Also, on sunny days, you will have more time to dry clothing and equipment.

(8) There should be a suitable spot downstream for safety swimmers.

2. **SAFETY PRECAUTIONS** (FMST.07.30b). The following two safety precautions must be taken while conducting stream crossings.

a. There must be a safety line at a 45-degree angle downstream across the stream. This is for anyone who slips and is swept downstream to grab so that he is capable to stopping himself.

b. There must be safety swimmers downstream. These are strong swimmers who are positioned downstream to help anyone who is swept downstream. The safety swimmers will use throw bags.

NOTE: The safety line, as well as any other lines that must be taken across the stream, will be taken across by using the lead swimmer method taught in ONE-ROPE BRIDGE.

3. **INDIVIDUAL CROSSING PREPARATIONS**. (FMST.07.30c) Prior to beginning to cross a stream, there are certain preparations that each individual should take. The five preparations are as follows:

a. Wear pack with shoulder straps fastened snugly. Waterproof the pack for buoyancy if possible.

b. Weapons will be slung diagonally over the shoulder with the weapon itself being between the pack and the individual's back.

c. Button all pockets and remove blousing garters. This prevents the water from creating added drag against the individual, which it would do if it could flow into open pockets.

d. Wear boots to protect the feet, but remove socks and insoles to keep them dry:

e. Wear the minimum amount of clothing. This reduces the amount of clothing that must be dried after the crossing.

f. Wear helmet in slow moving, shallow water.

NOTE: If you are in a tactical situation, the actual situation will dictate which of the above precautions will be taken.

4. **INDIVIDUAL CROSSING METHODS**. There are three individual methods that may be used.

a. Staff Method. (FMST.07.30d) A strong staff or pole about 6 feet long is used as a crossing aid. It should be strong enough to support the Marine's weight and trimmed clean of any branches. Placing both hands on the pole, the Marine should place the staff just upstream of his intended path. He should use the staff as the third leg of a tripod and should move only one leg or the staff at a time. He should face upstream using the staff to retain his balance. The staff is also used as a probe to discover bottom irregularities that could trip the Marine. The Marine should drag his feet instead of picking them up.

STAFF CROSSING METHOD

b. Swimming. This is an obvious method, if your Marines are good swimmers. This is not always the case, so usually this method is not a preferred one.

(1) In fast, shallow water, the Marine should angle across on his back with his feet downstream and his head up. He should use his hands to tread water and his feet to fend off obstructions.

(2) In fast, deep water, the Marine should angle across the stream on his stomach with his head upstream, to establish a proper ferry angle.

c. Belayed Method. In chest deep or water with a strong current, a rope can assist greatly. A rope will be secured from bank to bank, with the far anchor slightly downstream from the near anchor. The rope will be anchored off so that it lays at a minimum of 45 degrees. The Marine attaches himself to the rope by using a sling rope as safety line, tying a bowline around his waist and a figure-of-eight loop with a steel locking carabiner inserted. He then attaches the carabiner to the crossing line and crosses using the current to assist him.

5. **TEAM CROSSING METHODS** (FMST.07.30c)

a. Line Abreast. Small units (squad to platoon) can cross in moderate currents up to chest deep, by lining arms in a line abreast or chain method. The largest man of the chain is placed on the upstream side of the group. The group will enter the stream parallel to the flow of the stream. The middleman of the chain will control the group's movement and give the command when to step.

LINE ABREAST

b. Line Astern Method. Three or more men can line up facing the current. The upstream man, who should be the largest man in the group, breaks the current while the downstream men hold him steady. The upstream man may use a staff, similar to the individual staff method, to steady himself. All men side step at the same time with one man calling the cadence.

LINE ASTERN

c. Huddle Method. Between three and eight men can face inboard as in a football huddle. They will wrap their arms around each other's shoulders and cross the stream in this formation. The upstream man should change position as they cross because the entire formation will rotate. This prevents one man from becoming exhausted in the upstream position.

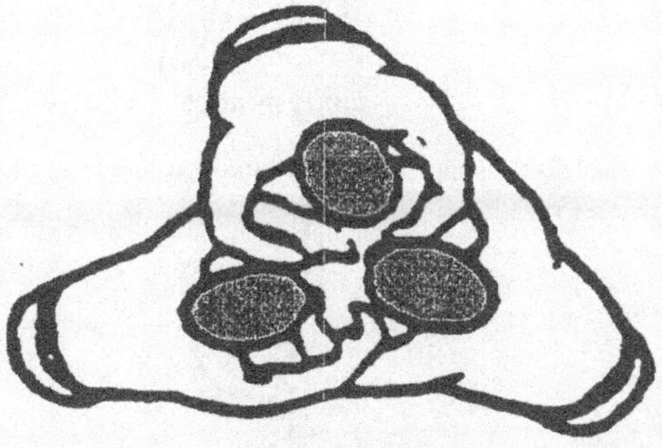

HUDDLE METHOD

UNITED STATES MARINE CORPS
Mountain Warfare Training Center
Bridgeport, California 93517-5001

WMC
04/09/02

STUDENT OUTLINE

ROPE MANAGEMENT

LESSON PURPOSE: To familiarize the student with some of the terms and procedures for working with a rope.

BODY

1. **TERMS USED IN ROPE WORK**

   a. Bight. A simple bend in the rope in which the rope does not cross itself.

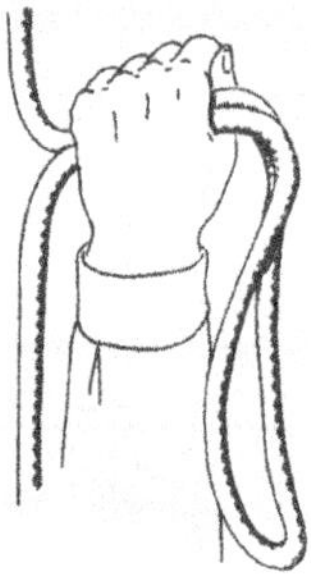

BIGHT OF ROPE

   b. Loop. A simple bend in the rope in which the rope does cross itself.

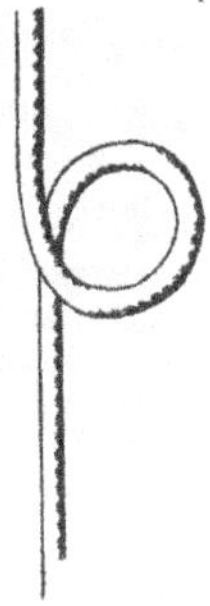

LOOP

c. Half Hitch. A loop which runs around an object in such a manner as to bind on itself.

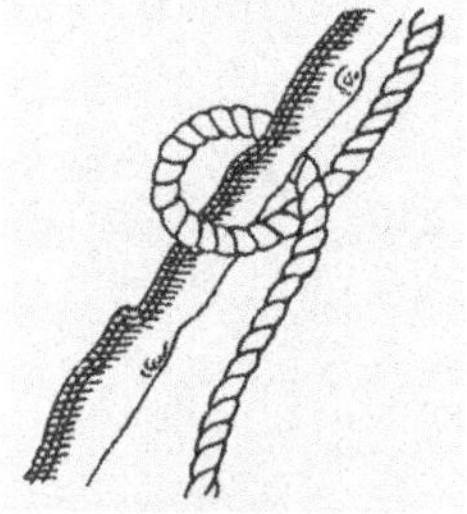

HALF HITCH

d. Standing End. The part of the rope which is anchored and cannot be used, also called the static end.

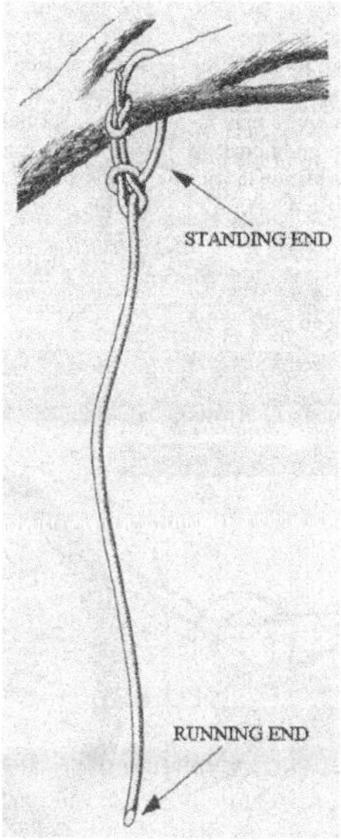

e. Running End. The free end of the rope which can be used.

f. Lay. The same as the twist of the rope. (Applies only to hawser laid ropes, such as manila.)

g. Pigtail. The short length left at the end of a rope after tying a knot or coiling a rope. It may or my not be tied off with a secondary knot, depending on the circumstance.

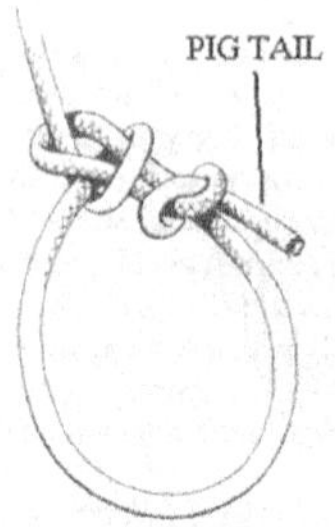

PIG TAIL

h. Stacking (or Flaking). Taking off one wrap at a time from a coil, and letting it fall naturally to the ground.

FLAKED OUT ROPEON THE GROUND

i. Dressing the knot. This involves the orientation of all of the knot parts so that they are properly aligned, straightened, or bundled, and so the parts of the knot look like the accompanying pictures. Neglecting this can result in an additional 50% reduction in knot strength.

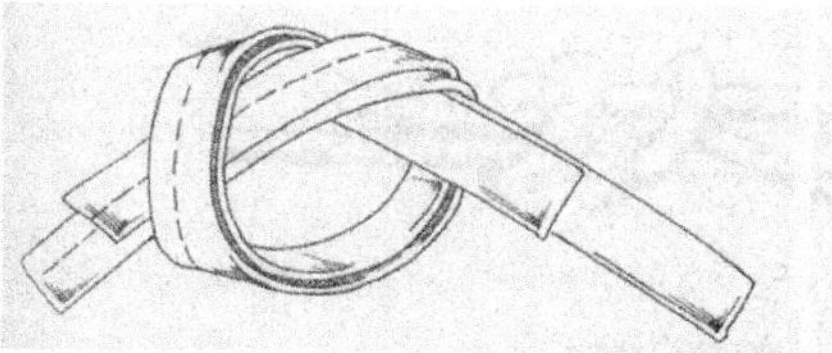

NON-DRESSED DOWN WATER TAPE KNOT

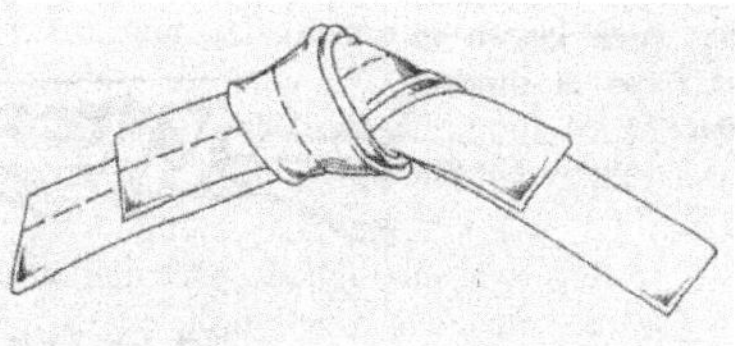

DRESSED DOWN WATER TAPE KNOT

j. Setting the knot. This involves tightening all parts of the knot so that all of the rope parts bind upon other parts of the knot so as to render it operational. A loosely tied knot can easily deform under strain and change character.

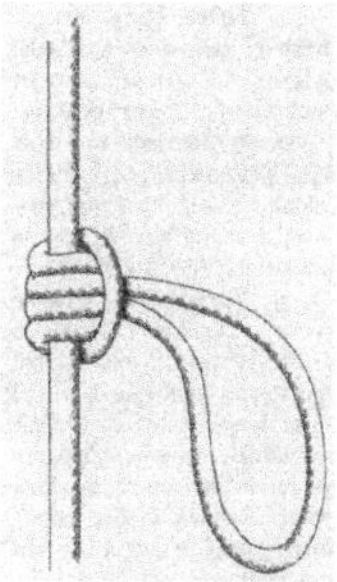

A FULLY SET PRUSIK KNOT

2. CONSIDERATIONS FOR THE CARE OF ROPE

a. The rope should not be stepped on or dragged on the ground unnecessarily. Small particles of dirt will get into and through the sheath causing unnecessary wear to the rope within.

b. The rope should never come in contact with sharp edges of any type. Nylon rope is easily cut, particularly when under tension. If a rope must be used around an edge which could cut it, then that edge must be padded or buffed using fire hose if available, or several small sticks.

c. Keep the rope as dry as possible. If it should become wet, hang it in large loops, above the ground, and allow it to dry. A rope should never be dried out by an open flame, or be hung to dry on metal pegs, as this will cause rust to get in the rope thus rendering it unserviceable.

d. Never leave a rope knotted or tightly stretched longer than necessary.

e. When using rope installations, never allow one rope to rub continually against another.

NOTE: With manila ropes this will cause the rope to fray, whereas nylon ropes can melt under the friction that this causes.

f. The rope should be inspected prior to each use for frayed or cut spots, mildew, rot or defects in construction.

g. Mark all climbing ropes at their midpoints to facilitate establishing the midpoint for a procedure requiring you to use the middle of the rope. The rope should be marked with a bright colored adhesive tape.

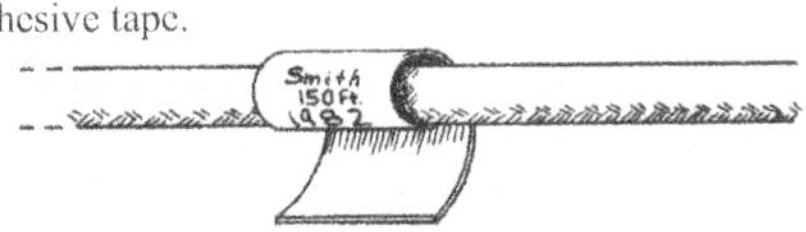

MARKING THE MIDDLE OF THE ROPE

h. The rope should not be marked with paints or allowed to come in contact with oils or petroleum products for these products will weaken it.

i. A climbing rope should NEVER be used for any other purpose except for mountaineering, i.e., towing vehicles.

j. The ends of a new rope or ends caused by a cut should be cut with the rope cutter contained in the MACK and marked with a serial number.

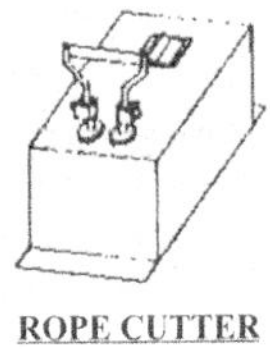

ROPE CUTTER

k. The rope should never be subjected to high heat or flame as this can significantly weaken it.

l. To clean rope use mild soap and rinse thoroughly with water. A rope washer can be used to clean or rinse the rope.

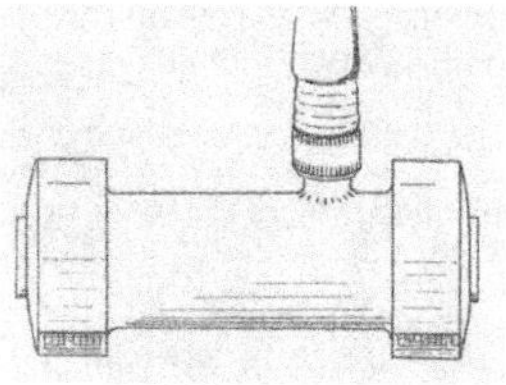

ROPE WASHER

m. When not in use, ropes should be coiled and hung on wooden pegs rather than on nails or any other metal object. They should be stored in a cool place out of the direct rays of the sun.

n. When in areas of loose rock, the rope must be inspected frequently for cuts and abrasions.

o. Always maintain an accurate Rope Log whenever using a rope.

p. Ropes 300-600 foot in length should be Mountain Coiled.

3. **INSPECTION OF ROPE**

a. All ropes have to be inspected before, during, and after all operations. Kernmantle rope is harder to inspect than a laid rope. I.e. green line. The Assault Climber must know what to look and feel for when inspecting a rope. Any of the below listed deficiencies can warrant the retirement of a rope.

(1) Excessive Fraying. Indicates broken sheath bundles or PIC breakage.

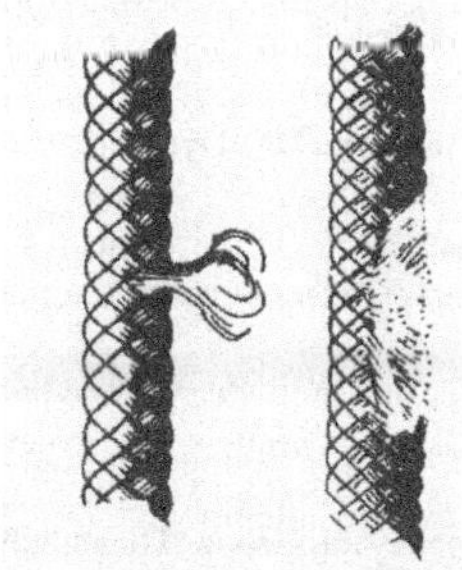

DAMAGED AND EXPOSED ROPE

(2) Exposed Core Fibers. Indicates severe sheath damage. (When you can see the inner core fibers)

(3) Glossy Marks. Signify heat fusion damage, also called a booger.

(4) Uniformity of Diameter / Size. May indicate core damage, noted by an obvious depression (hour glass) or exposure of white core fibers protruding from the sheath (puff).

(5) Discoloration. A drastic change from the ropes original color may indicate chemical change or damage.

(6) Stiffness or Soft Spots. Could signify core damage.

NOTE: Dynamic ropes measuring between 10mm and 12mm are marked at each end of its pigtails with a number "1" indicating that the rope is UIAA approved for single rope lead climbing. Dynamic ropes measuring between 8mm and 9mm are marked at each end of its pigtails with a "1/2" number indicating that two of these ropes are required to conduct a lead climb.

4. **ROPE LOG.** The purpose of the Rope Log is to maintain an accurate record for the use of each rope contained within the Marine Assault Climbers Kit (MACK). Due to the turnover of personnel, and the fact that no one person may have the same rope twice, the rope log is used to ensure the safe use, serviceability, and account of each rope.

a. Serial number. By assigning each rope a serial number, responsible units can determine information about a rope. As soon as the ropes are cut to the desired length for their intended purpose, each rope will be assigned a serial number by the responsible unit. That rope should then be labeled with that serial number in some permanent manner. The best method for this is as soon as the ends of the new rope have been whipped and fused, mark both ends of the rope with the serial number and then dip the ends of the rope, in effect, laminating the serial number to the rope ends.

(1) A rope serial number has five parts. The proper format for a rope serial number is as follows:

1. Type of rope: (S) - Static or (D) - Dynamic.

2. Last two digits of the year the rope was manufactured. Each rope has a shelf life of two years, after that it must not be used for any mountaineering purpose.

3. Four-digit number for that individual rope which is assigned by the responsible unit, and should be assigned sequentially as new ropes are issued out.

4. The length and diameter of the rope. The length may be recorded in feet, and the diameter in millimeters.

5. Responsible unit code. Example: S-96-0001-150/11mm-F2/8

Example meaning: This rope is a 150 foot, 11mm, static rope, manufactured in 1996. It is the first rope issued by Fox Co 2/8.

b. Recording Information in a Rope Log. Once a new rope has been serialized, the rope log for that rope should be started up. At a minimum, it should contain the following information:

(1) The rope serial number.

(2) Manufacturer. Depending on the manufacturer, ropes of the same type and diameter may vary in tensile strength, stretch factor, and durability.

(3) Date of manufacture. Five years from this shelf life date, ropes are considered to have reached their normal expiration date, and should be destroyed.

(4) Date in service. This is to be recorded for tracking purposes to establish how long a rope is in service. Two years is considered maximum rope life in service.

(5) Each time a rope is used, the using unit is responsible to record how that rope was used and how much use the rope received. Additionally, before checking out a rope and prior to turning it back in, the rope must be inspected by qualified personnel, and initial in the "Inspected By" block of the rope log.

NOTE: ANY TIME A DYNAMIC ROPE IS SUBJECTED TO A FALL FACTOR 2, THAT ROPE SHOULD NOT BE USED AGAIN FOR MOUNTAINEERING.

DYNAMIC					
SERIAL #					
MANUFACTURER					
DATE OF MANUFACTURE					
DATE IN SERVICE					
INDICATE THE DATE AND NUMBER OF EACH SPECIFIC USE					
CLIMBS	FALLS (FACTOR)	TOP ROPES	RAPPELS	REMARKS	INSPECTED BY

ROPE LOG

5. **COILING A ROPE**. There are two types of rope coils frequently used at MWTC. The Mountain Coil and the Butterfly Coil.

a. Mountain Coil. This coil is useful for carrying the rope over a pack or over a climber's shoulder and neck. It can be used for short time storage. The mountain coil can be tied in the following manner:

(1) Sit down with your leg bent at a 90-degree angle, heel on the deck. Starting at one end, the rope is looped around the leg in a clockwise fashion, going over the knee and under the boot sole until the entire rope is coiled.

(2) If coiling a 150-foot rope, use only one leg and offset the other, when coiling a 300-foot rope, use two legs and keep them together.

(3) With the starting end of the rope, form a 12-inch bight on the top of the coils.

(4) Uncoil the last loop and along the top of the coils, wrap 4-6 times towards the closed end of the bight.

(5) The end of the rope being wrapped is then placed through the closed end of the bight.

(6) The running end of the bight is then pulled snuggly to secure the coil.

(7) To prevent the coil from unraveling, the two pigtails are tied together with a square knot.

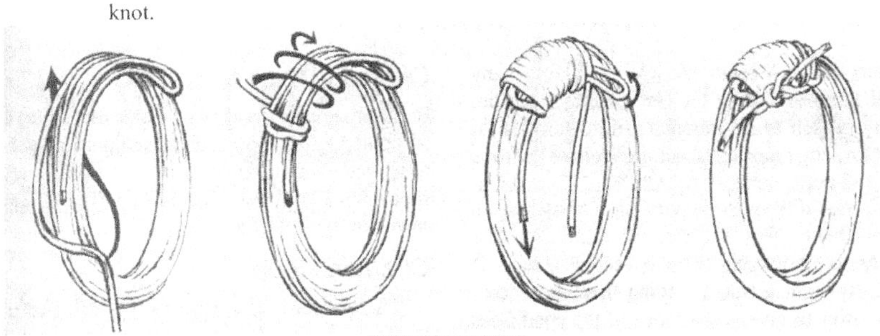

MOUNTAIN COIL

b. Butterfly Coil. This method is used for carrying a rope when the individual needs to have maximum use of his upper body, (i.e. while climbing), without the encumbrance of a large rope coil hanging across his chest.

(1) Coiling the Butterfly Coil

(a) Step 1: Find the middle of the rope, then form a three foot bight laying both ropes in the upraised palm at the two foot point.

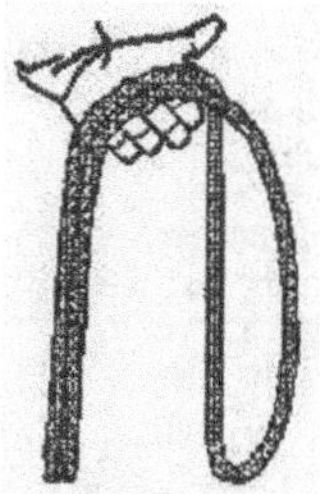

STARTING THE BUTTERLFY COIL

(b) Step 2: Form another two-foot bight with the running end. Place the rope at the two-foot bight along side on top of the original bight ensuring the running end is on the same side as the original bight.

(c) Step 3: Continue making two foot bights, laying them alternately into your palm until there is only six to eight feet remaining. At that point, begin wrapping the two pigtails horizontally four to six times at the mid way point of the ropes in a bight from bottom to top.

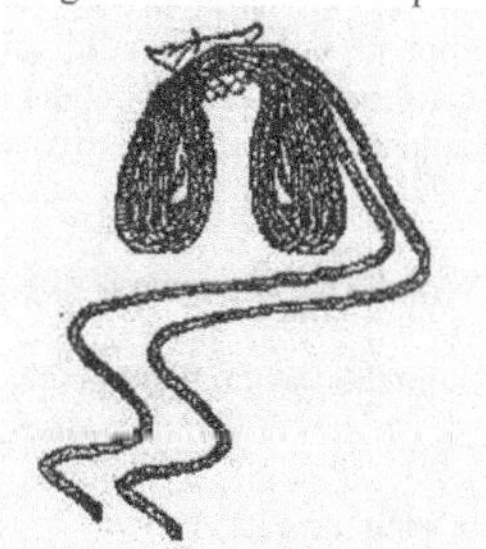

TWO FOOT BIGHTS ON BOTHS SIDES WITH 6 TO 8 FEET REMAINING

(d) Step 4: After completing your wraps, form a bight with the remaining pigtail and then thread it underneath your palm and upwards to one-foot above the coiled rope.

(e) Step 5: With the remaining pigtail, thread it through the one-foot bight in step four.

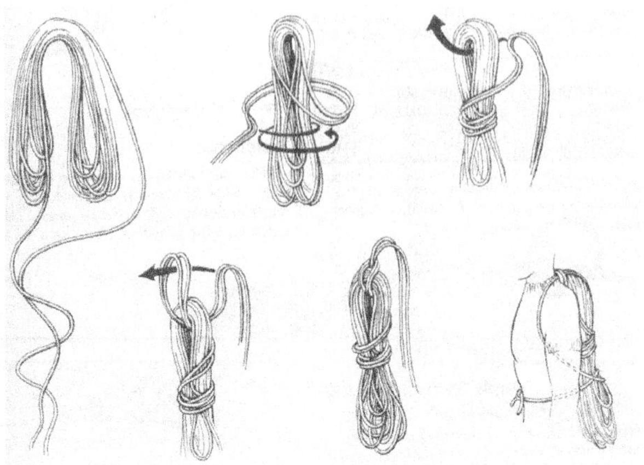

TYING AND CARRYING THE BUTTERFLY COIL

(2) Carrying the Butterfly coil. Separate the running ends, placing the coil in the center of the back of the carrier, then fun the two ends over his shoulders so as to form shoulder straps. The running ends are then brought under the arms, crossed in the back over the coil, brought around the body of the carrier and tied off with a square knot at his stomach.

6. **ROPE THROWING**. To insure that the rope will not get tangled when deployed, certain steps must be taken.

a. With a stacked rope, anchor off the standing end.

b. Take the opposite end of the rope and make 6-8 coils and place them in your strong arm. These wraps will serve as a throwing weight that you can aim.

c. 10-15 feet from the strong-arm coils create a second set of 6-8 wraps and place them in your weak arm.

d. From the edge of the cliff, sound off with the command "STAND-BY FOR ROPE". Just before you release the coils, sound off with the command "ROPE". At that time drop the weak arm coils from the cliff.

e. While taking aim, throw your strong-arm coils overhand or sidearm hard enough to hit your intended target.

f. If the throw was misdirected due to wind, tree, etc., reorganize and attempt to re-deploy the rope.

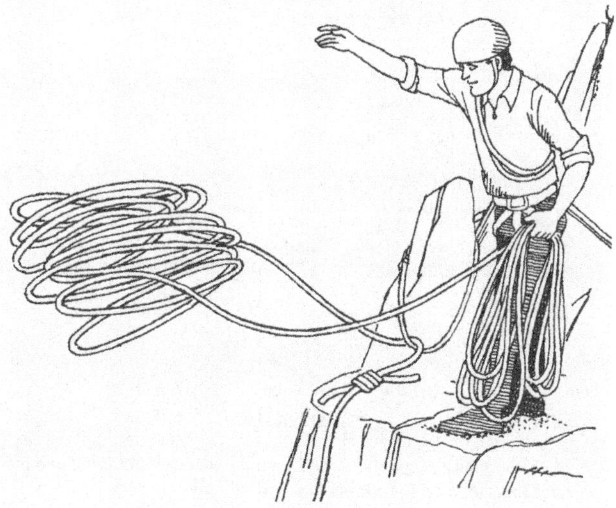

THROWING THE ROPE

6. MOUNTAINEERING KNOTS.

a. **Class I** – *End of the Rope Knots*

(1) Square Knot. Used to tie ends of two ropes of equal diameter together. It should be secured by overhand knots on both sides of the square knot.

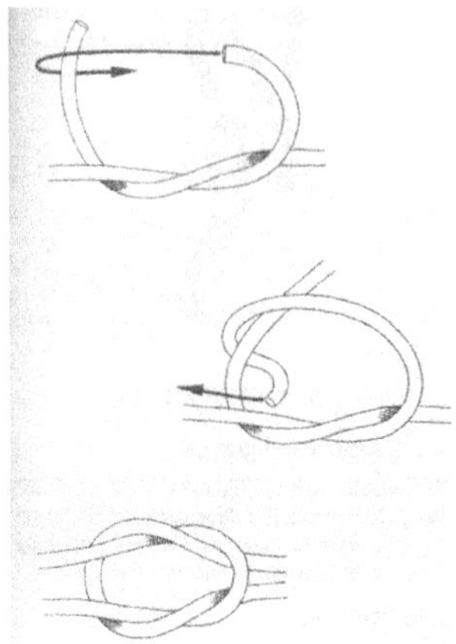

SQUARE KNOT

(2) Double Fisherman's Knot. It is a self-locking knot used for tying two ropes of equal diameter together. It can be tightened beyond untying.

DOUBLE FISHERMAN'S KNOT

(3) Water/Tape Knot. Used to secure webbing or tape runners. It is constructed by tying an overhand knot (without twists) in one end of the tape, and threading the other end of the tape through the knot from the opposite direction. After the knot is dressed down, each pigtail should be a minimum of two inches long.

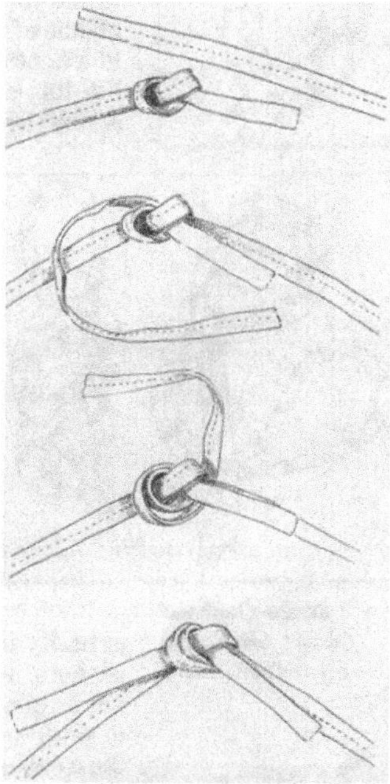

WATER TAPE KNOT

(4) Double Sheet Bend. Used to tie the ends of two or more ropes of equal or unequal diameter together.

SHEET BEND

b. **Class II** - *Anchor Knots*

(1) Bowline. Used to tie a fixed loop in the end of a rope. This knot is always tied with the pigtail on the inside and secured with an overhand knot.

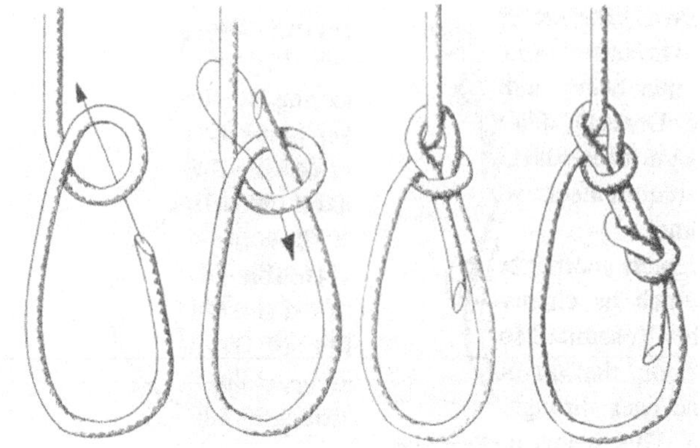

BOWLINE

(2) Round Turn with Two Half Hitches or a Bowline. A loop which runs around an object in such a manner as to provide 360 degree contact and may be used to distribute the load over a small diameter anchor. It will be secured with two half hitches or a bowline.

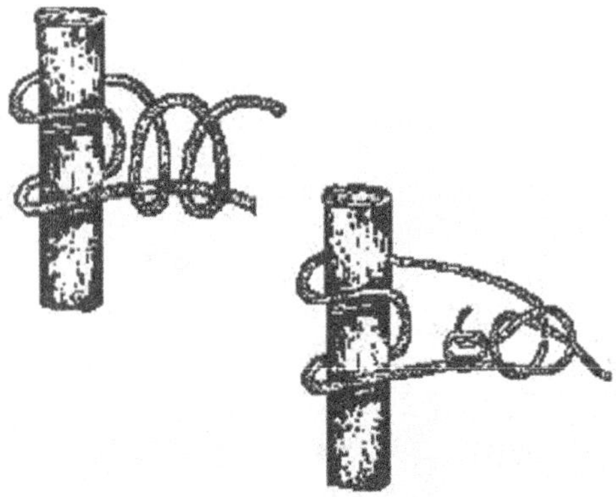

ROUND TURN AND TWO HALF HITCHES AND

ROUND TURN AND A BOWLINE

(3) Clove Hitch. This knot is an adjustable hitch. It could be considered a middle-of-the-rope anchor knot at the end-of-the-rope when used in conjunction with a bowline or round turn and two half hitches.

(1) Around-the-object Clove Hitch.

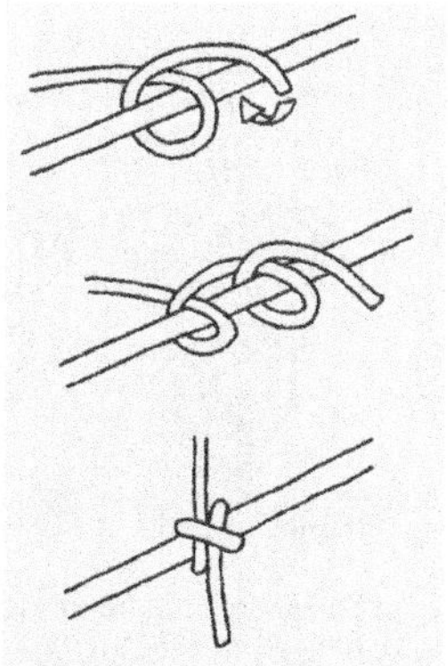

AROUND THE OBJECT CLOVE HITCH

(2) Over-the-object Clove Hitch.

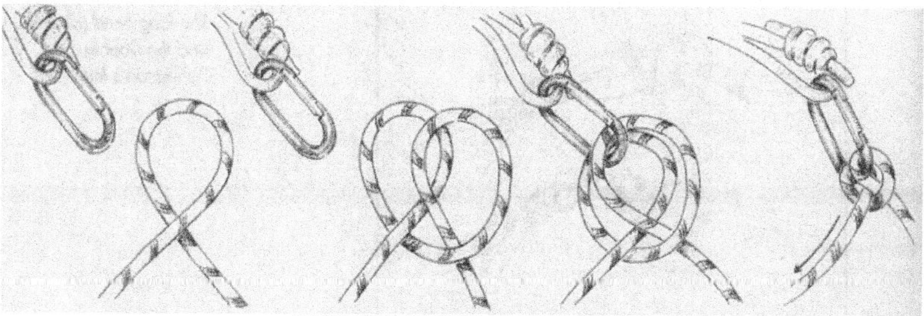

OVER THE OBJECT CLOVE HITCH

c. **Class III -** *Middle-of-the-rope*

(1) Figure-of-Eight-Loop. This is a strong knot that can be readily untied after being under load.

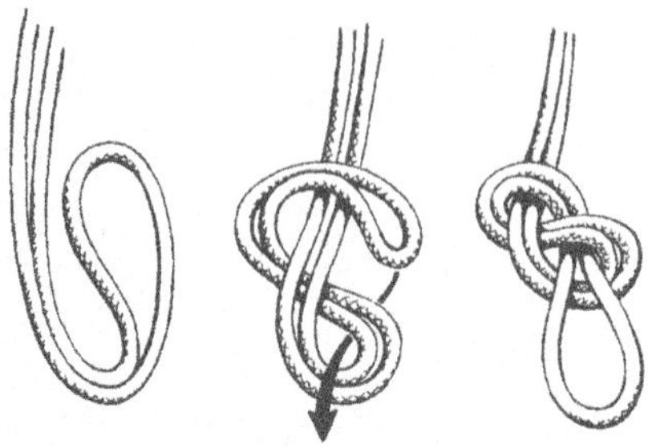

FIGURE 8 LOOP

(2) Double Figure of Eight Loop. The double Figure of Eight Loop is a strong knot and the double loop reduces the wear and strength loss from the rope bending around the carabiner by splitting the load between the two loops.

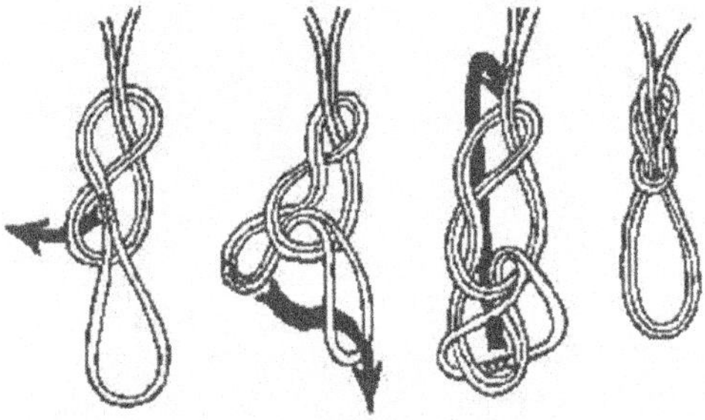

DOUBLE FIGURE 8 LOOP

(3) Two Loop Bowline. This knot can be used to construct a self-equalizing belay or to tie the middle person in on three people on a rope.

TWO LOOP BOWLINE

d. **Class IV** - *Special Knots*

(1) Prusik Knot. This knot functions by introducing friction that can be alternately set and released. For best results, tie the knot with a smaller diameter cord on a larger diameter cord. If slippage occurs, more wraps maybe used.

(a) Middle-of-the-Rope Prusik. This knot is created with an endless loop also known as a Prusik Cord. Do not tie this knot with tape due to less friction.

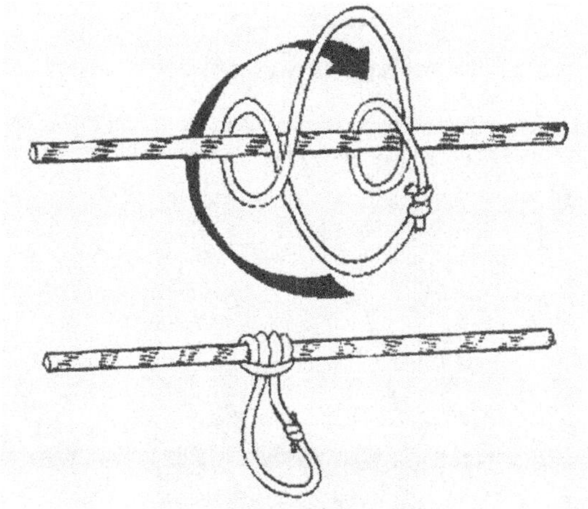

MIDDLE OF THE LINE PRUSIK

(b) End-of-the-Rope Prusik. This knot is always secured with a bowline. Do not tie this knot with tape due to less friction.

END OF THE LINE PRUSIK

(c) French Prusik. This can be constructed with a Prusik Cord or a single strand of cord with figure-of-eight loops on each end.

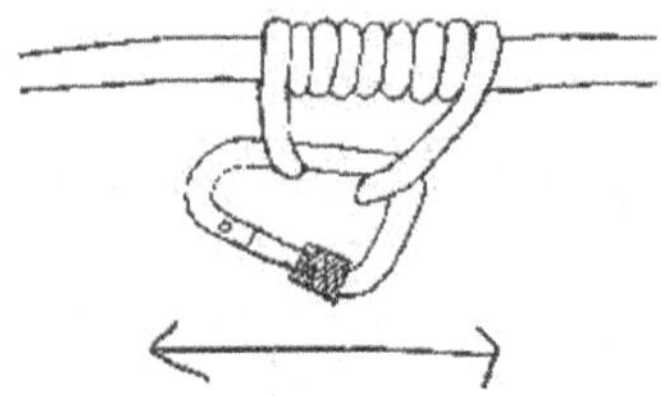

FRENCH PRUSIK

(2) Retraced Figure of Eight. Used to tie the end of the climbing rope into a harness or swammi wrap. The pigtail may or may not be secured with an overhand.

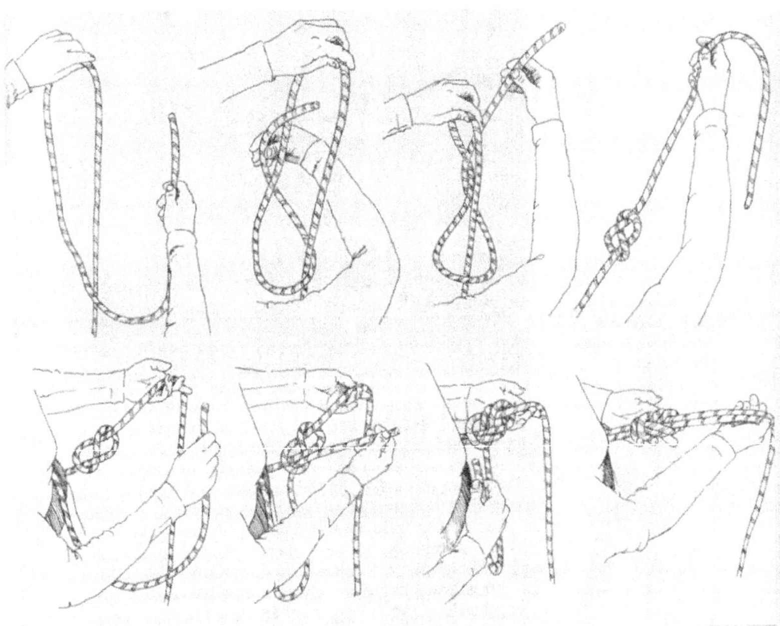

RETRACED FIGURE 8 IN HARNESS

(3) Directional Figure-Of-Eight. When tied and tensioned is applied to the standing and running ends of the rope, the knot will not pull apart. The loop will point toward the direction of pull.

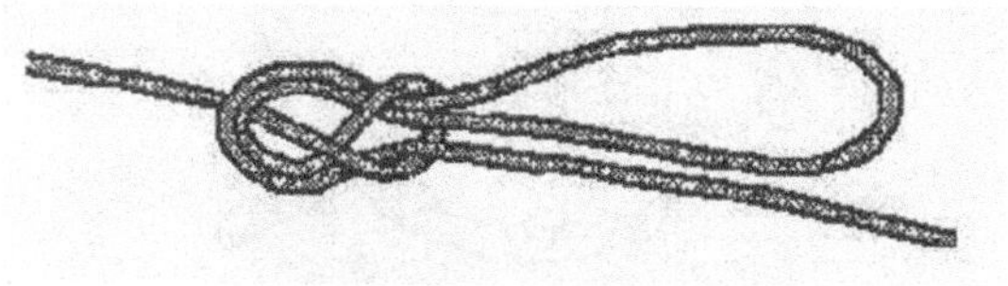

DIRECTIONAL FIGURE 8

(4) Slip Figure 8. This knot is used for retrievable anchors and fixed rope installations for the ease of untying the knot.

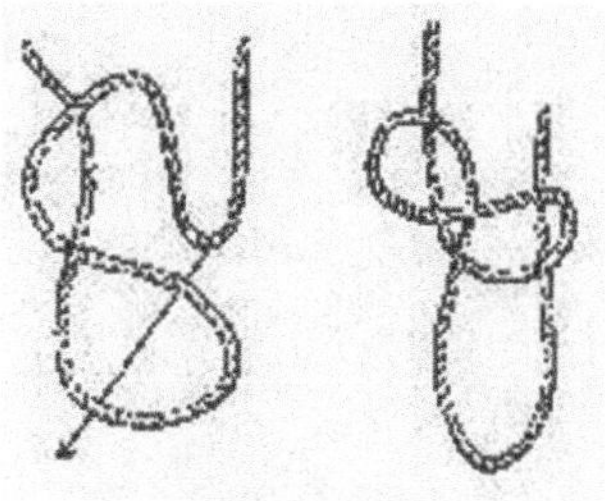

SLIP FIGURE 8

(5) Kliemheist. This is a friction knot.

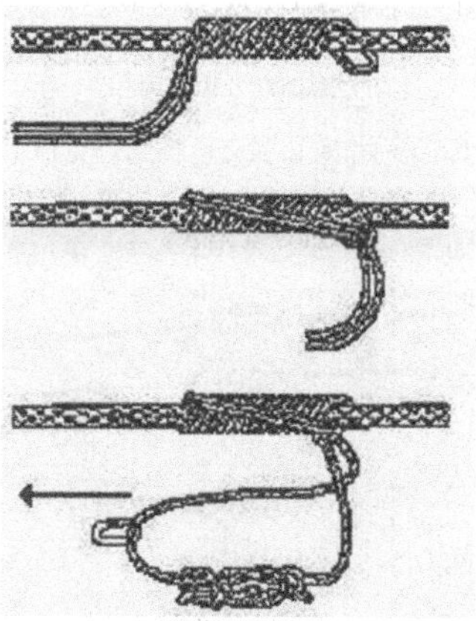

KLEIMHEIST

(6) Overhand Knot. Can be used to secure primary knots on itself.

OVERHAND

(7) M nter Hitch. This is a simple hitch in the rope that is clipped into a carabiner to put friction on the line.

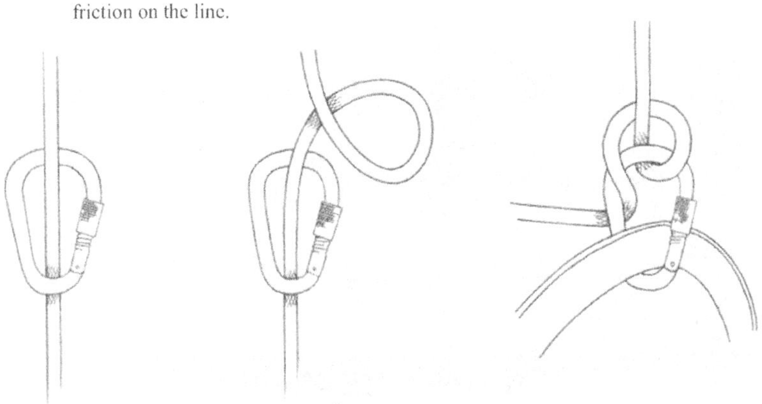

MÜNTER HITCH

(8) Timber Hitch. A Timber Hitch is used to fix a rope to a pole or equivalent for hoisting or towing purposes. It has the capability of casting off easily.

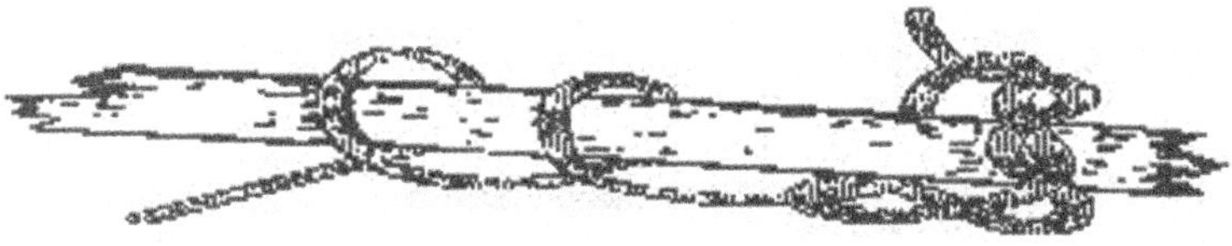

TIMBER HITCH

(9) Mariner's Hitch. This knot's advantage over the prusik is that it can be released under load.

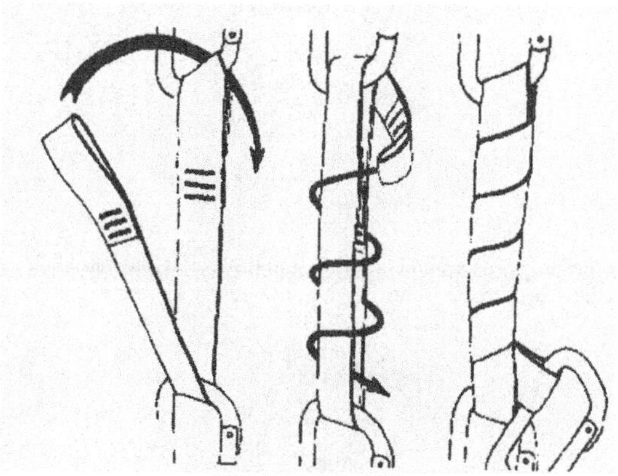

MARINER'S HITCH

(10) Krägur Knot. This friction knot is used to prevent unnecessary slippage when tied onto a rope of the same diameter.

KRÄGUR KNOT

(11) Rappel Seat. This is used as an expedient support harness for rappelling, crossing rope bridges, etc. It is constructed as follows:

(a) Center the sling rope on the left hip.

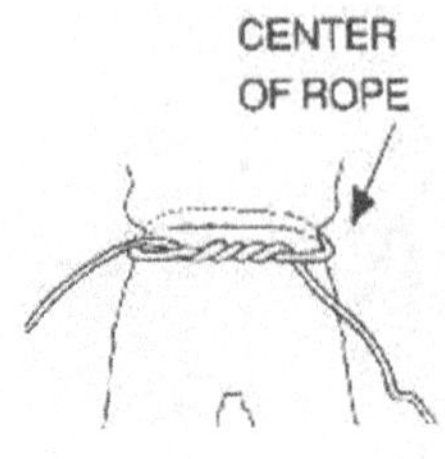

(b) Wrap the sling rope around the waist and tie at least one half hitch around itself (preferably two) in the front.

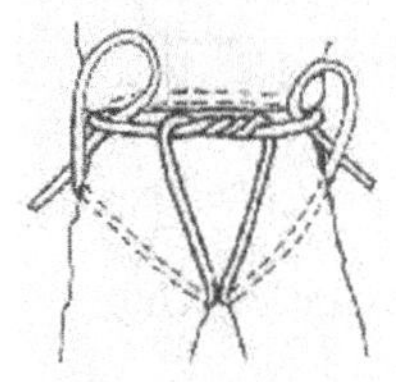

(c) Bring the running ends down through the legs, up over the buttocks, and over the original waist wrap, down between the waist wrap and the waist, and over itself, forming a bight. Cinch this up tightly.

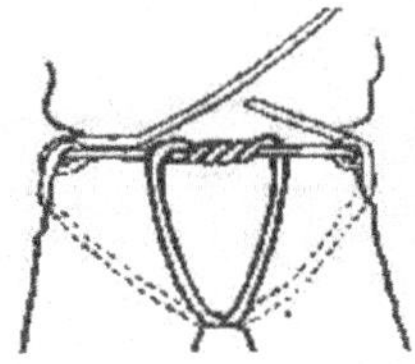

(d) Now tie a square knot with two overhand knots on the left hip ensuring that the overhand knots encompass all the wraps.

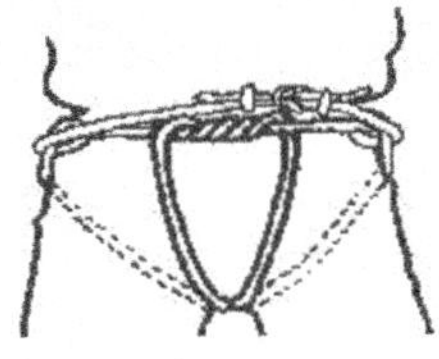

(e) Tuck any excess rope into a pocket.

(12) Bowline on a coil. Used by the first and last men on a climbing rope to tie into the rope. An overhand knot is used behind the knot. It distributes the force of a fall over a larger area of a climber's waist and is preferable to a single bowline around the waist. The bowline on a coil can also be used to take up excess rope. The bowline on a coil should have 4-6 wraps around the waist.

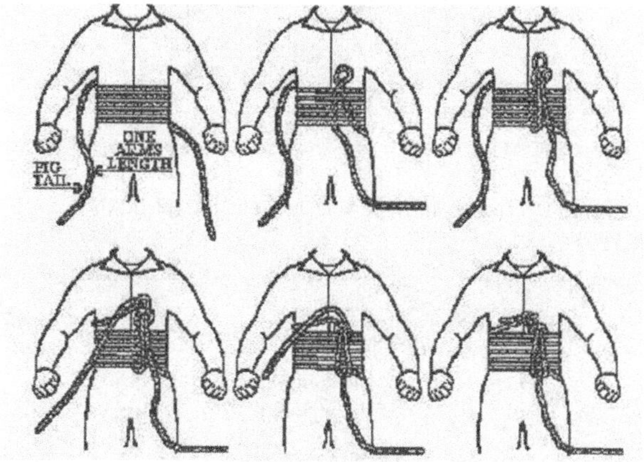

BOWLINE ON A COIL

(13) Swami wrap (Swami belt). The swami wrap (four to six wraps of rope or nylon webbing tied around the waist) popular with rock climbers who wanted to increase the working length of their climbing ropes. It also offers some improvement over the bowline on the coil in distributing the forces sustained in a fall over a larger area of the midsection of the body, but does not eliminate the problem of suffocation when hanging. The Swami wrap is constructed as follows:

(a) Take a sling rope and find the middle.

(b) Place the middle of the sling rope in the small of your back.

(c) Bring both pigtails to the front and continue to wrap around your body until both pigtails are about 15-20 inches long.

(d) Tie the pigtails on the left side with a square knot and two overhand knots.

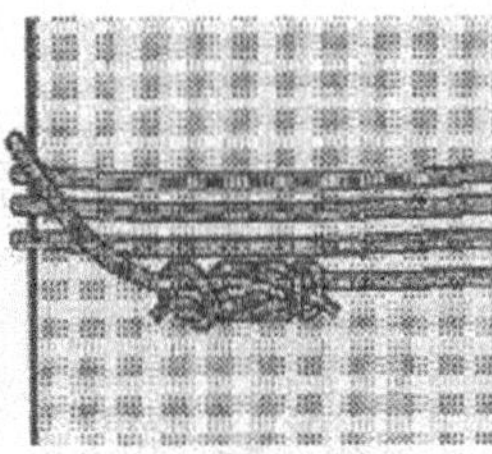

SWAMMI WRAP

(14) Three Loop Bowline. This knot will provide three bights, each of which can be adjusted against the others.

THREE LOOP BOWLINE

7 **KNOT STRENGTH.** Rope, cordage and webbing is strongest when loaded in a straight line. When bending a rope or web to create a knot, the strength of the rope is reduced. All knots should be dressed properly for maximum effective use.

KNOT	RELATIVE STRENGTH OF KNOT
No Knot	100%
Figure 8	75-80%
Bowline	70-75%
Double Bowline	70-75%

Double Fisherman	65-70%
Water Tape Knot	60-70%
Clove Hitch	60-65%
Overhand Knot	60-65%
Square Knot	45%

1. **KNOT TESTING TIME LIMIT.** The following times must be met to pass the knot tying portion of this course.

KNOT	UNIT OPERATIONS TIME LIMIT	SML, ACC TIME LIMIT (BLINDFOLDED)
Square knot	60 Seconds	30 Seconds
Double Fisherman's Knot	60 Seconds	30 Seconds
Water/Tape Knot	60 Seconds	30 Seconds
Round Turn and a Bowline	60 Seconds	30 Seconds
Round Turn and Two Half Hitches	60 Seconds	30 Seconds
Clove Hitch (around the object)	60 Seconds	30 Seconds
Munter Hitch	60 Seconds	30 Seconds
Slip Figure 8	60 Seconds	30 Seconds
Figure 8 Loop	60 Seconds	30 Seconds
Directional Figure 8	60 Seconds	30 Seconds
End of the Line Prusik	90 Seconds	45 Seconds
Retrace Figure 8	90 Seconds	45 Seconds
Rappel (Swiss) Seat	2 Minutes	60 Seconds

UNITED STATES MARINE CORPS
Mountain Warfare Training Center
Bridgeport, California 93517-5001

SMO
05/02/01

STUDENT OUTLINE

NOMENCLATURE AND CARE OF MOUNTAINEERING EQUIPMENT

LESSON PURPOSE. The purpose of this period of instruction is to introduce you to the types of equipment used here at MWTC and how to care for it so as to prevent its untimely failure. This lesson relates to all climbing and installation work that you perform here at MWTC.

OUTLINE

1. **ROPES.** All ropes used in the military must meet UIAA standards or U. S. Federal Test Standard 191A. Most ropes have a 5-year shelf life and maximum 2-year service life.

a. Static. Black in color.

(1) Construction. Kernmantle

(2) Minimum tensile strength. 7500 lbs.

(3) Maximum elongation. 1.5%

(4) Diameter. 11mm

(5) Sizes.

(a) 165 ft ± 5 ft

(b) 300 ft ± 10 ft

(6) Usage. Rescue operations and bridging where a low amount of elongation is desirable under a working load.

b. Dynamic. Olive Drab in color.

(1) Construction. Water-resistant treated Kernmantle to reduce friction.

(2) Minimum tensile strength. 6500 lbs.

(3) Maximum elongation. 6%

(4) Diameter. 10.5mm and 11mm

(5) Size. 165 ft ± 5 ft.

(6) Usage. For lead climbing/party climbing.

c. Maxim Dry Rope. Olive Drab or Multi-Colored.

(1) Construction. Water-repellent treated Kernmantle.

(2) Minimum tensile strength. 3472 lbs.

(3) Maximum elongation. 6%

(4) Diameter. 9mm

(5) Size. 150 ft ± 5 ft

(6) Usage. For glacier travel/ice climbing

d. Gold Line II.

(1) Construction. Eight strand braided nylon plymor.

(2) Minimum tensile strength. 4500 lbs.

(3) Maximum elongation. 20%

(4) Diameter. 11 mm

(5) Size. 300 ft or 600 ft spools

(6) Usage. Sling Ropes and litters only.

Three strand twisted or laid rope.

Static, kernmantle rope showing core (kern) and sheath (mantle).

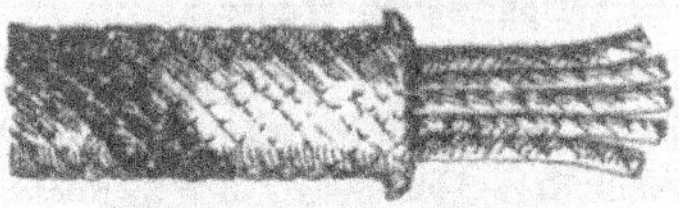

Dynamic, kernmantle rope used for rock climbing.

Double braid rope showing the braided sheath and braided core.

INSIDES AND OUTSIDE SHEATHS OF VARIOUS TYPES OF ROPES

NOTE: Sling ropes are made from 15 foot lengths of plymor or dynamic rope ONLY. Twenty-five foot practice coils should be constructed with static rope, but dynamic rope can be used.

2. ADVANTAGES/DISADVANTAGES

a. Advantages of Nylon Rope

(1) High strength to weight ratio.

(2) Good energy absorption in dynamic ropes.

(3) Flexible.

(4) Rot resistant, not affected by frost.

b. Disadvantages of Nylon Rope

(1) Low melting point. Nylon fuses at 400°F and melts at 480°F.

(2) Susceptible to abrasions and cuts.

(3) Affected by chemicals and light.

c. Advantages of Manila Rope

(1) Easily gripped.

(2) Hard wearing.

(3) Does not deteriorate in heat.

d. Disadvantages of Manila Rode

(1) Heavy, kinks, especially when wet. Absorbs water and swells.

(2) Burns at +300°F.

(3) Edible by rodents.

3. GENERAL INFORMATION

a. Nylon rope stretches under tension and will rupture at between 30% and 70% elongation, depending on construction.

b. Nylon rope loses as much as 30% strength when wet.

c. Temperatures as low as 250°F will damage a nylon rope.

4. NYLON WEBBING

a. The type of nylon webbing available is tubular. Tubular nylon webbing is very strong and flexible. All rules that apply to nylon rope apply to tubular nylon webbing. The size of nylon webbing used is:

(1) 1 inch tubular nylon. Tensile strength approximately 4,000 - 4,500 lbs., depending on the manufacturer.

b. Pre-sewn Spectra Runners. Tensile strength approximately 5,500 lbs.

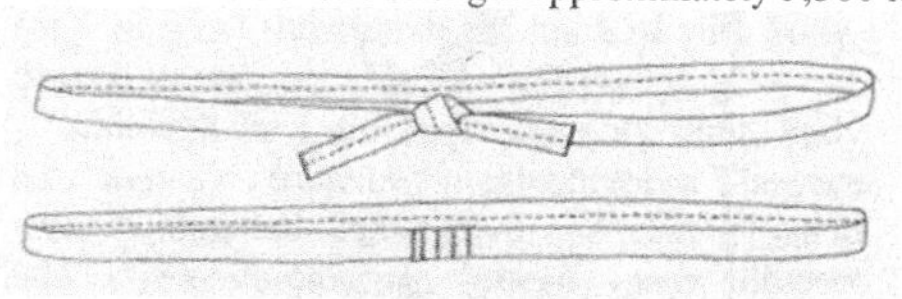

NYLON WEBBING

NOTE: These are minimum strengths. Some manufactures make even stronger webbing.

1. **CARABINERS**. Also commonly known as snap links. There are two different types of carabiners frequently used at MWTC, the steel locking carbine and the non-locking aluminum carabiners.

a. Steel Locking Carabiner. There are two different sizes of carabiners:

(1) Steel Locking Stubai 82.

(2) Large Locking "D" Stubai 85.

b. Aluminum Non-Locking Carabiner.

(1) The standard "D" non-locking carabiner with a tensile strength of 2000 lbs. or more.

2. **NOMENCLATURE OF A CARABINER**.

a. There are four parts to a non-locking carabiner. They are as follows:

(1) Gate.

(2) Gate pivot pin.

(3) Locking pin

(4) Body.

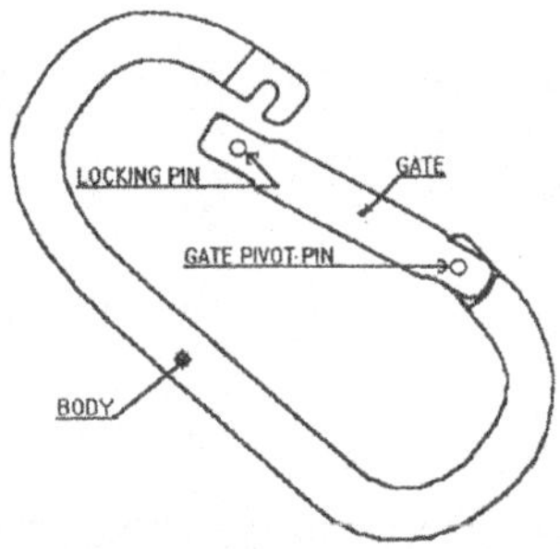

STEEL NON-LOCKING CARABINER

b. There are five parts to a locking carabiner. They are as follows:

(1) Gate

(2) Gate pivot pin

(3) Locking notch

(4) Locking nut

(5) Body

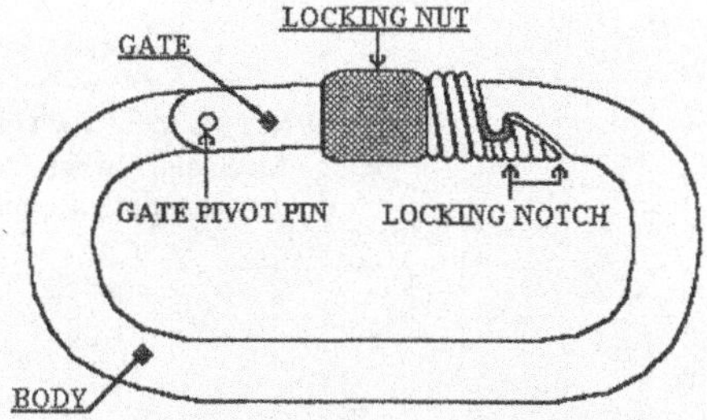

STEEL LOCKING CARABINER

3. **<u>CARE, MAINTENANCE AND SERVICEABILITY OF A CARABINER</u>**. The following inspections and maintenance should be performed on a routine basis:

a. <u>Care</u>.

(1) Avoid dropping of a carabiner onto a hard surface, hairline factures could occur.

(2) Keep the carabiners off of the bare ground because dirt and grit can get into the working parts.

(3) If the carabiner has fallen from a significant height, it should be retired.

b. <u>Maintenance</u>.

(1) Remove all dirt, moisture and grime after use.

(2) Lubricate with tri-flow graphite and wipe off with a clean cloth.

c. <u>Serviceability Checks</u>.

(1) The gate snaps shut with no friction and with no gap between the locking pin and locking notch.

(2) There is no excessive side to side movement of the gate

(3) The pivot pin is tight.

(4) That the spring-loaded gate functions properly.

(5) The locking pin is tight.

(6) The locking nut travels freely and locks securely.

(7) There are no cracks or flaws in the metal.

NOTE: The weakest part of a carabiner is the gate. If an engraver is used to mark a carabiner, it should be applied to the gate and not the load bearing side.

UNITED STATES MARINE CORPS
Mountain Warfare Training Center
Bridgeport, California 93517-5001

WMC
04/11/02

LESSON PLAN

BALANCE CLIMBING

1. LEARNING OBJECTIVES

a. TERMINAL LEARNING OBJECTIVE. In a summer mountainous environment, execute balance climbing, in accordance with the references. (WMC)

b. ENABLING LEARNING OBJECTIVES.
(1) With the aid of references, state orally the safety requirements for balance climbing, in accordance with the references. (WMC.C3a)

(2) With the aid of references, state orally the individual preparations for a balance climb, in accordance with the references. (WMC.C3b)

(3) In a summer mountainous environment, execute the duties of a spotter for a balance climb, in accordance with the references. (WMC.C3c)

(4) In a summer mountainous environment, execute the commands used between the climber and the spotter during a balance climb, in accordance with the references. (WMC.C3d)

(5) In a summer mountainous environment, execute each type of hold used in balance climbing, in accordance with the references. (WMC.C3e)

(6) Without the aid of references and given the acronym "CASHWORTH", describe orally the nine considerations for proper body position while climbing, in accordance with the references. (WMC.C3f)

OUTLINE

1. **SAFETY PRECAUTIONS**. There are two safety precautions that always apply to balance climbing. They are as follows:

C3

a. Never climb more than 10 feet above the ground. By this it is meant that the climber's feet are never more than 10 feet above the ground.

b. A spotter is required for all balance climbs.

SPOTTER WATCHING THE CLIMBER

2. **INDIVIDUAL PREPARATIONS**. Prior to beginning a balance climb there are seven things that the climber must do to prepare himself. They are as follows:

a. Helmet with a serviceable chinstrap must be worn.

b. Sleeves rolled down_to give hand and arm freedom of movement. Blouse tucked in to your trousers. In case of fall, it may catch on a rock and cause you to flip over sideways.

c. All watches, rings, and jewelry must be removed before climbing.

d. Gloves will not be worn, as they can slip, and also give a false feel for the rock.

e. Unblouse trousers, if they restrict movement.

f. Soles of boots clean and dry.

g. Select route where vegetation is minimal. Never use vegetation for hand or foot holds.

3. **DUTIES OF THE SPOTTER**. The spotter is the balance climber's partner who, rather than climbing himself, acts as the safety man for the climber during the climb. The five duties of the spotter are as follows:

a. Positions himself directly behind the climber before the climb starts.

b. Maintains his position facing the cliff, directly below the climber and approximately 3-4 feet away from the base of the cliff, for the duration of the climb. He will move diagonally as necessary to remain below the climber.

c. The spotter will stand with his feet shoulder width apart and arms ready to stop the climber if he falls.

d. If the climber falls, the spotter will not "catch" him; he will prevent the climber from falling further down the hill. He will do this by pushing the climber towards the base of the cliff, thereby preventing him from tumbling backwards.

e. At no time will the spotter allow anyone to come between himself and the face of the cliff while a balance climb is taking place. He will require anyone who wants to pass by his position to go behind him.

4. **SPOTTING AND CLIMBING COMMANDS**. The following are the commands used by both the spotter and climber.

COMMAND	GIVEN BY	MEANING
"Last Name, climbing."	Climber	I am ready to climb.
"Climb Climber."	Spotter	I am ready to spot you.
"Last Name, off Climb."	Climber	I am off the climb.
"ROCK!"	Climber or Spotter.	A rock has been knocked off the rock face and is falling.
"FALLING!"	Climber	I am Falling.

a. If the command "ROCK" is given, all personnel in the vicinity will take the following action:

(1) If close to the cliff face, move against the cliff face with your face against the cliff face and your hands between you and the cliff face.

(2) If not close to the cliff face, look up to locate the rock and avoid it.

5. **ACTIONS IF FALLING**. If, while making a balance climb, the climber feels himself slipping and beginning to fall, he will take the following action:

a. Sound the command "falling".

b. Push himself away from the rock face.

c. Maintain proper body position as follows:

(1) Head up.

(2) Hands out toward the rock.

(3) Body relaxed.

(4) Feet kept below the body, slightly apart.

(5) Ensure you face the cliff face as you fall.

6. **TYPES OF HOLDS**. There are five basic holds that are used in balance climbing. They are as follows:

a. Push Holds.

(1) Most effective when hands are kept low.

(2) Often used in combination with a pull hold.

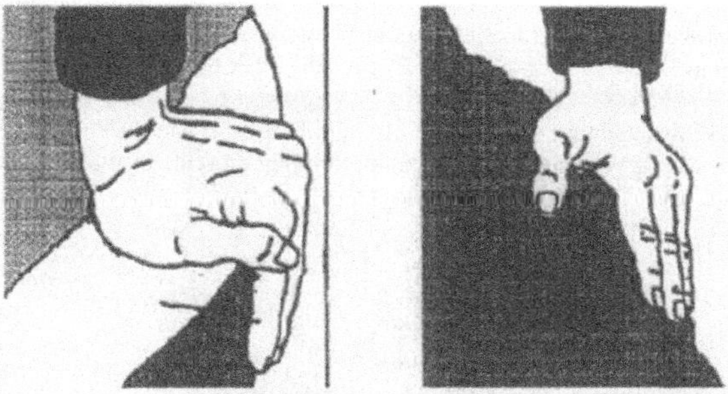

TYPES OF PUSH HOLDS

b. Pull Holds.

(1) The easiest hold to use and, consequently, often overused.

(2) Can be effective on small projections.

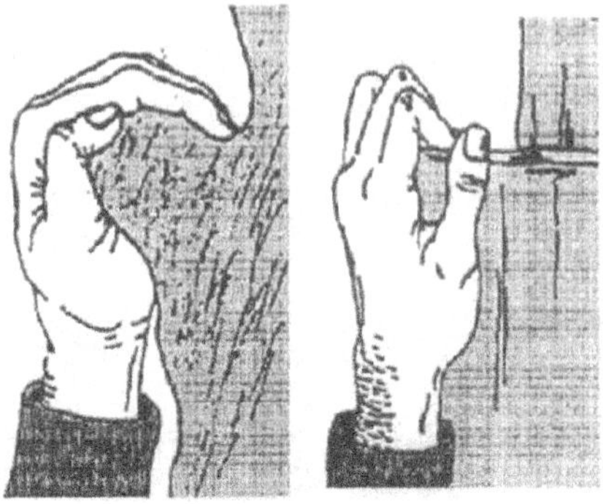

TYPES OF PULL HOLDS

c. Foot Holds

(1) Feet should be positioned with the inside of the foot to the rock.

(2) Use full sole contact as much as possible.

(3) Avoid crossing your feet. If you must cross your feet use a change step. A change step is a method of substituting one foot for the other foot on the same foothold.

(4) Making maximum use of footholds, climbing with your feet, is an effective means of conserving your body strength, since your leg muscles are stronger than your arm muscles.

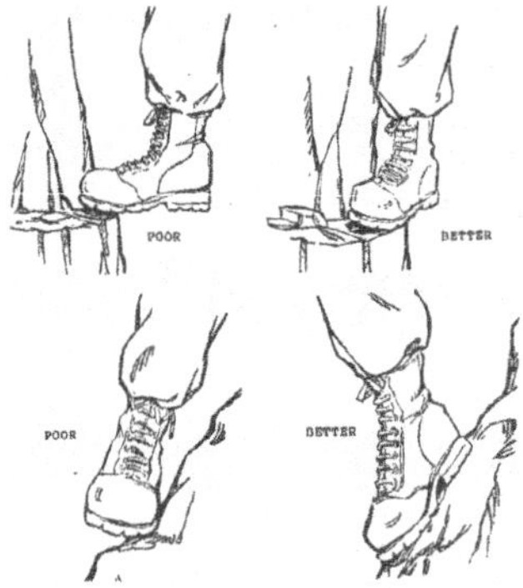

USE OF FOOT HOLDS

d. Friction Holds. A friction hold is anytime you are relying on the friction of your foot or hand against the face of the rock for traction, rather than pushing/pulling against a projection on the face of the rock.

(1) It is a type of hold that feels very insecure to an inexperienced climber.

(2) The effectiveness of this type of hold is dependent upon many things i.e., type, condition and angle of the rock face, type of boot soles, confidence, etc.

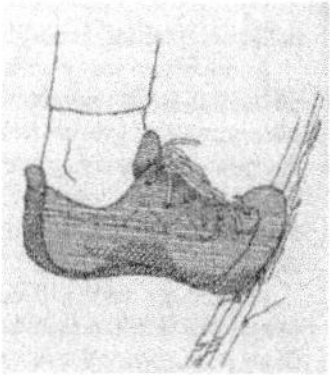

FRICTION USED IN SMEARING

e. Jam Holds. This type of hold involves jamming/wedging any part of your body or your entire body into a crack/opening in the rock.

(1) An important consideration is that you do not jam such that you cannot free that portion of your body after you complete the move. This sounds like a ridiculous statement; however, you must remember that after you complete your move, you may be withdrawing the portion of your body that you used from a different angle than you inserted it.

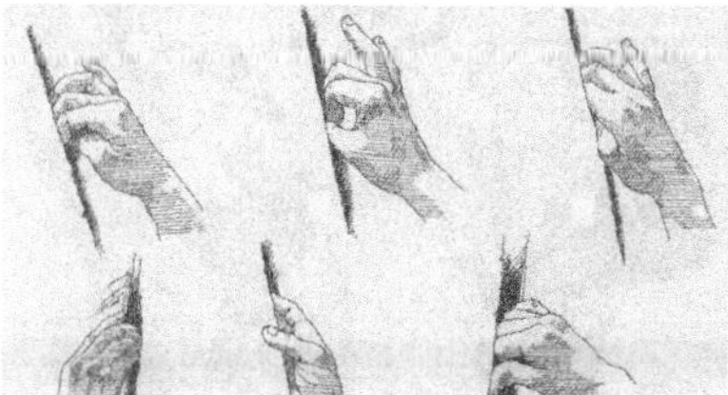

FINGER JAMS

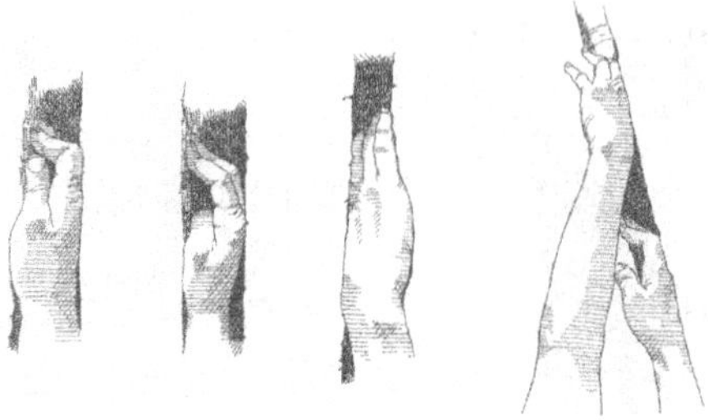

HAND JAMS

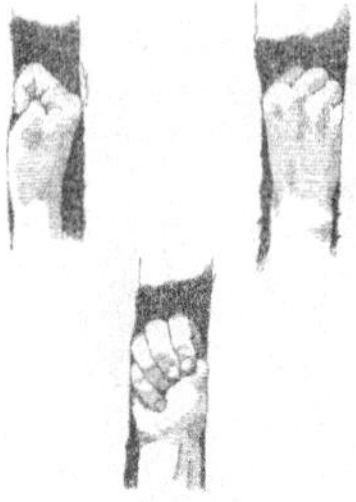

FIST JAMS

7.) **COMBINATION HOLDS.** The five types of holds just mentioned above are not just used individually. They are most often used in combinations with each other. Some examples are:

a. Chimney Climbing. This is when you insert your entire body into a crack in the rock and by using both sides of the opening, and possibly all five types of basic holds, move up' the crack.

CHIMNEY CLIMBING TECHNIQUES

b. Lie-back. This is a combination of both pull holds with your hands and friction holds with your feet.

LIE BACK TECHNIQUE

c. Push-Pull. As the name implies, this is when you use a push hold and a pull hold together.

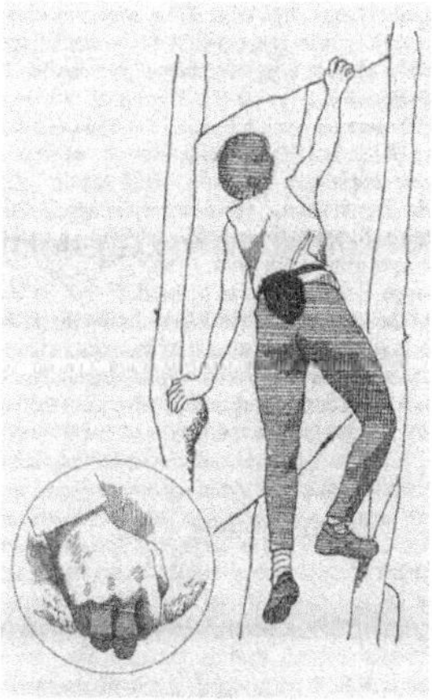

PUSH-PULL

d. Mantling. This is a technique where you continue to climb without moving your hands off a projection by pulling yourself up until your hands are at chest level and then invert your hands and push on the same projection.

MANTELING

MANTELING WITH FACE HOLD

e. Cross-Pressure in Cracks. This is a technique of putting both hands in the same crack and pulling your hands apart to hold/raise yourself.

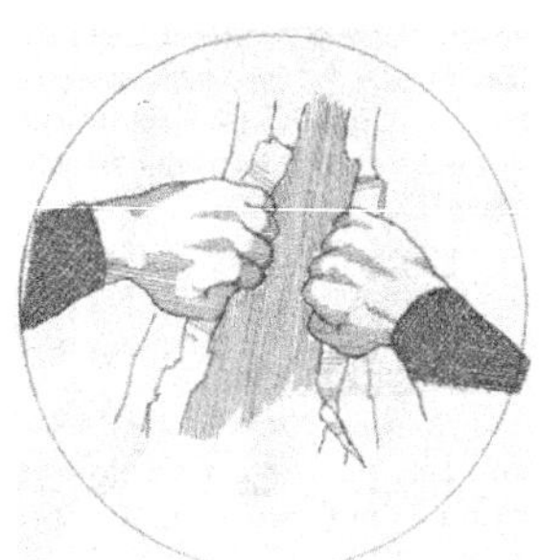

CROSS-PRESSURE METHOD

f. Inverted. Pull or push.

INVERTED TECHNIQUE USING COUNTERFORCE PRESSURE BETWEEN FEET AND HANDS

g. Pinch. As the name implies this is a grip used on tiny little nubbins.

PINCH

h. Stemming. The spreading of arms or legs to maintain a proper body position. (i.e. usually used in a book or chimney.)

STEMMING

8. **GENERAL USE OF HOLDS**. How you use an individual hold is dependent on your experience level, or sometimes, your imagination. Here are some general guidelines.

a. Most handholds can be used as foot holds as you move up the rock.

b. Use all holds possible in order to conserve energy.

c. Even small projections may be used as holds.

d. Do not make use of your knees or elbows due to the reason that it is skin on bone and a slip could occur if pressure is exerted on them. Knees and elbows can be used with the extension of a limb jam.

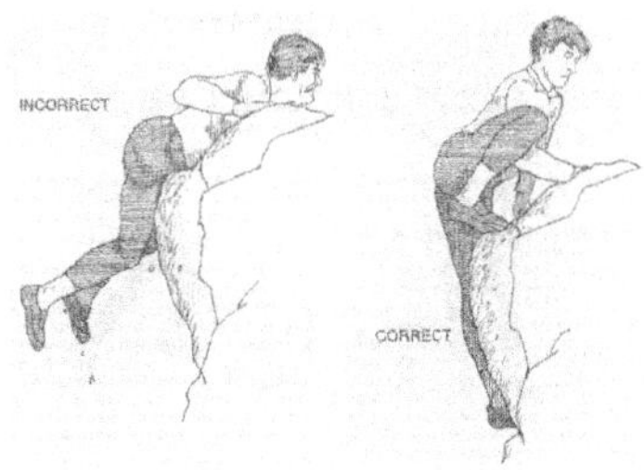

PROPER AND IMPROPER

9. **MOVEMENT ON SLAB**. Movement on slab is based on friction holds.

a. Use any and all irregularities in the slope to gain additional friction.

b. Traversing requires both hands and feet.

TRAVERSING

c. Descending steep slab may require turning inboard to face the slab and backing down.

d. The biggest mistake in slab climbing is leaning into the rock. Maintain maximum friction by keeping weight centered.

SLAB CLIMBING

10. BODY POSITION

a. The climber should climb with his body in balance by keeping his weight centered over and between his feet. Don't hug the rock. Don't over extend and become "spread-eagled". While climbing, keep in mind the acronym "CASHWORTH" for proper body position and movement. (WMC.C3F)

C - Conserve energy.

A - Always test holds.

S - Stand upright, on flexed joints.

H - Hands kept low; handholds should be waist to shoulder high.

W - Watch your feet.

O - On three points of contact.

R - Rhythmic movement.

T - Think ahead.

H - Heels kept low, lower than the toes.

UNITED STATES MARINE CORPS
Mountain Warfare Training Center
Bridgeport, California 93517-5001

WMC.C4
04/11/02

STUDENT OUTLINE

NATURAL AND ARTIFICIAL ANCHORS

TERMINAL LEARNING OBJECTIVE In a summer mountainous environment, establish an anchor system in accordance with the references. (WMC.C4)

ENABLING LEARNING OBJECTIVES.

(1) Without the aid of references and from a given list, choose the two types of anchors, in accordance with the references. (WMC.C4a)

(2) In a summer mountainous environment, establish a natural anchor point, in accordance with the references. (WMC.C4b)

OUTLINE

1. **TYPES OF ANCHORS.** (WMC.C4a)There are two types of anchors that we use. These two types are:

1) NATURAL
2) ARTIFICIAL

C4

2. **GENERAL CONSIDERATIONS FOR ANCHORS**. Anytime you employ natural or artificial anchors; there are special considerations that you must apply. These considerations apply both to the anchor itself and to the material being used to build the anchor. Some examples of these considerations are:

a. Whether using natural or artificial anchors, the installing unit must insure that the anchor is suitable for the load.

b. The anchor position must be relative to the direction of pull on the anchor.

c. The angle between the anchor points should not exceed 90 degrees. This is to ensure that no added stress is put upon the anchors, as well as the equipment being used to construct the anchor.

(1) To decrease the angle between anchor points, materials (i.e. sling ropes, web runners, Prusik cord, etc..) could be used to extend the anchor which will decrease the angle between anchor points.

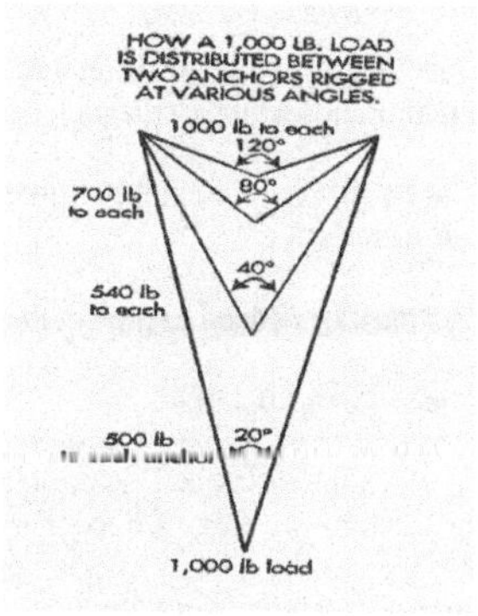

3. **NATURAL ANCHORS.** A natural feature is the preferred type of anchor point. Some examples and considerations are as followed:

a. Types of natural anchors.

(1) Trees.

(a) Select a tree that has not been chopped, burned or is rotten.

(b) The tree should be at least 6" in diameter and strong enough to support the intended load.

(c) Trees growing on rocky terrain should be treated with suspicion, since the roots normally are shallow and spread out along a relatively flat surface.

(2) Shrubs and Bushes.

(a) Select a shrub or bush that is alive and is not brittle, charred or loose.

(b) To avoid leverage, locate the central root and construct the anchor as near to the base as possible.

(3) Rocks and Boulders.

(a) Stability is of prime importance when considering a rock or boulder for an anchor. It must be strong enough and secure enough for the intended load.

(b) All surfaces of the rock or boulder should be inspected for any rough or sharp points. These areas must be padded to protect the rope from being abraded or cut.

(4) Spikes and Flakes.

(a) To check stability of a spike or flake, thump it with the heel of your hand. Anything that sounds hallow is suspicious.

(b) They should be checked for cracks or other signs of weathering that may impair their firmness.

(c) Sharp edges must be padded to protect the rope against cuts and abrasion.

(5) Threads and Chockstones.

(a) A thread is when the rock weathers or cracks to form a hole in the main wall. A chockstone is a rock wedged in a crack.

(b) Check a thread by thumping it with the heel of your hand. Make a visual inspection for cracks and weather. Common sense will prevail for choosing this type of anchor.

(c) When choosing a chockstone, ensure that it has substantial contact with the crack and that the stone's symmetry corresponds with the intended direction of pull.

b. Types of Natural High Tension Anchors. The following are types of Natural High Tension Anchor Systems used to construct installations. These can be tied on any suitable natural anchor point:

(1) Tree Wrap Anchor. The tree wrap is an anchor used to relieve tension on the actual knot itself. This system requires more rope when tied around a large anchor point.

(a) Tie a Figure 8 Loop on the standing end of the rope and wrap the 4-6 times around an anchor point.

(b) Attach a locking carabiner through the knot's loop and clip it onto the running end of the rope.

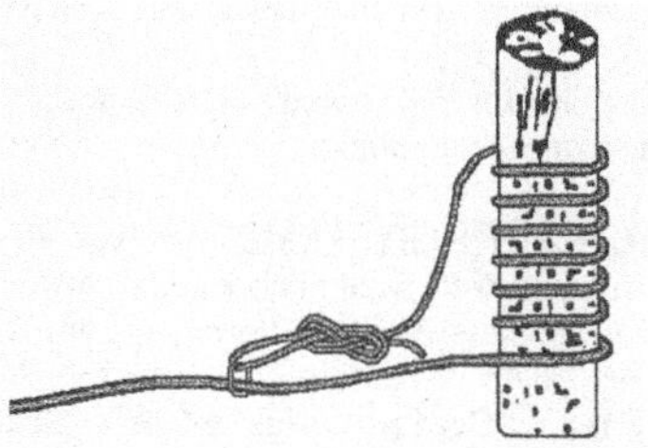

NOTE: If the anchor point to be wrapped is larger then 18' around, then three wraps will suffice.

(2) Swami Wrap Anchor. When tying the Swami Wrap, ensure that the joining knot is kept on the side of the anchor point. The Swami Wrap can be loosened under load if need be. This system can be substituted for a tree wrap when the installation's rope length is an issue.

(a) Select a suitable anchor point and tie a Swami Wrap around it.

(b) Clip a locking carabiner into as many wraps as possible. This will serve as the attachment point for the anchor system.

(3) Figure of 8 Anchor. The Figure of 8 Anchor is a quick and efficient system. This system cannot be loosened under load. It is tied in the following manner:

(a) Tie a Figure of 8 Loop on the standing end of the rope and wrap the knot around a suitable anchor point. Attach a locking carabiner through the knot's loop and clip it onto the running end of the rope.

(b) Before tensioning the anchor system, adjust the rope so the running end of the rope runs smoothly through the carabiner towards the direction of pull. This will prevent any lateral tension.

(b) If two ropes are used, the upper rope's carabiner has its gate upwards, and the lower rope's carabiner has it's gate downward.

c. Primary and Secondary Natural Anchor Point System. The two anchor points should be in line with the direction of pull. The primary anchor is the point nearest the running end while the secondary anchor point is directly behind the primary. It is constructed in the following manner:

(1) The primary anchor knot should be an around the object clove hitch. This is chosen for ease of untying the system after tension has been placed on the rope.

(2) The secondary knot is tied around a suitable anchor point ensuring that the rope is taut between the two anchor points.

4. **ARTIFICIAL ANCHOR SYSTEMS**. Any time we use anything other then a natural feature, we are using an artificial anchor point. Artificial anchors can be constructed in the ground, or on the rock itself. The following are artificial anchor systems:

a. Single Timber Deadman. This system is constructed in the ground and it requires considerable time and effort. The steps of it's construction are as follows:

(1) Dig a trench 6 feet long and 3 feet deep and wide enough to work in at a 90 degree angle to the direction of the pull.

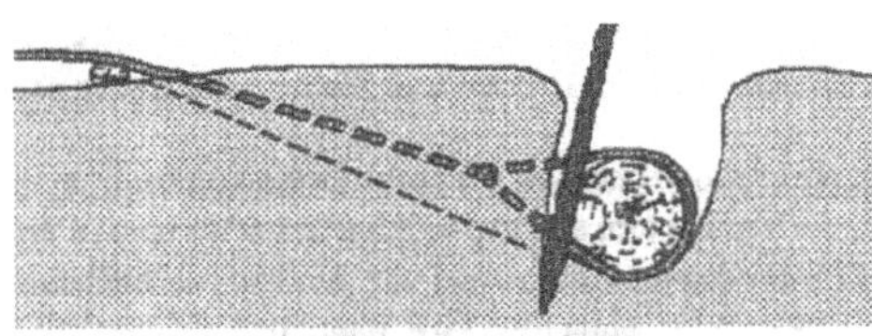

(2) Dig another trench about 12 inches wide. This trench is dug so that it intersects the main trench at a right angle in the middle. The bottom of this trench should be parallel to the direction of pull and should join the bottom of the main trench.

(3) Take an anchoring device (i.e. log, engineer stakes, bundled up branches etc.) that is strong enough to support the intended load. The anchor is placed into the main trench and covered with dirt with the exception of that part of the anchor that joins the second trench. Stakes approximately 3 feet long should be driven approximately 1-1/2 feet into the ground between the dead-man and the slanted side of the trench to assist in holding the dead-man in place if the soil is soft.

b. Picket Hold Fast. The picket hold fast is an easier anchor to construct than the Deadman, and can be used almost anywhere. The strength of the picket system depends on the pickets and the soil or snow conditions. The picket system can be used for both high and non-tension systems. Construction is as follows:

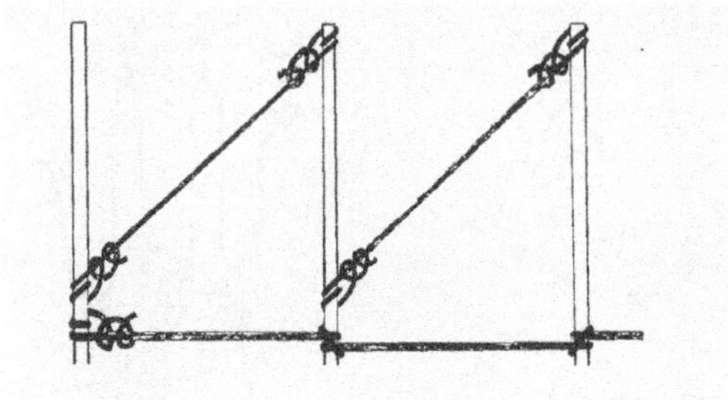

(1) Three stakes (i.e. logs, engineer stakes, etc.) are driven into the ground at a 30 degree angle against the direction of pull.

(2) The line of pickets should be driven in line with the direction of pull. The distance between the pickets can be anywhere from 3-12 feet apart depending on the terrain and soil conditions.

(3) Before tying any rope to the picket anchors, the base of the pickets must be buffed or padded.

(4) To tie a rope to the pickets, go to the furthest picket away from the cliff edge. Tie a round turn and two half hitches at the base of the picket. Take the running end of the rope and tie an over the object clove hitch to the base of the middle picket. Then with the running end of the rope tie an over the object clove hitch to the picket closest to the edge of the cliff face. Ensure that there is tension on the rope in between each picket.

(5) To tie off the pickets to themselves, go to the furthest picket from the cliff edge. And with a sling rope, tie a round turn and two half hitches at the base of that picket. With the running end of that sling rope tie a round turn and two half hitches at the top of the middle picket. Using a second sling rope tie a round turn and two half hitches to the base of the middle picket. Then with the running end of that sling rope tie a round turn and two half hitches to the top of the picket closest to the cliff edge.

c. Equalized Anchor. This system is built with a minimum of three pieces of protection. It can be tied with the standing end of a rope or by utilizing a practice coil.

(1) Construction with the Standing End of the Rope:

(a) Tie a Figure of 8 Loop in the standing end of the rope.

(b) Place a carabiner in each artificial anchor point and attach the knot in either of the outside carabiners.

(c) Clip the rope into the remaining carabiners.

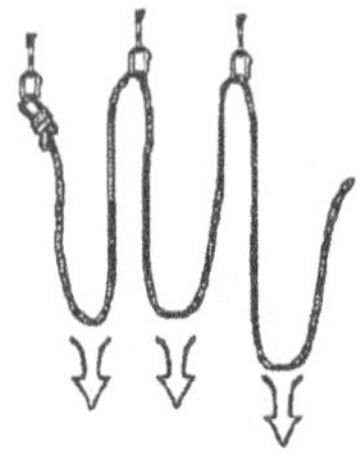

(d) A bight of rope is then pulled down after each carabiner into the anticipated direction of pull. With all three bights, tie an overhand knot around itself to include the running end of the rope.

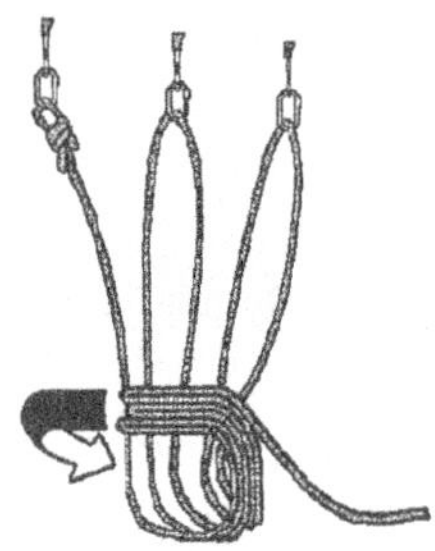

(e) With the running end of the rope coming from the overhand knot, tie a Figure of 8 Loop and attach it to the overhand knot's loop with a locking carabiner.

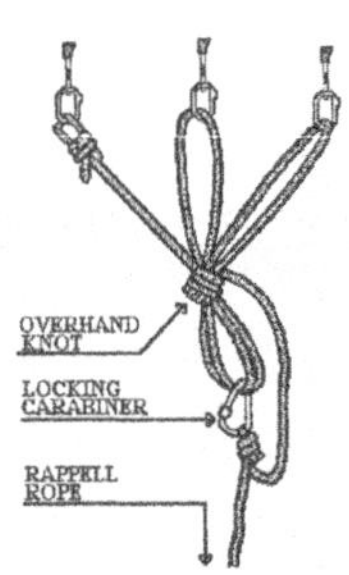

(f) Ensure that there is sufficient slack in the dead rope to prevent disfiguring the knot thus ruining the structural integrity of the knot.

(2) Construction with the Cordalette Method:

(a) Create an endless loop with a practice coil and clip it into each of the artificial anchor points.

(b) A bight of rope is then pulled down between each carabiner into the intended direction of pull.

(c) With all three bights, tie an overhand knot on itself.

(d) Place a locking carabiner into the overhand knot's loop. This will serve as the attachment point for the anchor system.

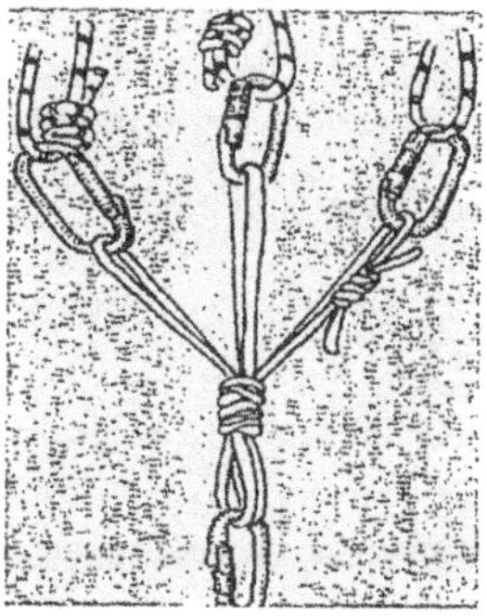

NOTE: If constructing a two-rope high-tension installation, a minimum of three artificial anchors per rope will be used.

UNITED STATES MARINE CORPS
Mountain Warfare Training Center
Bridgeport, California 93517-5001

SML
ACC
02/11/02

LESSON PLAN

SIT HARNESS

LEARNING OBJECTIVES

TERMINAL LEARNING OBJECTIVE: Without the aid of references, properly wear and maintain the MWTC sit harness, in accordance with the references.

ENABLING LEARNING OBJECTIVES

(1) Without the aid of references and given a diagram, name in writing the parts of a sit harness, in accordance with the references.

(2) Without the aid of references, wear a sit harness, in accordance with the references.

BODY

1. (5 Min) **NOMENCLATURE**

   a. Waist belt.

   b. Leg loops (adjustable).

   c. Buttocks straps (adjustable).

   d. Fastex buckle - 2.

   e. Doughnut.

   f. Equipment loops.

   g. D-ring (older models).

   h. Waist belt tie-in point.

   i. Crotch strap.

WEARING OF THE SIT

C5

HARNESS

a. First, disconnect the fastex buckle at the rear of the harness.

b. Hold the harness in front of you, put your feet through the leg loops ensuring that the buckles on the leg loops are outboard, on your thighs.

c. Fasten the waist belt into the buckle, ensuring that it is a tight but comfortable fit. **You must ensure that the waist belt is threaded back through the buckle, this action locks the waist belt to the buckle. Failure to do this will cause the waist belt to slip through the buckle when it is under load.**

d. Adjust the leg loops so that they are high on your thigh, once you have them adjusted, get your buddy to clip the fastex buckle male to the most comfortable female fastex. Then adjust the buttock straps so as the leg loops are held up.

e. If all the above are done correctly, the harness should now be a comfortable but snug fit, after you have fitted it for the first time there is no need to go through the same procedure each time you put it on. Simply hold the harness in front of you, step into it and attach the waist belt to the buckle in the approved manner.

2. TYING IN THE ROPE (END OF ROPE)

a. Pass the end of the climbing rope up through the crotch strap, then through the doughnut and through the waist belt tie-in point.

b. Tie the rope using a retraceable figure eight, adjusting the knot to get it as close as possible to the body.

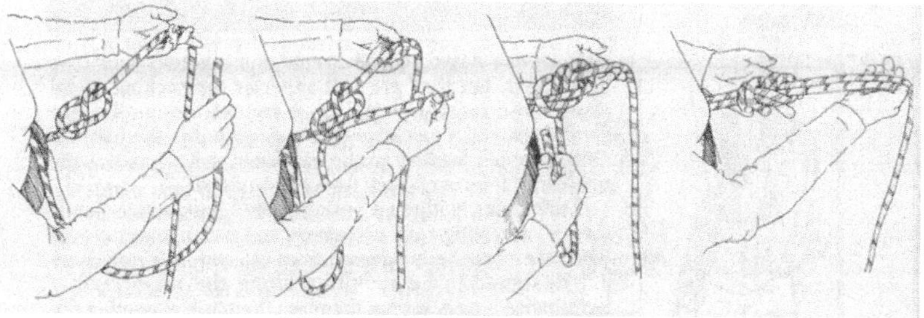

RETRACE FIGURE 8 IN SIT HARNESS

3. TYING IN THE ROPE (MIDDLE OF THE ROPE)

a. Take up a bight of rope and tie a figure eight loop, then take a steel locking carabiner and attach the carabiner to the harness by securing the crotch strap, doughnut, and the waist belt tie-in loop. Now attach the figure eight loop to the carabiner.

4. (5 Min) **CARE AND MAINTENANCE**

a. Avoid contact with chemicals, as this will damage the nylon.

b. Regularly inspect for signs of abrasions and normal wear. Pay particular attention to wear points such as the tie-in loops, buckles, and sewn joints.

c. Keep away from heat such as open flames, cigarettes, etc.

d. If soiled by grit and seawater, wash in lukewarm water with pure soap and allow to dry in a warm room away from direct heat.

e. Two to three years of life can be expected during normal climbing use.

f. It is recommended that a harness that has experienced a serious fall should be discarded.

g. Under no circumstances will you ever tie into the equipment rack of any sit harness as a belay, anchor point, etc.

UNITED STATES MARINE CORPS
Mountain Warfare Training Center
Bridgeport, California 93517-5001

FMST.07.33
05/02/01

STUDENT OUTLINE

ESTABLISHMENT OF RAPPEL POINTS AND RAPPELLING

TERMINAL LEARNING OBJECTIVE Given a mountainous environment, mountaineering equipment, helmet, sit harness, rappel gloves, 165 ft static rope, rescue 8, prusic cord, and one or more assistants, conduct rappelling operations to support mission requirements per the reference. (FMST.07.33)

ENABLING LEARNING OBJECTIVES.

(1) Without the aid of references, select from a given list the three criteria involved in site selection for a rappel site, in accordance with the references. (FMST.07.33a)

(2) In a summer mountainous environment, execute the commands used between a rappeller and his belay man when conducting a rappel, in accordance with the references. (FMST.07.33b)

(3) Without the aid of references, select from a given list three of the nine duties of the rappel point NCOIC, in accordance with the references. (FMST.07.33c)

(4) Without the aid of references, select from a given list three of the six duties of the first man down a rappel rope, in accordance with the references. (FMST.07.33d)

(5) Without the aid of references, select from a given list the three types of rappels, in accordance with the references. (FMST.07.33e)

(6) In a summer mountainous environment, rappel down a cliff face using a seat-hip rappel, in accordance with the references. (FMST.07.33g)

C6

OUTLINE

1. **INHERENT DANGER OF RAPPELLING**. Rappelling is inherently dangerous because rappellers rely totally on the equipment.

   a. To ensure a safe training evolution, two ropes will be utilized at MWTC.

   b. All the rappels taught at MWTC can be completed with a single rope if the situation arises.

   c. If utilizing a one-rope system with the carabiner wrap, the rope will be attached to the carabiner with two wraps vice one.

2. **SITE SELECTION**. (FMST.07.33a) when selecting a rappel site consider these three factors.

   a. There must be a good anchor. As previously taught in *NATURAL* and *ARTIFICIAL ANCHORS*, natural anchors are preferred.

   b. The rappel route down should be as free of obstacles (i.e., vegetation, debris) as possible.

   c. There must be suitable loading and unloading platforms.

NOTE: The evaluation of a site for the above factors should be made by the rappel point NCOIC, who should be the most experienced rappeller in the unit.

3. **DUTIES OF THE RAPPEL POINT NCOIC**. (FMST.07.33c)Once a rappelling site has been selected, one person will be appointed to each rappel lane as a rappel point NCOIC. These individuals should have experience as rappellers. The rappel point NCOIC has ten duties and responsibilities, which are:

   a. Ensures that the anchor points are sound and that the knots are properly tied.

   b. Ensures that loose rock and debris is cleared from the loading platform.

   c. Allows only one man on the loading platform at a time and ensures that the rappel point is run in an orderly manner.

   d. Ensures that each man is properly prepared for the particular rappel; i.e. gloves on, sleeves down, helmet secured, rappel seat tied correctly and secured properly.

   e. Attaches the rappeller to the rope and ensures the rappeller knows the proper braking position for that particular rappel.

   f. Ensures that the proper commands or signals are used.

   g. Dispatches each man down the rope.

h. The rappel point NCOIC will be the last man down the rope.

i. The rappel point NCOIC will ensure that the ropes are inspected after every 50 rappels.

j. The rappel point NCOIC will maintain a rope log.

4. **TYPES/USE**. (FMST.07.33e) The three types of rappels and when they are preferred to be used are:

a. Hasty Rappel: It is used when carrying loads down moderate slopes.

b. Seat-Shoulder Rappel: It is used for heavily laden troops over vertical to near vertical cliff faces.

c. Seat-Hip Rappel: It is used when carrying loads over vertical to near vertical faces.

5. **HASTY RAPPEL**. The hasty rappel is the easiest type of rappel to prepare for. It requires no equipment other than a rope and gloves.

a. Conduct. A hasty rappel is conducted in the following manner:

(1) Sleeves will be rolled down and gloves will be put on.

(2) Face slightly sideways.

(3) Place the rappel rope across your back, grasping it with both hands, palms forward, and arms extended.

(4) The hand nearest the anchor is the guide hand. The hand farthest from the anchor is the brake hand.

(5) Lean out at a moderate angle to the slope.

(6) Descend down the hill facing half-sideways, taking small steps and continually looking downhill while leading with the brake hand.

(7) Feet should not cross and the downhill foot should lead at all times.

b. Braking. The steps for braking during a hasty rappel are as follows:

(1) Bring the lower (brake) hand across the front of the chest to brake.

(2) At the same time, turn to face up toward the anchor point.

HASTY RAPPEL

6. **SEAT-SHOULDER RAPPEL**. The seat-shoulder rappel relies on friction as the main effort of controlling the descent. It is very efficient for men with heavy packs because it provides support for heavy loads on the back.

a. Conduct. A seat-shoulder rappel is conducted in the following manner:

(1) Put on your rappel-seat, roll down your sleeves and put on your gloves.

NOTE: To avoid causing a possible injury, it is advised that the rappel seat be constructed at the rear of the rappel site loading platform. If the rappel seat is worn for too long, the rappel seat could loosen up enough to cause you to slip out of your seat while rappelling.

(2) The steel locking carabiner is placed on the rappel-seat so that the gate opens down and away, to prevent the gate from opening once the wraps are placed into the carabiner.

(3) Step-up to the rope with you're left shoulder facing the anchor.

(4) The rappel rope is attached to the rappeller's hard point carabiner as follows:

(a) Snap the rope into the locking carabiner.

(b) Taking slack from the standing (anchor) end of the rope, make one wrap with the rope around the body of the carabiner and back through the gate.

(c) Ensure that the locking nut of the carabiner is fastened to lock the carabiner closed.

NOTE*:* If you are using only one rope to rappel with due to the tactical situation or equipment availability, the procedures are the same, EXCEPT the individual will make two wraps around the body of the carabiner instead of one.

(5) Take the rope across your chest, over your left shoulder, diagonally across your pack and down to the right (brake) hand.

(6) Descend by walking down the cliff using the braking procedure to control the rate of descent. Look under your brake arm for possible obstacles to avoid.

b. Braking. The steps for braking during a seat-shoulder rappel are as follows:

(1) Lean back.

(2) Face directly uphill while bringing the brake hand across the chest.

7. **SEAT-HIP RAPPEL**. The seat-hip rappel is the most commonly used rappel.

a. Conduct. A seat-hip rappel is conducted in the following manner:

(1) Construct the rappel seat; roll down sleeves and put gloves on.

(2) The steel locking carabiner is placed on the rappel-seat so that the gate opens up and away.

(3) Step up to the rope with your left shoulder facing the anchor.

(4) The rappel rope is snapped into the carabiner as follows:

(a) Snap the rope into the locking carabiner.

(b) Taking slack from the standing (anchor) end of the rope, make one wrap with the rope around the body of the carabiner and through the gate again.

(c) Ensure that the locking nut of the carabiner is fastened to lock the carabiner closed

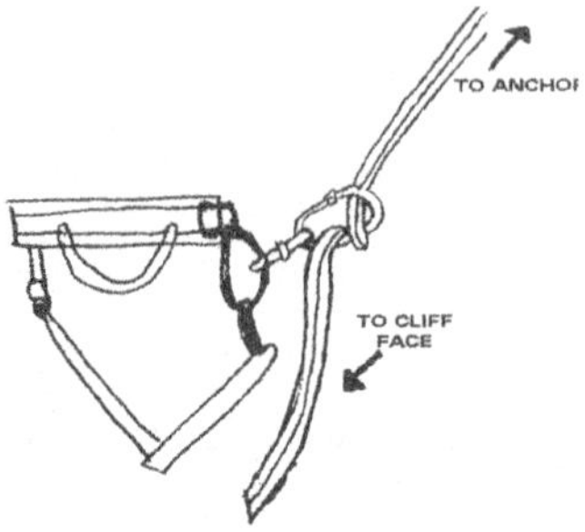

SEAT-HIP RAPPEL

NOTE: If you are using only one rope to rappel with due to the tactical situation or equipment availability, the procedures are the same, EXCEPT the individual will make two wraps around the body of the carabiner instead of one.

(5) The rappeller will grasp the running end of the rope with the brake (right) hand, palm down and turned slightly inboard, near the hip.

b. Braking. The steps in braking for a seat-hip rappel are as follows:

(1) Grasp the rope tightly with the brake hand.

(2) Take the brake hand and place it in the small of the back. This will create enough friction to stop all momentum.

NOTE: At no time will you bound or jump while you are descending. You "walk down" the cliff face using the proper braking procedure to control your rate of descent.

8. **SAFETY OF THE FIRST MAN DOWN**. To ensure the safety of the first man down, the rappeller will:

a. Before deploying the ropes, tie the two ends of the ropes together with an overhand knot.

b. Tie a friction knot on the standing end of the rappel rope with a Prusik cord.

c. Attach the rappel ropes to the harness's hard point with a locking carabiner.

d. Attach the Prusik cord to the harness's hard point with a locking carabiner. This cord is also referred to as a safety Prusik.

NOTE: If the Safety Prusik is too short, it can be extended with a web runner.

9. **DUTIES OF THE FIRST MAN DOWN**. (FMST.07.33d) The first man down the rope has specific duties. They are as follows:

a. Selects a smooth route down for the ropes.

b. Clears the route of loose rocks and debris.

c. The first man down will untie the overhand knot and straighten the ropes out once he reaches the bottom.

d. The first man down belays the next man down the rope.

(1) To belay correctly, the belay man will stand facing the cliff face, arms up with palms facing upward and over lapping. The rope will pass through the opening of the over lapping hand's thumb and index fingers.

(2) To stop a fallen rappeller, the belay man will grab the rope with closed fists and pull straight down bringing the forearms parallel to the deck.

e. Take charge of personnel as they arrive at the bottom to include appointing a belay man.

10. **RAPPELLING COMMANDS.** (FMST.07.33b) In order to conduct rappelling operations safely, it is essential that everyone understands the sequence of events. The following voice commands or rope tugs will be utilized:

VOICE COMMANDS	GIVEN BY	MEANING
"Lane # on Rappel"	Rappeller	I am ready to begin rappelling
"Lane # on Belay"	Belayer	I am on belay and you may begin to rappel
"Lane # off Rappel"	Rappeller	I have completed the rappel and am off the rope
"Lane # off Belay"	Belayer	I have completed the belaying of the rappeller

ROPE TUGS	GIVEN BY	MEANING
Three Tugs	Rappel Point NCOIC	The rappeller is ready to begin rappelling
Three Tugs	Belayer	I am on belay and the rappeller may begin to rappel
Three Tugs	Belayer	The rappeller is off the rappel rope

11. **TYING-OFF**. Occasionally, it may be necessary to stop during a rappel before reaching the bottom of a cliff. The following sequence is used:

a. The commands used in tying-off are as listed below:

COMMAND	GIVEN BY	MEANING
"Lane # Tying-off"	Rappeller	I am ready to tie-off, give me some slack.
"Lane # Tying-off"	Belayer	I have given you enough slack and you may tie-off.
"Lane # On rappel"	Rappeller	I have completed tying-off and I am ready to resume rappelling.
"Lane # On rappel"	Belayer	I am on belay, you may rappel.

b. The procedure used when tying-off is as follows:

(1) The rappeller gives the command, "Lane #, tying-off."

(2) The Belayer gives the rappeller slack and gives the command "Lane # Tying-off". He continues to hold the rope in the belay position, remaining alert and watching the rappeller.

(3) The rappeller quickly brings his brake hand to the twelve o'clock position, so that the running end and standing end are parallel.

(4) The rappeller grasps all ropes with his guide hand as close to the carabiner as possible.

(5) The rappeller steps over the rope so that the running end is going between his legs.

(6) The rappeller releases the rope with his brake hand then reaches under the running end, over the standing end. He then takes up a bight from the running end about two feet long and pulls it over the standing end and under the running end, forming a half hitch.

(7) He pulls the half hitch tightly against the guide hand.

(8) He works the half hitch down snugly against the carabiner while maintaining contact with the guide hand as long as possible.

NOTE*:* A loose half hitch could bind into the carabiner causing difficulties in clearing the knot out. Make sure that the first half hitch is dressed down tightly before moving it against the carabiner.

(9) Place another half hitch above the one already tied.

(10) To untie, reverse the steps. Remove the safety half hitch, then shrink the first half hitch down to a small loop by grabbing the running end of both ropes and pulling them straight out to the left. Place your right hand in the middle of the two bights until they are snug on all four fingers. Then remove one finger and make the bights snug on three fingers. Repeat this process until it's down to one finger and both the bights are equal.

(11) Now, grasp the running end with both hands and smartly jerk the running end of the ropes upward to pop the small loop out from the first half hitch. From this position, keep the guide hand around both of the ropes next to the carabiner; step back over the rope so that the running end is to your right side. Grab just the running end with your brake hand and quickly set the brake behind the small of your back then, readjust you guide hand onto the standing end. You now should be in the seat-hip rappel position.

NOTE: The wraps may bind up some after untying, where no further movement down is possible. To alleviate this, keep the brake on and force your body weight down to pop the wraps in the carabiner back to their intended position and then continue on with the rappel.

(12) The rappeller gives the command "Lane #, on rappel."

(13) The Belayer takes up the slack and gives the command "Lane #, on belay."

(14) The rappeller resumes to rappel down.

12. **RESCUE TECHNIQUES**. When conducting rappelling operations, the possibility of a rappeller getting caught on the rope due to either his clothing or equipment can occur. There are two types of rescues that can be performed to free the rappeller from this situation.

a. The Self-Rescue Technique.

(1) After realizing that you are caught up on the rope, check with your belay man to ensure that he has a solid brake set.

(2) Using a safety Prusik, tie a friction knot onto the rappel rope approximately an arms length above the area that is fouled. Anchor the other end of your safety Prusik to your rappel seats hard point. This is the same technique used as discussed earlier with the first man down the rope.

(3) Work the friction knot up until there is no tension on the rappel device. This will give you the necessary slack to free the malfunction.

(4) Once you have corrected the problem, continue to rappel down using the first man down method.

b. The Buddy Rescue Technique. Himself uses this method when the rappeller is unable to correct the problem.

(1) A rescuer will rappel down on another rope to the disabled rappeller and tie-off.

(2) The rescuer will then establish a safety Prusik onto the victim as taught in the self-rescue technique.

(3) After clearing the malfunction, continue with rappelling operations.

c. Rappelling Casualty Rescue. There are three different methods in which to get an injured rappeller to the bottom.

(1) The first method is to allow the belay man to lower the casualty by slowly releasing the tension on the rappel rope.

(2) The second method is used for critical injuries or when the belay man cannot properly control the casualty's descent.

(a) The rescuer will rappel down using another rope. Once the rescuer gets to the casualty, the rescuer will have his belay man brake him off, enabling the rescuer to use both hands.

(b) If necessary, the rescuer will perform the basic first aid needed. Once the casualty is ready to be lowered, the rescuer will call down to both of the belay men to simultaneously lower both men. The rescuer will hold onto the casualty the entire way down so that he doesn't bounce off the rock face.

(3) The third method is called a tandem rappel. Its used to rescue a casualty who has sustained an injury serious enough that he cannot operate the rappel device himself and requires the assistance of a second rappeller. This type of rescue can be conducted from the top of the cliff face or while on rappel, depending on where the injury occurred.

(a) If the injury occurs while on top of the cliff face, the following steps should be taken:

1. Take either a long sling rope or Prusik cord and tie a figure-eight knot offset so that one length of the cord is longer then the other.

2. Tie a figure-eight knot at each end.

3. Take a rappel device and attach it to the offset figure-eight loop with a carabiner and attach it to the rappel rope.

4. Next, attach the casualty's hard point to the short end of the sling rope/Prusik cord and the rescuer to the other end.

5. The rescuer will then utilize a safety Prusik in the same method as the first man down technique, except that the friction knot will be tied below the rappelling device and controlled with the brake hand.

6. The rescuer will then maneuver himself under the casualty to provide assistance and support.

7. At this point you are ready to rappel both rescuer and casualty at the same time.

NOTE: Your rappel device should be far enough in front of you so that the casualty will not be able to reach it.

(b) If the injury occurs during the rappel, the following steps are taken:

1. The rescuer will preset the same system on a separate rappel rope and rappel down to the casualty.

2. The rescuer will then take the short end of the sling rope and attach it to the casualty's hard point.

3. With a knife, the rescuer will cut the casualty's rappel rope away.

4. Rappel down to the bottom of the cliff face.

NOTE: THIS IS THE ONLY TIME THAT A KNIFE WILL BE USED DURING A RAPPEL. EXTREME CAUTION SHOULD BE USED. IT IS NOT TO BE DONE IN TRAINING, BUT ONLY DURING AN ACTUAL RESCUE AND AS A LAST RESORT.

13. **RETRIEVABLE RAPPELS**. Once a unit has rappelled down a vertical obstacle, it may be necessary to retrieve the rope(s). Depending on the height of the obstacle either one or two ropes will used to construct the rappel lane.

a. One-Rope Retrievable Rappel:

(1) Find the middle of the rope and place it directly behind a suitable anchor point.

(2) Join the pigtails of the rope with an overhand knot and deploy the rope down the obstacle.

(3) On one side of the rope in front of the anchor point, tie an over-the-object clove hitch onto a locking carabiner.

(4) On the other side of the rope, tie another over-the-object clove hitch and attach it to the same carabiner.

(5) The first man down will utilize a safety Prusik and untie the overhand knot after he reaches the bottom.

(6) All others conduct proper rappel procedures.

(7) The last man down will disconnect the carabiner from the rope, and ensuring that the middle of the rope is directly behind the anchor point, rappel down the rope.

(8) Pulling on either end of the rope then retrieves the rope.

b. Two-Rope Retrievable Rappel. This type of system is identical to the one rope retrievable rappel except for a few considerations.

(1) The reason for using two ropes vice one rope is that the height of the rappel is greater.

(2) When two ropes are used they should be joined together using a square knot. This knot will be placed directly behind the anchor point.

(3) All else remains the same as the one rope retrievable rappel except:

(a) The last man down will disconnect the carabiner from the rope and will move the joining knot as close to the vertical obstacle's edge as possible. This will prevent the knot from possibly getting caught up while retrieving.

(b) The last man down will then place a carabiner on the rope below the knot. This will enable him to know which line to pull for retrieval.

14. **EQUIPMENT**. Equipment should be worn in accordance with unit SOP. The unit has the responsibility to determine which methods it feels are most beneficial to the mission. Weapons should be worn across shoulder; muzzle down and away from the brake hand with a tight sling securely attached to the weapon. Weapons should also be dummy corded to the individuals.

UNITED STATES MARINE CORPS
Mountain Warfare Training Center
Bridgeport California, 93517-5001

WMC
04/11/02

STUDENT OUTLINE

TOP ROPING

LEARNING OBJECTIVES.

TERMINAL LEARNING OBJECTIVE. In a summer mountainous environment, and given a rock face and a belay man, conduct top roping with cartridge belt and rifle, in accordance with the references. (WMC)

ENABLING LEARNING OBJECTIVE

(1) In a summer mountainous environment, establish a belay point from the top of a cliff for use in a top rope climb, in accordance with the references. (WMC.a)

(2) In a summer mountainous environment, demonstrate the use of climbing commands while conducting a top rope, in accordance with the references. (WMC.b)

OUTLINE

1. **ESTABLISHING A BELAY STANCE FROM THE TOP**.

a. The Belayer will establish a sitting belay stance on the cliff head by:

(1) Constructing a suitable anchor with the standing end of the rope.

(2) With the direction of pull away from the anchor, tie a directional figure 8 loop near enough to the cliff's edge so that the climber can be observed, if possible.

(3) Tying a swami Wrap around oneself, clip a locking carabiner through all the rear wraps and clip a large locking carabiner through all the front wraps.

(4) Secure the rear locking carabiner into the directional figure of 8 loop.

(5) Secure the running end of the rope to the large locking carabiner utilizing a suitable friction belay (i.e., Munter hitch, stitch plate, etc.).

NOTE: Gloves **will not** be worn while belaying a climber during a top rope.

2. **SECURING THE CLIMBER TO THE ROPE**.

a. The climber will tie into the end of the top rope by:

(1) Constructing a Swami wrap around oneself.

(2) Tying a retrace figure 8 loop through all of the Swami's wraps or;

(3) A figure of 8 loop clipped into a locking carabiner secured through all the Swami's wraps.

b. Alternative methods.

(1) A bowline on a coil as discussed in ROPE MANAGEMENT or;

(2) Tie into the hard point of the harness as discussed in SIT HARNESS.

3. **ESTABLISHING A BELAY STANCE FROM THE BOTTOM**.

a. Belaying from the bottom is commonly referred to as a "yo-yo" or "Sling Shot" belay. This system can be constructed for either a direct or indirect belay.

(1) Construct a suitable top anchor utilizing either a pulley or one steel locking or two non-locking carabiners as the attachment point for the anchor.

(2) The rope will travel up from the bottom and pass through the attachment point then return to the bottom.

(3) The climber will be secured to one end of the rope and the Belayer from the other end.

4. **CLIMBER'S RESPONSIBILITIES**.

a. The climber ensures that the Belayer is anchored and on belay, by use of commands or prearranged signals prior to beginning the climb.

b. The climber will not out climb the belayer; this will cause slack in the rope between the belayer and the climber.

c. Avoid placing excess pressure/weight on belay man.

d. Weapons will be worn across shoulder, muzzle down and to the left, with a tight sling securely attached to the weapon.

5. **CLIMBING COMMANDS AND SIGNALS.** In order to conduct top rope operations safely, it is essential that everyone understands the sequence of events. The following voice commands or rope tugs will be utilized:

VOICE COMMANDS	GIVEN BY	MEANING
"Lane #, Up Rope"	Climber	Belayer needs to take in the slack.
"That's Me"	Climber	Excess slack has been taken up between us.
"Lane #, On Belay"	Belayer	I am on belay.
"Lane #, On Climb"	Climber	I am ready to climb.
"Climb Climber"	Belayer	You may begin to climb.

ROPE TUGS	GIVEN BY	MEANING
Three Tugs	Climber	Belayer needs to take in the slack.
Three Tugs	Belayer	I have taken up the slack and you may begin to climb.
Three Tugs	Climber	Climbing.

UNITED STATES MARINE CORPS
Mountain Warfare Training Center
Bridgeport, California 93517-5001

FMST.07.40
03/02/02

STUDENT OUTLINE

MOUNTAIN CASUALTY EVACUATIONS

TERMINAL LEARNING OBJECTIVES. In a summer mountainous environment and given a simulated/actual casualty, evacuate casualties, in accordance with the references. (FMST.07.40)

ENABLING LEARNING OBJECTIVES

(1) With the aid of references, state orally the general considerations for medevac procedures, in accordance with the references. (FMST.07.40a)

(2) In a summer mountainous environment, and with the aid of references, construct an expedient litter, in accordance with the references. (FMST.07.40b)

(3) In a summer mountainous environment, and with the aid of references, secure a simulated/actual casualty in a SKEDCO litter, in accordance with the references. (FMST.07.40c)

(4) In a summer mountainous environment, and with the aid of references, attach a belay line to a SKEDCO litter for movement up/down moderate slopes, in accordance with the references. (FMST.07.40d)

(5) In a summer mountainous environment, with the aid of references, and given a simulated/actual casualty, conduct a medevac using the barrow-boy method, in accordance with the references. (FMST.07.40e)

C8

OUTLINE

1. **GENERAL CONSIDERATIONS** (FMST.07.40a) the general considerations are a set of guidelines that can be used no matter how serious the casualty is. They are remembered by a simple acronym, **APASSNGG.**

   a. Apply Essential First Aid. (i.e. splints, pressure bandage, etc.)

   b. Protect the Patient from the Elements. Provide the casualty with proper insulation from the ground. Ensure that he is warm and dry. If there are any natural hazards (i.e. rock fall, lighting, etc.) either move the casualty as quickly as possible or ensure that he is well protected.

   c. Avoid Unnecessary Handling of Patient.

   d. Select Easiest Route. Send scouts ahead if possible, to break trails.

   e. Set Up Relay Points and Warming Stations. If the route is long and arduous, set up relay points and warming stations with minimum amount of medical personnel at warming stations to:

      (1) Permit emergency treatment. Treat for shock, hemorrhage, or other conditions that may arise.

      (2) Reevaluate the patient constantly. If patient develops increased signs of shock or other symptoms during the evacuation, he may be retained at an emergency station until stable.

   f. Normal Litter Teams Must Be Augmented in Arduous Terrain.

   g. Give Litter Teams Specific Goals. This litter teams job is extremely tiring, both physically and mentally. The litter teams must be given realistic goals to work towards.

   h. Gear. Ensure all of the patient's gear is kept with him throughout the evacuation.

2. **ONE -MAN CARRIES**

   a. Sling Rope Carry. The sling rope carry requires two men and a 15-foot sling rope. One as the bearer, and an assistant to help in securing the casualty to the bearer. Conscious or unconscious casualties my be transported in this manner:

(1) Bearer kneels on all fours and the assistant places casualty face down on bearer's back ensuring the casualty's armpits are even with the bearer's shoulders.

(2) He then finds the middle of the sling rope and places it between the casualty's shoulders and the ends of the sling rope are run under the casualties armpits, crossed and over the bearer's shoulders and under his arms.

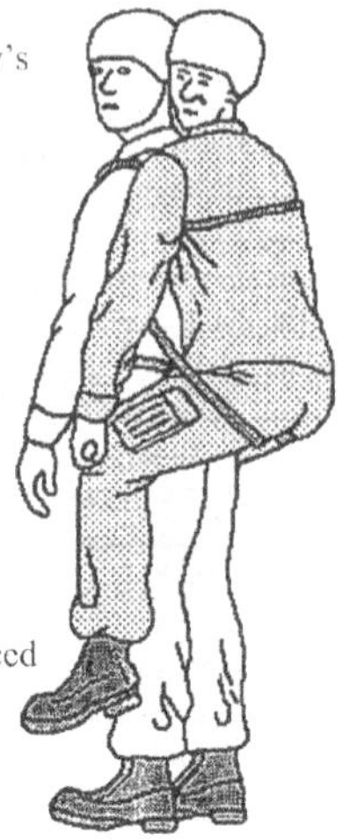

(3) Then the ropes are run between the casualty's legs, around his thighs, and tied with a square knot with two overhands just above the bearer's belt buckle.

(4) Ensure the rope is tight. Padding, when available, should be placed where the rope passes over the bearer's shoulders and under the casualties thighs.

b. The rope coil carry. This requires a bearer and a rope coil. It can be used to carry a conscious or unconscious casualty.

(1) Position the casualty on his back.

(2) Separate the loops of the mountain coil into two approximately equal groups.

(3) Slip ½ of the coil over the casualty's left leg and ½ over his right leg so that the wraps holding the coil are in the casualty's crotch, the loops extending upward the armpits.

(4) The bearer lies on his back between the casualty's leg and slips his arms the loops. He then moves forward until the coil is extended. When using the rope coil the bearer ties the coil to himself vice slipping his arms through the loops.

(5) Grasping the casualty's right or left arm, the bearer rolls over, rolling to the casualty's uninjured side, pulling casualty onto the bearer's back.

(6) Holding the casualty's wrists, the bearer carefully stands, using his legs to lift up and keeping his back as straight as possible.

NOTE: The length of the coils on the rope coil and the height of the bearer are to be considered. If the coils are too long and the bearer happens to be a shorter person, it will require the coils to be uncoiled and shortened. If this is not done, then the casualty will hang too low on the bearer's back and make it a very cumbersome evacuation. A sling rope harness can be used around the victim's back and the bearer's chest, which will free the bearer's hands.

3. **TWO-MAN CARRY**

a. Pole Carry. The pole carry method is a field expedient method and should be considered as a last resort only, when narrow ledges must be traversed; vegetation limits the bearers to a narrow trail. This method is difficult for the bearers and uncomfortable for the casualty. Two bearers, four sling ropes and a 12 foot pole - 3 inches in diameter, are required for this carry.

(1) The casualty is placed on his back in a sleeping bag or wrapped in a poncho or blanket, then placed on an insulated pad.

(2) One sling rope is placed under the casualty below the armpits and tied with a square knot across the casualty's chest.

(3) The second sling rope is tied in the same manner at the casualty's waist.

(4) The third sling rope is placed at the casualty's legs below the knee.

(5) The fourth sling rope is tied around the ankles.

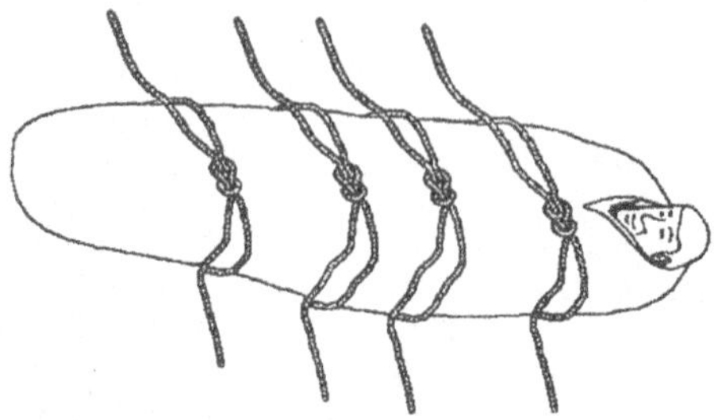

(6) The pole is placed along the casualty's length and secured using square knots with two overhands with the ends of the sling ropes. The square knots should be so tight that the overhands are tied onto themselves

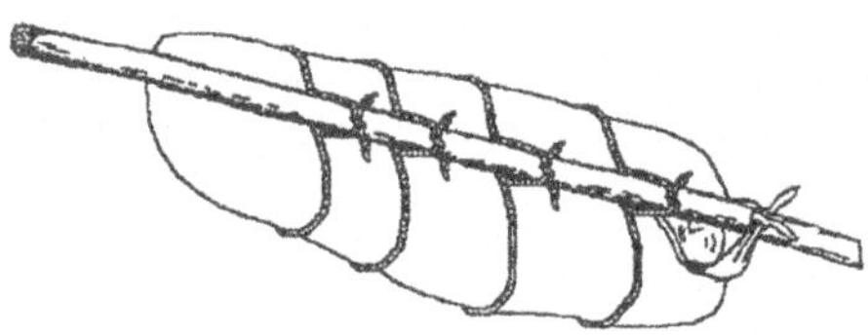

(7) The casualty should hang below the pole, as close to the pole as possible, to prevent swinging during movement.

(8) The casualty's head may be supported using a triangular bandage or a cartridge belt passed around the pole.

(9) For additional support and of movement, two additional bearers may be required, as well as a mountain coil.

(a) Mountain coil is split into two equal coils.

(b) Place knot of mountain coil under casualty's lower back.

(c) Additional bearers slip into each half of the hasty coil one on each side of casualty, aiding in support and movement of the casualty.

4. **EXPEDIENT LITTERS.** (FMST.07.40b) The two types of expedient litters that we will talk about are the alpine basket and the poncho litter.

a. Alpine Basket. To belay the alpine basket, the pre-rigs are attached to the bights formed coming through the loops.

(1) If barrow boy is to be used, the procedure previously discussed will be adhered to.

(2) If barrow boy is not used, then a tag line from the bottom must be implemented to keep the casualty away from the cliff face on the decent.

(3) Construction of the Alpine Basket:

(a) Start by making the same amount of bights as in the rope litter, but start from one end and tie a figure-of-eight loop to run the first bight through.

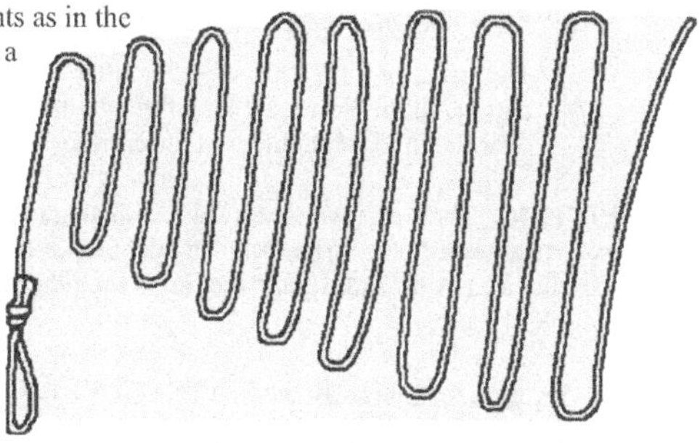

(b) Place padding i.e., isopor mat, on top of the bights and then lay the casualty on the padding and bights.

(c) Start at the casualty's feet and pull the first bight up around the casualty's ankles and through the figure-of-eight loop tied into the starting end of the rope.

(d) Go to the opposite side of the casualty and pull up the second bight and pull it through the loop formed by the bight that was pulled through the figure-of-eight.

(e) Continue until you get to the casualty's armpits, bring the second to the last bight up over the casualty's shoulder and into the bight and then bring the last bight up over the casualty's other shoulder and into the last bight formed.

(f) Secure the end of the bight with an appropriate anchor knot.

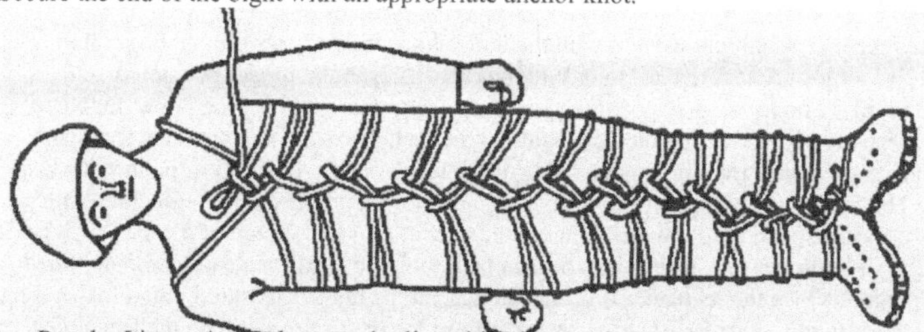

b. Poncho Litter. A poncho, poncho liner, bivy bag or similar piece of material may be used. In addition you will need six individuals with sling ropes.

(1) Lay poncho litter flat on the ground.

(2) Select six rocks about the size of a golf ball. Place one rock in each of the corners and one in the middle on each side in the middle of the litter. The rocks are placed on the under side of the poncho or like material. If a bivy bag is used the casualty should be zipped inside the bivy bag. The rocks should then be arranged in the same manner only on the inside below the zipper.

(3) Tie the sling rope together with an overhand knot. Take the middle of the rope and secure it around the rock with a clove hitch.

(4) An isomat may be laid on the poncho to help make the litter firmer.

(5) The casualty is then placed in the litter. The sling ropes are adjusted by feeding the pigtails of the over hand knot through itself to adjust for length. The loop is then put over the inboard shoulder of the carriers. Insure that the casualty is carried level.

5. **LITTERS.** There are two kinds of litters that are used for casualty evacuations in moderate to vertical terrain. The SKEDCO and the Stokes litter. Each has a variation of procedures for securing a casualty and rigging the litters for either raising or lowering. Let's first look at the SKEDCO litter.

(1) Securing a Casualty to the SKEDCO (FMST.07.40c)

(a) First unroll the litter. The litter must be re-rolled the opposite way to allow the litter to lay flat. Then lay the litter next to the casualty.

(b) If the casualty has any possibility of a spinal injury the Oregon Spine Splint must be used. Secure the splint to the casualty by use of the color-coded buckles. Experienced medical personal are recommended if spinal immobilization is necessary.

(c) Once the casualty is on the SKEDCO use the four body straps to secure the casualty to the litter. Unless injuries prevent, the casualty's arms should be at his sides to prevent further injuries to himself or the rescuers.

(d) Once the casualty is secured with the body straps the feet straps must be secured. The feet straps are secured last to ensure the casualty is in the proper position on the SKED. The feet can be positioned in three ways. The first position is feet together with the straps running on the outside of the feet. The second position is feet apart with the straps running on the inside of the feet. The last position is the feet stacked. This is the most uncomfortable position and not recommended for casualties with possible spinal cord injuries. Placing the heel of one foot on top of the toes of the other forms this position. This position will only be used for a casualty in confined spaces. Once the feet are positioned the feet straps must be secured. Start by bending the feet end of the SKED to form a plat form for the feet. Then loop the feet straps through the second grommets on each side.

(e) The last thing to do is to form the head end to protect the casualty's head. If possible the casualty should wear a helmet. Form the head end tying the pull strap up and secure it to the first body straps.

(2) Medevacing a Casualty. There are many ways to move a casualty once in the SKEDCO. However the medevac team must keep the general considerations in mind. The two methods that we will talk about next are the simplest and require the least amount of additional rigging

(a) The first way is to drag the casualty by the drag strap located at the head end of the SKED. You can also use the SKEDCO's carrying bag as a harness in conjunction with the pull strap and towing harness. If additional people are required, cordage can be added to the pull strap or the front carrying handles.

(b) The second way is to carry the casualty using the carrying handles. By using the set of four removable webbing the litter team can be augmented. To do this each pieces of webbing is tied to make an endless loop. Then pass a bight of the loop through one of the grommets to create additional handles.

(3) Rigging the SKED for Vertical Terrain and Helicopter Lift. In this type of terrain special requirements must be taken to raise or lower a casualty in the SKEDCO.

(a) Rigging the SKED for a Vertical Employment. A vertical raise or lower is used when moving a casualty on steep earth to avoid any further injury to the casualty. On vertical terrain the vertical raise or lower can be used if the terrain is not uniformed or there is a chance of rock fall. Ensure that the casualty's head is always above his feet.

(1) Identify the 30-foot piece of cordage that comes with the SKEDCO. Then tie a figure eight in the middle of the rope. If the rope is worn or missing, the same process can be done with two sling ropes

(2) Next pass each end of the rope through the grommets at the head of the SKEDCO. Leaving approximately 1-2 feet of rope between the stretcher and the knot.

(3) Continue to feed each end through the grommets and the carrying handles towards the foot end of the SKEDCO. Pass the ends of the rope through the last grommets at the foot end and secure the two ends with a square knot without over hands.

(4) Bring the pigtails up and over the casualty's feet and pass the ends through the
a carrying handles towards the middle of the casualty. Then tie a square knot
b with two over hands.

c

(4) Rigging the SKEDCO for a Horizontal Employment. A horizontal raise or lower is preferred on uniformed vertical terrain. The horizontal employment allows the rescuer to assist the casualty on either a raise or lower. It also allows the rescuer the ability to monitor the casualty's condition and can easily treat the casualty if the need arises.

(a) Identify the two 4 inch nylon straps. They should be two lengths; one four inches shorter than the other. The shorter strap should be marked HEAD STRAPS.

(b) Insert one end of the head strap into a slot near the head of the litter. Then wrap the rest of the straps under the SKEDCO and pass the other end through the opposing slot. Do the same at the foot end of the SKEDCO with the other strap. Ensure that the strap runs smoothly under the SKEDCO.

(c) Connect the strap ends with the large locking caribiner that comes with the SKEDCO. If the caribiner is worn or missing, opposing Stubai 85 locking carabiners will suffice.

6. **STOKES LITTER**. The stokes litter is a litter that is constructed of metal tubing with a plastic covering. The litter is formed in a rectangular basket shape with mesh attached to the frame. Using the stokes for an evacuation (as with any evacuation) it should be padded for the casualty.

a. Securing the casualty to the stokes litter. In the event that the "seat belts" are missing from the stokes litter, sling ropes can be used to the lash the casualty. The steps involved are:

(1) Tie two sling ropes together using square knots and two overhands.

(2) Tie a stirrup hitch around ankles and feet, feed the two pigtails through the right angles of the stokes. Do not cross the ropes at the ankles.

(3) Lace the sling rope towards the casualty's head by passing the rope through the right angles (not over the top of the rails of the stokes).

(4) Secure the ends of the sling ropes by tying a clove hitch with two half hitches on the thick vertical bar located by the victim's shoulder.

7. **YOSEMITE PRE-RIG.** The title of this method, pre-rig implies the meaning ready to use, but in this case the rig must be constructed. Its purpose is for attaching a belay line to a litter and to make the litter easily adjusted. This is the one method that is used to secure a litter to a belay line.

a. Construction: This method normally requires four sling ropes. The steps are: (1) using one sling rope, tie a figure-of-eight loop with one tail.

(1) Take the remaining tail and run it through the window of the stokes litter or in the stirrups of the collapsible litter.

(2) Tie a Kragur knot onto the same sling rope.

(3) Repeat steps (1), (2), and (3) with the three remaining sling ropes.

(4) Suspend the litter to ensure that the comers are balanced.

8. **ASCENTS OR DESCENT OVER STEEP TO MODERATE SLOPES.** When the litter team is ascending or descending a slope they must consider the potential for further injury to the casualty or to themselves. If the risk of injury is high a belay line may be used to prevent injury to the casualty and the rescuers.

a. Preparing Casualty for Ascents or Decent over Steep to Moderate Terrain. This procedure will be depending on several things. Initially, site selection should contain the following features.

(1) Suitable anchor points.
(2) Clearance for casualty along the route
(3) Loading and unloading points.

b. Additional considerations.

(1) The casualty will always be rigged for vertical employment when on steep to moderate terrain.
(2) The smoothest possible route must be selected.
(3) Ensure that the casualty's head is above his feet.

c. Rescuers positions. There are two methods that can be used for the rescuers for moving a casualty in steep to moderate terrain.

(1) Two to four men will position themselves on each side of the litter. They can then carry the litter by the carrying handles. In steep terrain a second belay line may be used to assist the rescuers. We will discuses the belay line later in the chapter.

(2) The Caterpillar method will require as many personnel as possible. The personnel will split in half and position themselves on each side of the litter forming a tunnel. As the litter is raised or lowered each member will hand the litter to the next member in the tunnel. As the litter passes each person in the tunnel he will peel off and assume the lead either at the top or bottom of the tunnel. This will continue until the litter reaches its desired destination.

d. Belay Line. (FMST.07.40d) For belaying of a casualty, one rope will be used and from the top using one of two methods depending on the application.

(1) Body Belay. This method should only be used over moderate terrain. The belay man will establish a sitting position behind a suitable anchor (i.e., rock, tree, etc.) and pass the standing end of the rope behind his back. The running end of the rope will feed out from the belay man's right side. A figure of eight loop is tied to the end of the running end of the rope. It is then attached to the litter's figure eight loop with a locking carabiner. The belay man will then remove all of the slack between himself and the litter. The standing end of the rope should be stacked on the belay man's left side and run through his left side. As the casualty is lowered, the belay man will feed the rope from behind his back allowing it to run through his right hand. If the belay man needs to stop the casualty, he will clench the rope in his left hand, and

bring the rope to the center of his chest.

(2) Direct Belay. This method is the safest for either raising or lowering a casualty in either moderate to steep terrain.

(a) To Lower a Casualty. First a swami wrap will be tied around a suitable anchor point. Two locking carabiners will be clipped into all of the wraps of the swami wraps, gates up. A figure of eight loop is tied into the end of the static rope and attached to the litter with a locking carabiner. After all the slack has been taken up between the litter and the anchor, the rope must be tied through an appropriate belay device. The belay device is attached to the anchor through one of the two locking carabiners on the anchor. A safety (French) prussic will be tied to the running end of the rope and clipped into the second locking carabiner on the anchor. While the casualty is being lowered, one person will control the rope running through the belay device. The safety prussic will be controlled by a second person. Should the primary belay man lose control, the person operating the safety prussic simply lets go and the prussic will bind onto the rope, stopping the casualty.

(b) To Raise a Casualty. The anchors will be established in the same manner as discussed in lowering the casualty with one minor change. The one change is that instead of running the rope through a belay device, the rope will only run through a locking carabiner. The load will be raised by the use of a mule team. The mule must consist of as many people as possible. The mule team will raise the load in as straight a line from the anchor as possible. If the space does not permit a ninety-degree angle away from the anchor is also an option. The mule team will walk backward until the last man reaches his limit of advance. Once he reaches that limit he will peel off the end and return to the front of the mule team. This process is continued until the casualty reaches the top. If the load becomes unmanageable, the safety prussic will be allowed to bind on the rope while the mule team repositions themselves. If the person operating the safety prussic can not see the casualty a Point NCO will be in charge of communicating with the mule team.

(c) To Belay the Rescuers. If the route is too steep or the footing is poor the rescuers may need some assistance either on the raising or lowering of a casualty. If this is the case a separate belay line will be established for the rescuers. The anchor and the belay line are established in the same manner. The same anchor can be used if it is suitable for the load. The rescuers will then tie either around the chest bowlines or swami wraps. A figure eight loop will be tied into the end of the static rope and connected to the bottom rescuer with a locking carabiner. The other rescuers will connect themselves to the same rope with middle of the line prussic. They will be connected in this manner so that they can adjust their position to the casualty.

9. **BARROW BOY.** (FMST.07.40e) A barrow boy is no more than an assistant to the litter on vertical to near vertical cliff faces. The barrow boy can be used for either the Stokes or the

SKED litters. For this situation the Stokes litter should only be used in the horizontal position. However the SKED can be employed in either the horizontal or the vertical positions.

NOTE: Static load, lowered slowly, little friction, so don't need 2 ropes.

a. Rigging the Barrow Boy.

(1) First the rescuer must ensure that a suitable anchor has been established, a proper belay has been constructed, and that a safety prussic has been constructed.
(2) Then the rescuer must ensure that if an A-frame is used that it has been constructed and anchored properly.

(3) Next the rescuer will tie a rappel seat on. (a sit harness can also be used) Then he will ensure that an around the body bowline is tied onto the casualty. A figure eight will be tied on to the end to act as the casualty's safety.

(4) Then after running the running end of the rope though the carabiners or pulley of the A-frame he will tie a figure eight loop at the end of the static rope. He will then attach the figure eight loop to his hard point with a locking carabiner. Then he will tie a middle of the line prussic above the figure eight and attach it to the same locking carabiner in his hard point. This is called the adjustment prussic. It is used to adjust the position of the Barrow Boy in relation to the litter. Next he will take six to eight feet of slack from the end of the line figure eight and tie a directional figure of eight with direction of pull down. (Note: the prussic should be between the end of the line figure of eight and the directional figure of eight.) The directional figure of eight is the attaching point for the litter and the casualty's safety.

(5) Once the rescuer and the litter are secured, the belay man must take all the slack out of the system. The rescuers will the maneuver the litter through the apex of the A-frame with the help of the point NCO.

(6) Once onto the cliff face the rescuer will then position him self with his adjustment prussic so that he can be of most assistance to the litter on the raise or lower. The rescuer will pull the litter out away from the cliff face so that the casualty rides smoothly up or down the cliff face.

(7) The Point NCO will be in charge of the belay men or the mule team. He will also communicate with the rescuer about the rate of speed, if the rescuer needs to be stopped along the route, and when he reaches the top or bottom of the cliff.

10. **TANDEM LOWERING.** The tandem lowering system can be used for the walking wounded, POW's, or more serious casualties when situation would not permit using the barrow boy.

a. The assistant to the casualty should first tie a rappel seat on himself and then assist the casualty with his.

b. The assistant will take only one belay line and tie a end of the line figure 8 loop, and clip this into his rappel seat.

c. A directional figure 8 will be tied approximately 12 inches up the rope from the figure 8 loop (with the loop pointed down) and this will be clipped to the casualty's rappel seat.

d. If needed, safety prussic cords should be tied above the casualty's directional figure 8. These safety cords should be attached in the same manner as the first man down a rappel.

e. The casualty and the assistant will lower as one, with the assistant helping on the way down the cliff.

11. **OTHER CONSIDERATIONS.** All of the techniques we have discussed for the evacuation of a casualty from top to bottom can also be used on a suspension traverse or rope bridge, with a slight variation in the belay line. Two belay lines may be used for rope bridges and the suspension traverse, if they are available. No matter what type of litter is used, the individuals involved in the evacuation must ensure that the head is always uphill or not lower than the feet.

UNITED STATES MARINE CORPS
Mountain Warfare Training Center
Bridgeport, California 93517-5001

FMST.0
7.34
04/12/02

STUDENT HANDOUT

ONE-ROPE BRIDGE

LEARNING OBJECTIVES

TERMINAL LEARNING OBJECTIVES. In a summer mountainous environment, conduct bridging, in accordance with the references. (FMST.07.34)

ENABLING LEARNING OBJECTIVES

(1) Without the aid of references, list in writing the criteria for site selection for a one-rope bridge, in accordance with the reference. (FMST.07.34a)

(2) In a summer mountainous environment with an assault load , cross a one-rope bridge using the rappel seat method, in accordance with the reference. (FMST.07.34b)

(3) Without the aid of references, list in writing the rescue techniques, in accordance with the reference. (FMST.07.34c)

(4) In a summer mountainous environment transport a SKED stretcher across a one rope bridge, in accordance with the reference. (FMST.07.34d)

BODY

1. **SITE SELECTION**

a. The two criteria for site selection for a one-rope bridge are: (FMST.07.34a)

1) There must be suitable anchors on both sides of the stream.

2) The anchors must offer good loading and unloading platforms.

b. Other considerations involved are:

1) The site chosen for the initial crossing does not have to be at the location for the construction of the bridge, just as long as the rope can be taken to the selected site the

C9

crossing.

2) The site chosen for the lead swimmer to cross should be as free as possible from obstacles in the water, such as large boulders, stumps or logs as discussed in *STREAM CROSSING*.

3) The anchors must be close enough for the 150 foot coil to reach both near side and far side anchors. Keep in mind that it will take approximately 1/3 of the 150 foot rope for tightening and anchoring of the bridge.

2. **DISTANCE ESTIMATION**. The follow methods can be used to determine the distance between anchor points:

a. Azimuth Method. Shoot an azimuth to a point on the far side of the intended obstacle to cross. Then move LEFT or RIGHT (perpendicular to the azimuth) until you get a 15 degree offset of your previous azimuth. Next, measure the distance that you paced in feet from your first azimuth to your second azimuth, and multiply that distance by three. This total will give you the approximate distance across the obstacle in feet.

b. Unit Average Method. Get three Marines to judge (best guess) the distance across the intended obstacle. Add up the total accumulated distance, then divide by three. This will give an estimation of the distance across the obstacle.

Example:	Marine 1	240 ft
	Marine 2	230 ft
	Marine 3	250 ft
		720 ft

Divided by 3 = 240 ft

3. **ORGANIZATION**. The organization of construction is broken down into four groups. They are as follow:

a. The Bridging Team - consisting of the bridge NCO and another Marine.

b. The Safety Line Team - consisting of the lead swimmer and his belay man.

c. The Mule Team - consisting 3 Marines for a 9:1, and 6 Marines for a 3:1.

d. The Security Team -- consisting of the rest of the party.

4. **CONSTRUCTION**. This bridge is ideal for squad and platoon sized units, for its quick construction and minimal amount of equipment required.

a. Once the site for the bridge has been designated by the bridge NCO, the Safety Line Team will move up stream from this site to enter the water. Their first step is to flake out the rope. Once this is accomplished, the lead swimmer's will take a bight of the rope and tie a figure 8 loop at the end that will be going to the far side, ensuring the knot has a 18"-24" loop. The lead swimmer's upstream arm will go in it, with the belay man tending his rope the lead swimmer will cross the never using the flow to assist him. (Ferry angle as taught in STREAM CROSSING)

b. At the same time as the lead swimmer is crossing, the bridging team will be preparing the near side anchor. The bridging team will tie a swammi wrap around the anchor, using a sling rope or practice coil, and ensuring the square knot is behind or on the side of the anchor. Once the anchor is secure, flake out the bridging line in order to send it across.

c. Once the lead swimmer is across he will move to the site where the far side anchor will be established and secure his rope. The belay man will then move down to the near side anchor, and as close to the waters edge as possible. He will attach the bridging line to the safety by tying a middle of the rope in a figure 8 knot to the safety line and an end of the rope figure 8 knot to the bridging line connecting the two ropes with an 85 carabiner. The lead swimmer will then pull the rope across. The belay man with the help of another Marine will belay the bridging line across keeping tension on the rope to keep the rope out of the water.

d. After the bridging line is across the lead swimmer will detach the bridging line from the safety line and secure it to the far side anchor using a tree wrap, wrapping from right to left, ensuring there are 3-5 wraps.

e. Once the bridging line is secure. The lead swimmer and the belay man will secure the safety line on both banks, ensuring the rope is creating a ferry angle (minimum 45 degrees). This will be the safety line for the Marines crossing the bridge.

NOTE: If a fall from a rope bridge would result in death or serious injury, then utilize a safety line that will be anchored parallel with the bridge and hand tightened. The Marine crossing will clip into both ropes prior crossing.

f. On the near side bank:

1. The bridging team will need the following gear to build the mechanical advantage:

    (a) Four steel locking carabiners & 3 -Three foot prusiks

2. Once the bridging line is secured on the far side, the bridging team will take one 85 and clip it into the near side anchor (swammi wrap). Then they will take the bridging line and clip into that 85, then using one three foot long prusik (16" loop) the bridging line will be secured to the 85 by tying a French Prusik, which will act as a braking knot. The bridging team will now pull the rope taut, and begin construction of the mechanical advantage.

3. The first step is to take one three foot prusik (16" loop) and tie a French prusik on the bridging line as far away from the anchor as possible. Then take one 85 and clip it to the loops of the French prusik.

4. Taking a bight from the running end of the rope clip it into the 85 hanging from the tails of the French prusik.

5. Then bring a bight back from the clipped in rope to the anchor, and taking one 85 clip it into the 85 created in step (2), take the bight of rope and clip into this 85.

6. Take one three foot prusik and tie a French prusik onto the bottom end of the bight clipped into the 85 created in step (3), take one 85 and clip it into the tail loops of the French prusik.

7. From the anchor, take a bight of rope from the 85 created in step (6) and clip it into the 85 created in step (6).

8. You have made approximately a 9:1 mechanical advantage.

NOTE: If the manpower is available the squad/platoon has the option of building a 3:1 mechanical advantage vise a 9: 1, the mule team will just have to add another body having no more then 3 bodies on a 9:1 and 6 bodies on a 3:1

g. Tensioning of the Bridge

1. Now that the bridge is built, the bridge NCO can call up the mule team to tension the bridge.

2. This procedure is facilitated by the mechanical advantage system that was just put into place. The braking knot (French prusik) is used to hold tension on the bridging line while the bridge is being tensioned.

3. The mule team will begin to pull on the running end of the rope coming out of the mechanical advantage. They will pull the rope straight back, trying to keep it in line with the bridge as best as possible.

4. The mule team will tension the rope as much as possible, the bridge NCO will monitor the system. Once the mule team cannot pull anymore tension, the bridge NCO will then have them hold the rope in place, and will reset the braking knot. Once the braking knot is set the bridge NCO will ask for slack from the mule team and cycle the system out. This process will continue until the bridge is tight.

5. On the last cycle the bridge NCO will set the brake knot and with the help of another Marine they will make a bight out of the running end and bring it around the tree while keeping tension.

6. Take a bight and make a complete round turn on the body of the 85 created in step (2). Last, the bridge NCO will tie two half hitches encompassing all the ropes just behind the anchor.

7. At this stage the bridge is tight and secured, the bridge NCO will now call up the remainder of the squad/platoon to cross. The bridge NCO will monitor the crossing of all Marines, and will be the last one to cross.

TRANSITION: Now that we have discussed the construction of the bridge, are there any questions over the material? Next we will discuss the crossing of the bridge.

5. **<u>CROSSING</u>.** (FMST.07.34b) The method used to cross is known as a horizontal traverse. This traverse can be accomplished in the following ways.

a. <u>Rappel Seat Method</u>

1.) The Marine ties himself into a rappel seat and inserts a large steel locking carabiner with the gate facing down and away.

2.) The Marine faces the bridge with his right hand towards the near anchor to snap into the bridge, once the 85 is locked onto the bridge rope, a helper flips the 85 over so that the locking nut screws down. A carrier rope with a girth hitch may be added to assist the Marine in mounting the rope.

3.) The Marine hangs below the bridge from his rappel seat with his head pointing in the direction of the far anchor and allows his legs to hang free.

4) Pulling with his arms makes progress.

5) This method is the safest and therefore the preferred method.

6) If the Marine must take a pack across, he may wear it with the waistband secured: However, the preferred method is to have another carabiner (aluminum non-locking is sufficient) attached to the pack frame at the top and attach this to the bridge behind the Marine putting his legs through the shoulder straps and pulling the pack across with him.

7) One man at a time will cross, although one can load and another can unload concurrently.

8) Weapons will be worn across the shoulder; muzzle down with a tight sling securely attached to the weapon.

b. <u>Pulley Method.</u> This method is used when the one rope bridge is long, uphill, or speed is vital and the Marines crossing it have a lot of heavy personal equipment; M24OG's, radios, etc.

1) Equipment required. One pulley, four steel locking carabiners, and a hauling line twice the length of the obstacle.

2) Setting up the system. Construction of the first suspension point is done by attaching a pulley to the one rope bridge, one 85 carabiner is attached to the pulley, gate down, and a second 85 is attached to the first 85 with the gate facing the near side bank. Into this second 85, attach the hauling rope with a figure-8 loop. The figure-8 is placed halfway along the hauling rope and attached to the 85, which is then locked. Four feet down the line another figure-8 on a bight is placed into the second suspension point.

3) Connection. The Marine clips his 85 into the lower 85, his equipment is clipped into the second 85. The mule team starts to pull.

c. Helmets will be worn for all methods of crossing the bridge. Gloves are optional.

NOTE: There are additional methods of crossing a one rope bridge such as the Commando crawl, Monkey crawl and hand over hand techniques, however, they are not taught at MWTC due to safety reasons.

6. **<u>RESCUE TECHNIQUES</u>** If an individual is unable to complete the crossing on the one rope bridge using the rappel seat method of crossing (i.e. from a fall or exhaustion), a rescue will have to be made in the following manner: (FMST.07.34c)

a. Reach. First try to reach the victim by using an object such as a pole, your hand, if the victim is close enough, etc ...

b. Throw. If reaching the victim is impossible, try throwing a rope to the victim and have him attach the rope to himself, preferably his seat, and pull him back to the desired side of the installation.

c. Tow. If the victim is unable to catch or reach a rope being thrown to him or the victim is unconscious, tie a figure-of-eight loop into the middle of the safety line and then connect the rope to the rope bridge with a steel locking carabiner. The mule team will pull the carabiner up against the victim's seat and begin towing the victim to the desired side of the installation. The rope used to tow the victim should be twice the length of the span of the rope bridge (If necessary, ropes can be tied together to accommodate the span). This will allow towing from either side without having to throw or have a Marine carry the safety line back across the bridge each time a towing rescue is performed.

d. Go. If all else fails, the last option will be to go after the victim. The rescuer will move out onto the rope bridge with a safety line attached to himself at approximately eight feet from the end of the safety line. In the end of the safety line tie a figure-of-eight loop and insert a steel locking carabiner in the end of it. Once the rescuer has made contact with the victim. He will attach the steel locking carabiner to either the victim's seat (preferably the victim's carabiner that is attached to the rope bridge), or to the bridge itself. To ensuring that the carabiner is placed so that it pulls against the victim's carabiner. The mule team now starts to pull both the victim and the rescuer to the desired side of the installation.

e. Cut. If while crossing the rope over water the individual goes underwater and no other rescue technique can be employed, the rope will be cut.

7. RETRIEVING THE ONE ROPE BRIDGE

a. Before the bridge NCO sends the last Marine to cross, they must make the bridge retrievable.

b. The first step is to break down the mechanical advantage, ensuring the braking knot is set before doing so. The system will be broken down until the bridging line is attached only to the brake knot.

c. They will now take the rope around to the back of the anchor, and tie a slip figure 8, with the loop of the slip 8, take it and attach it to the bridging line just in back of the braking knot with a 85.

d. Now pull the running end coming out of the slip 8 until it tightens, and bites on the rope. Once the rope bites, tie a thumb knot behind the slip 8 as close to the slip figure 8 as possible.

e. The bridge NCO, with the help of another Marine, will now untie the swammi wrap, until all of the tension is on the bridging line.

f. The safety line will now be brought up, and attached to the end of the bridging line with an overhand knot.

g. The bridge NCO and the other. Marine will now cross the bridge.

NOTE: It is possible that the bridge may lose tension after breaking the bridge down. The packs of the last two Marines across can be sent over with other Marines, prior to the slip figure being tied.

h. After both Marines have crossed, the safely line will be brought up from down stream.

i. The tree wrap will now be taken off. Once the tree wrap is off, all knots and carabiners will be taken out.

j. The safety line will now be pulled, the bridge is retrieved.

UNITED STATES MARINE CORPS
Mountain Warfare Training Center
Bridgeport, California 93517-5001

WMC
04/12/02

STUDENT OUTLINE

SUMMER MOUNTAIN WARFIGHTING LOAD REQUIREMENTS

LESSON PURPOSE. The purpose of this period of instruction is to introduce the student to the different levels of gear requirements in accordance with MWTC's uniform codes. This lesson relates to mountain walking and tactical evolutions

OUTLINE

1. **BASIC UNIFORM REQUIREMENTS.**

a. Marines operating in mountainous terrain will generally wear the standard issued utility uniform. However, due to the wide range of temperatures, and sudden changes in weather in the mountains, Marines will find the need to be continually adding and removing layers. The amount of clothing worn may vary depending upon the severity of the weather, the activity level of the Marine, and the individual metabolism of the Marine. Every person in that unit should still maintain the same outer camouflage layer.

b. As part of the basic uniform, each man will be required to have in his possession, at all times, seven required pocket items. These seven items should be carried in the pockets of your utility uniform:

(1) Pocketknife

(2) Whistle

(3) Pressure Bandage

(4) Chapstick and sunscreen

(5) Sunglasses

C10

(6) Survival Kit and rations (fire starting material, signaling material, food gathering material, water procuring material, 1st aid, shelter material, and some high energy, lightweight snacks.)

(7) Notebook with pen/pencil

c. Some additional items that should be carried in your pockets at all times:

(1) Contact gloves

(2) Paracord (10 meters)

(3) Flashlight with tactical lens and spare batteries

(4) Chemlights

2. **ASSAULT LOAD.**

a. The Assault Load is equipment in addition to the basic uniform requirements. It is carried in the load-bearing vest (LBV), butt pack and the pack system. This is the equipment carried for short duration missions such as security patrols or during the assault. It is also carried at all times when away from the bivouac site.

(1) An extra insulating layer (Field jacket liner, wooly pully, etc.)

(2) Protective layer (ECWCS parka and trousers)

(3) LBV with 2 quarts of water and first aid kit

(4) Helmet

(5) Rations for the time away from your bivouac site

(6) Cold weather hat or balaclava

(7) Individual mountaineering gear

a. Sling rope

b. Carabiners

c. Rappelling Gloves

(8) Specialized mountaineering equipment

(9) Mission essential gear

3. **COMBAT LOAD**. The Combat Load is the equipment carried for longer duration missions.

The following items are in addition to the items already being carried in the Assault Load:

a. Sleeping bag with bivy bag

b. Isopor mat

c. Stove

d. Fuel bottle

e. Poncho (for expedient shelters or medevac purposes)

4. **EXISTENCE LOAD**. The Existence Load is to be packed for longer duration combat missions, where it is necessary to replace worn out items. These items could be uniforms, boots, personal hygiene gear, etc. The Existence Load should be brought to the forward elements once the situation allows.

5. **PACKING CONSIDERATIONS.** Because most Marines are familiar with how to pack a pack, these are general guidelines only.

a. When not wearing your protective layer or insulating layers, keep it handy at the top of the pack. When taking a break during a movement, you will be able to quickly don a layer to prevent getting chilled.

b. Keep your stove and fuel bottle in the outside pockets of your pack. They may leak and soak your equipment with fuel if stored inside your pack.

c. Keep the climbing rope in the rope bag. This will reduce the rope's exposure to petroleum products, ultra-violet light, cuts, and abrasions.

d. Ice ax is placed through the ice ax loops with spike up, adze outboard, and secured in place.

e. Crampons are attached to the outside of the pack with the tips covered and facing outboard.

Mountain Operations

Department of Defense

Pentagon
Publishing

Preface

FM 3-97.6 describes the tactics, techniques, and procedures that the United States (US) Army uses to fight in mountainous regions. It is directly linked to doctrinal principles found in FM 3-0 and FM 3-100.40 and should be used in conjunction with them. It provides key information and considerations for commanders and staffs regarding how mountains affect personnel, equipment, and operations, it also assists them in planning, preparing, and executing operations, battles, and engagements in a mountainous environment.

Army units do not routinely train for operations in a mountainous environment. Therefore, commanders and trainers at all levels should use this manual in conjunction with TC 90-6-1. Army Training and Evaluation Program (ARTEP) mission training plans, and the training principles in FM 7-0 and FM 7-10 when preparing to conduct operations in mountainous terrain.

The proponent of this publication is Headquarters TRADOC. Send comments and recommendations on DA Form 2028 directly to Commander, US Army Combined Arms Center and Fort Leavenworth, ATTN: ATZL-SWW. Fort Leavenworth, Kansas 66027-6900.

Unless this publication states otherwise, masculine nouns and pronouns do not refer exclusively to men.

***FM 3-97.6**

Field Manual
No. 3-97.6

Headquarters
Department of the Army
Washington, DC, 28 November 2000

Mountain Operations

Contents

Page

DISTRIBUTION RESTRICTION: Approved for public release; distribution unlimited

This publication supersedes FM 90-6, 30 June 1980.

Introduction

The US Army has a global area of responsibility and deploys to accomplish missions in both violent and nonviolent environments. The contemporary strategic environment and the scope of US commitment dictate that the US Army be prepared for a wide range of contingencies anywhere in the world, from the deserts of southwest Asia and the jungles of South America and southeast Asia to the Korean Peninsula and central and northern Europe. The multiplicity of possible missions makes the likelihood of US involvement in mountain operations extremely high. With approximately 38 percent of the world's landmass classified as mountains, the Army must be prepared to deter conflict, resist coercion, and defeat aggression in mountains as in other areas.

Throughout the course of history, armies have been significantly affected by the requirement to fight in mountains. During the 1982 Falkland Islands (Malvinas) War, the first British soldier to set foot on enemy-held territory on the island of South Georgia did so on a glacier. A 3.000-meter (10,000-foot) peak crowns the island, and great glaciers descend from the mountain spine. In southwest Asia, the borders of Iraq, Iran, and Turkey come together in mountainous terrain with elevations of up to 3,000 meters (10,000 feet).

Mountainous terrain influenced the outcome of many battles during the Iran-Iraq war of the 1980s. In the mountains of Kurdistan, small Kurdish formations took advantage of the terrain in an attempt to survive the Iraqi Army's attempt to eliminate them. In the wake of the successful United Nations (UN) coalition effort against Iraq, US forces provided humanitarian assistance to Kurdish people suffering from the effects of the harsh mountain climate.

Major mountain ranges, which are found in desert regions, jungles, and cold climate zones, present many challenges to military operations. Mountain operations may require special equipment, special training, and acclimatization. Historically, the focus of mountain operations has been to control the heights or passes. Changes in weaponry, equipment, and technology have not significantly shifted this focus. Commanders should understand a broad range of different requirements imposed by mountain terrain, including two key characteristics addressed in this manual: (1) the significant impact of severe environmental conditions on the capabilities of units and their equipment, and (2) the extreme difficulty of ground mobility in mountainous terrain.

Field Manual
No. 3-97.6

***FM 3-97.6**

Headquarters
Department of the Army
Washington, DC, 28 November 2000

Mountain Operations

Contents

Page

DISTRIBUTION RESTRICTION: Approved for public release; distribution unlimited

This publication supersedes FM 90-6, 30 June 1980.

Chapter 1

Intelligence

Before they can understand how to fight in mountainous environment, commanders must analyze the area of operations (AO), understand its distinct characteristics, and understand how these characteristics affect personnel and equipment. This chapter provides detailed information on terrain and weather necessary to conduct a thorough intelligence preparation of the battlefield (IPB). however, the IPB process remains unaffected by mountains (see FM 2-01.3 for detailed information on how to conduct IPB).

SECTION I - THE PHYSICAL ENVIRONMENT

CONTENTS

1-1. The requirement to conduct military operations in mountainous regions presents commanders with challenges distinct from those encountered in less rugged environments and demands increased perseverance, strength. will, and courage. Terrain characterized by steep slopes, great variations in local relief, natural obstacles, and lack of accessible routes restricts mobility, drastically increases movement times, limits the effectiveness of some weapons, and complicates supply operations. The weather, variable with the season and time of day, combined with the terrain, can greatly affect mobility and tactical operations. Even under non-violent conditions, operations in a mountainous environment may pose significant risks and dangers.

TERRAIN

1-2. Mountains may rise abruptly from the plains to form a giant barrier or ascend gradually as a series of parallel ridges extending unbroken for great distances. They may consist of varying combinations of isolated peaks, rounded crests, eroded ridges, high plains cut by valleys, gorges, and deep ravines. Some mountains, such as those found in desert regions, are dry and barren, with temperatures ranging from extreme heal in the summer to extreme cold in thewinter. In tropical regions, tush jungleswith heavy seasonal rains and little temperature variation frequently cover mountains. High, rocky crags with glaciated peaks and year-round snow cover exist in mountain ranges at most latitudes along the western portion of the Americas and in Asia. No matter what form mountains take, their common denominator is rugged terrain.

MOUNTAINOUS REGIONS

1-3. The principal mountain ranges of the world lie along the broad belts shown in Figure 1-1. Called Cordillera, after the Spanish word for rope, they encircle the Pacific basin and then lead westward across Eurasia into North Africa. Secondary, though less rugged, chains of mountains lie along the Atlantic margins of America and Europe.

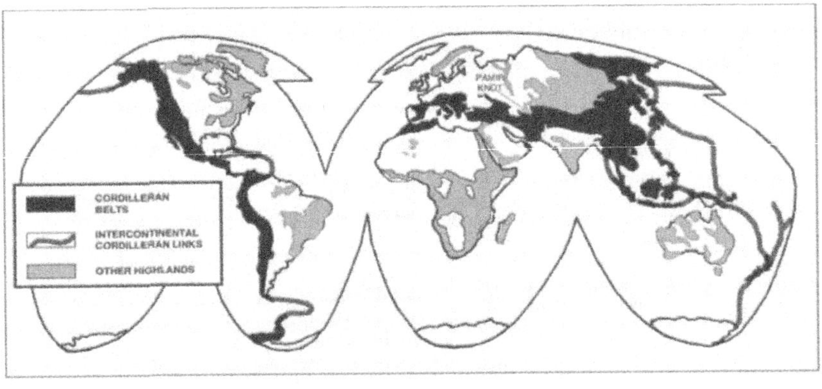

Figure 1-1. Mountain Regions of the World

1-4. A broad mountainous region approximateiy 1,600 kilometers wide dominates northwestern North America. It occupies much of Alaska, more than a quarter of Canada and the US, and all but a small portion of Mexico and Central America The Rocky Mountain Range includes extensive high plains and basins. Numerous peaks in this belt rise above 3,000 meters (10,000 feet). Its climate varies from arctic cold to tropical heat, with the full range of seasonal and local extremes.

1-5, Farther south, the Andes stretch as a continuous narrow band along the western region of South America, Narrower than its counterpart in the north, this range is less than 800 kilometers wide. However, it continuously exceeds an elevation of 3,000 meters (10,000 feet) for a distance of 3,200 kilometers.

1-6. I n its western extreme, the Eurasian mountain belt includes the Pyrenees, Alps, Balkans, and Carpathian ranges of Europe. These loosely linked systems are separated by broad low basins and are cut by numerous valleys. The Atlas Mountains of North Africa are also a part of this belt Moving eastward into Asia, this system becomes more complex as it reaches the extreme heights of the Hindu Kush and the Himalayas. Just beyond the Pamir Knot on the Russian-Afghan frontier, it begins to fan out across all parts of eastern Asia. Branches of this belt continue south along the rugged island chains to New Zealand and northeast through the Bering Sea to Alaska.

MOUNTAIN CHARACTERISTICS

1-7. Mountain slopes generally vary between 15 and 45 degrees. Cliffs and other rocky precipices may be near vertical, or even overhanging. Aside from obvious rock formations and other local vegetation characteristics, actual slope surfaces are usually found as some type of relatively firm earth or grass. Grassy slopes may include grassy clumps known as tussocks, short alpine grasses, or tundra (the latter more common at higher elevations and latitudes). Many slopeswill be scattered with rocky debris deposited from the higher peaks and ridges. Extensive rock or boulder fields are known as talus. Slopes covered with smaller rocks, usually fist-si zed or smaller, are called scree fields. Slopes covered in talus often prove to be a relatively easy ascent route. On the other hand, climbing a scree slope can be extremely difficult, as the small rocks tend to loosen easi ly and giveway. H owever, this characteristic often makes scree fields excel lent descent routes. Before attempt i ng to descend scree slopes, commanders should carefully analyze the potential for creati ng dangerous rock fall and take necessary avoidance measures.

1-8. In winter, and at higher elevations throughout the year, snow may blanket slopes, creating an environment with its own distinct affects. Some snow conditions can aid travel by covering rough terrain with a consistent surface. Deep snow, however, greatly impedes movement and requires soldiers well-trained in using snowshoes, skis, and over-snow vehicles. Steep snow covered terrain presents the risk of snow avalanches as well. Snow can pose a serious threat to soldiers not properly trained and equipped for movement under such conditions. Avalanches have taken the lives of more soldiers engaged in mountain warfare than all other terrain hazards combined,

1-9. Commanders operating in arctic and subarctic mountain regions, as well as the upper elevations of the world's high mountains, may be confronted with vast areas of glaciation. Valleys in these areas are frequently buried under massive glaciers and present additional hazards, such as hidden crevices and ice and snow avalanches. The mountain slopes of these peaks are often glacialed and their surfaces are generally composed of varying combinations of rock, snow, and ice. Although glaciers have their own peculiar hazards requiring special training and equipment, movement over valley glaciers is often the safest route through these areas (TC 90-6-1 contains more information on avalanches and glaciers, and their effects on operations).

MOUNTAIN CLASSIFICATIONS

1-10. There is no simple system available to classify mountain environments. Soil composition, surface configuration, elevation, latitude, and climatic patterns determine the specific characteristics of each major mountain range. When alerted to the potential requirement to conduct mountain operations, commanders must carefully analyze each of these characteristics for the specific mountain region in which their forces will operate. However, mountains are generally classified or described according to their local relief; for military purposes, they may be classified according to operational terrain levels and dismounted mobility and skill requirements.

Local Relief

1-11. Mountains are commonly classified as low or high, depending on their local relief and, to some extent, elevation. Low mountains have a local relief of 300 to 900 meters (1,000 to 3,000 feet) with summits usually below the timberline. High mountains have a local relief usually exceeding 900 meters (3,000 feet) and are characterized by barren alpine zones above the timberline. Glaciers and perennial snow cover are common in high mountains and usually present commanders with more obstacles and hazards to movement than do low mountains.

Operational Terrain Levels

1-12. Mountain operations are generally carried out at three different operational terrain levels (see Figure 1-2). Level I terrain is located at the bottom of valleys and along the main lines of communications. At this level, heavy forces can operate, but maneuver space is often restricted. Light and heavy forces are normally combined, since vital lines of communication usually follow the valley highways, roads, and trails.

Level	Description
I	The bottoms of valleys and main lines of communications
II	The ridges, slopes, and passes that overlook valleys
III	The dominant terrain of the summit region

Figure 1-2. Operational Terrain Levels

1-13. Level II terrain lies between valleys and shoulders of mountains. Generally, narrow roads and trails, which serve as secondary lines of communication, cross this ridge system. Ground mobility is difficult and light forces will expend great effort on these ridges, since they can easily influence operations at Level I. Similarly, enemy positions at the next level can threaten operations on these ridges.

1-14. Level MI includes the dominant terrain of summit regions. Although summit regions may contain relatively gentleterrain, mobility in Level III is usually the most difficult to achieve and maintain Level III terrain, however, can provide opportunities for well-trained units to attack the enemy from the flanks and rear At this terrain level, acclimatized soldiers with advanced mountaineering training can infiltrate to attack lines of communication, logistics bases, air defense sites, and command infrastructures

Dismounted Mobility Classification

1-15. When conducting mountain operations, commanders must clearly understand the effect the operational terrain level has on dismounted movement. Therefore, in addition to the general mobility classification contained in FM 2-01.3 (unrestricted, restricted, severely restricted), mountainous terrain may be categorized into five classes based on the type of individual movement skill required (see Figure 1-3), Operations conducted in class 1 and 2 terrain require little to no mountaineering skills. Operations in class 3, 4, and 5 terrain require a higher level of mountaineering skills for safe and efficient movement. Commanders should plan and prepare for mountain operations based, in large part, on this type of terrain analysis.

Class	Terrain	Mobility Requirements	Skill Level Required*
1	Gentler slopes/ trails	Walking techniques	Unskilled (with some assistance) and Basic Mountaineers
2	Steeper/rugged terrain	Some use of hands	
3	Easy climbing	Fixed ropes where exposed	Basic Mountaineers (with assistance from assault climbers)
4	Steep/exposed climbing	Fixed ropes required	
5	Near vertical	Technical climbing required	Assault Climbers

* See Chapter 2 for a discussion of mountaineering skill levels

Figure 1-3. Dismounted Mobility Classification

WEATHER

1-16. In general, mountain climates tend to be cooler, wetter versions of the climates of the surrounding lowlands. Most mountainous regions exhibit at least two different climatic zones — a zone at low elevations and another at elevations nearer the summit regions. In some areas, an almost endless variety of local dimates may exist within a given mountainous region. Conditions change markedly with elevation, latitude, and exposure to atmospheric winds and air masses. In addition, the climatic patterns of two ranges located at the same latitude may differ radically.

1-17. Like most other landforms, oceans influence mountain dimates. Mountain ranges in dose proximity to oceans and other large bodies of water usually exhibit a maritime dimate. Maritime dimates generally produce milder temperatures and much larger amounts of rain and snow. Their relatively mild winters produce heavy snowfalls, while their summer temperatures rarely get excessively hot. Mountains farther inland usually display a more continental climate Winters in this type climate are often bitterly cold, while summers can be extremely hot. Annual rain- and snowfall here is far less than in a maritime climate and may be quite scarce for long periods. Relatively shallow snow-packs are normal during a continental dimale's winter season.

1-18. Major mountain ranges force air masses and storm systems to drop significant amounts of rain and snow on the windward side of the range. Asair masses pass over mountains, the leeward slopes receive far less precipitation than the windward slopes. It is not uncommon for the dimate on the windward side of a mountain range to be humid and the dimate on the leeward side arid. This phenomenon affeds coastal mountains, as well as mountains farther inland. The deepest winter snow-packs will almost always be found on the windward side of mountain ranges. As a result, vegetation and forest characteristics may be markedly different between these two areas. Prevailing winds and storm patterns normally determine the severity of these effect s.

1-19. Mountain weather can be erratic, varying from strong winds to calm, and from extreme cold to relative warmth within a short time or a minor shift in locality. The severity and variance of the weather require soldiers to be prepared for alternating periods of heat and cold, as well as conditions ranging from dry to extremely wet. At higher elevations, noticeable temperature differences may exist between sunny and shady areas or between areas exposed to wind and those protected from it. This greatly increases every soldier's clothing load and a unit's overall logistical requirements. Figure 1-4 summarizes the effects of mountain weather discussed below. FM 2-33,201 and FM 3-97 22 contain additional information on how weather affects operation s,

TEMPERATURE

1-20. Normally, soldiers encounter a temperature drop of three to five degrees Fahrenheit per 300-meter (1,000-foot) gain in elevation. In an atmosphere containi ng considerable water vapor, the temperature drops about one degree Fahrenheit for every 100-meter (300-foot) increase. In very dry air. it drops about one degree Fahrenheit for every 50 meters (150 feet). However, on cold, dear, and calm mornings, when a troop movement or dimb begins from a valley, soldiers may encounter higher temperatures as they gain elevation This reversal of the normal situation is called temperature inversion Additionally, during winter months, the temperature is often higher during a storm than during periods of dear weather. However, the dampness of predpitation and penetration of the wind may still cause soldiers to chill faster This is compounded by the fad that the cover afforded by vegetation often does not exist abovethe trt?e-line. Under these conditions, com ma riders must weigh the tactical advantage of retaining positions on high ground against seeking shelter and warmth at lower elevations with reduced visibility.

1-21, At high elevations, there may be differences of 40 to 50 degrees Fahrenheit between the temperature in the sun and that in the shade. This is similar in magnitude to the day-to-night temperature fluctuations experienced in some deserts (see FM 3-97.3). Besides permitting rapid heating, the dear air at high altitudes also results in rapid cooling at night. Consequently. temperatures ri se swiftly after sunrise and drop quickly after sunset. M uch of the chilled air drains downward so that the differences between day and night temperatures ere greater in valleys than on slopes.

Weather Condition	Flat to Moderate Terrain Effects	Added Mountain Effects
Sunshine	Sunburn Snow blindness Temperature differences between sun and shade	Increased risk of sunburn and snow blindness Severe, unexpected temperature variations between sun and shade Avalanches
Wind	Windchill	Increased risk and severity of windchill Blowing debris or driven snow causing reduced visibility Avalanches
Rain	Reduced visibility Cooler temperatures	Landslides Flash floods Avalanches
Snow	Cold weather injuries Reduced mobility and visibility Snow blindness Blowing snow	Increased risk and severity of common effects Avalanches
Storms	Rain/snow Reduced visibility Lightning	Extended duration and intensity greatly affecting visibility and mobility Extremely high winds Avalanches
Fog	Reduced mobility/visibility	Increased frequency and duration
Cloudiness	Reduced visibility	Greatly decreased visibility at higher elevations

Figure 1-4. Comparison of Weather Effects

WIND

1-22. In high mountains, the ridges and passes are seldom calm. By contrast, strong winds in protected valleys are rare. Normally, wind velocity increases with altitude and is intensified by mountainous terrain. Valley breezes moving up-slope are more common in the morning, while descending mountain breezes are more common in the evening. Wind speed increases when winds are forced over ridges and peaks (orographic lifting), or when they funnel through narrowing mountain valleys, passes, and canyons (Venturi effect). Wind may blow with great force on an exposed mountainside or summit. As wind speed doubles, its force on an object nearly quadruples

1-23, Mountain winds cause rapid temperature changes and may result in blowing snow, sand, or debris that can impair movement and observation. Commanders should routinely consider the combined cooling effect of ambient temperature and wind (windchill) experienced by their soldiers (see Figure 1-5 on page 1-8). At higher elevations, air is considerably dryer than air at sea level. Due to this increased dryness, soldiers must increase their fluid intake by approximately one-third. However, equipment will not rust as quickly, and organic matter will decompose more slowly.

WIND SPEED		COOLING POWER OF WIND EXPRESSED AS "EQUIVALENT CHILL TEMPERATURE"																				
KNOTS	MPH	TEMPERATURE (° F)																				
CALM	CALM	40	35	30	25	20	15	10	5	0	-5	-10	-15	-20	-25	-30	-35	-40	-45	-50	-55	-60
		EQUIVALENT CHILL TEMPERATURE																				
3-6	5	35	30	25	20	15	10	5	0	-5	-10	-15	-20	-25	-30	-35	-40	-45	-50	-55	-60	-70
7-10	10	30	20	15	10	5	0	-10	-15	-20	-25	-35	-40	-45	-50	-60	-65	-70	-75	-80	-90	-95
11-15	15	25	15	10	0	-5	-10	-20	-25	-30	-40	-45	-50	-60	-65	-70	-80	-85	-90	-100	-105	-110
16-19	20	20	10	5	0	-10	-15	-25	-30	-35	-45	-50	-60	-65	-75	-80	-85	-95	-100	-110	-115	-120
20-23	25	15	10	0	-5	-15	-20	-30	-35	-45	-50	-60	-65	-75	-80	-90	-95	-105	-110	-120	-125	-135
24-28	30	10	5	0	-10	-20	-25	-30	-40	-50	-55	-65	-70	-80	-85	-95	-100	-110	-115	-125	-130	-140
29-32	35	10	5	-5	-10	-20	-30	-35	-40	-50	-60	-65	-75	-80	-90	-100	-105	-115	-120	-130	-135	-145
33-36	40	10	0	-5	-15	-20	-30	-35	-45	-55	-60	-70	-75	-85	-95	-100	-110	-115	-125	-130	-140	-150
WINDS ABOVE 40 HAVE LITTLE ADDITIONAL EFFECT		LITTLE DANGER					INCREASING DANGER (Flesh may freeze within 1 minute)							GREAT DANGER (Flesh may freeze within 30 secs)								

Figure 1-5. Windchill Chart

PRECIPITATION

1-24. The rapid rise of air masses over mountains creates distinct local weather patterns. Precipitation in mountains increases with elevation and occurs more often on the windward than on the leeward side of ranges. Maximum cloudiness and precipitation generally occur near 1,800 meters (6,000 feet) elevation in the middle latitudes and at lower levels in the higher latitudes. Usually, a heavily wooded belt marks the zone of maximum precipitation.

Rain and Snow

1-25. Both rain and snow are common in mountainous regions. Rain presents the same challenges as at lower elevations, but snow has a more significant influence on all operations. Depending on the specific region, snow may occur at anytime during the year at elevations above 1,500 meters (5,000 feet). Heavy snowfall greatly increases avalanche hazards and can force changes to previously selected movement routes. In certain regions, the intensity of snowfall may delay major operations for several months. Dry, flat riverbeds may initially seem to be excellent locations for assembly areas and support activities, however, heavy rains and rapidly thawing snow and ice may create flash floods many miles downstream from the actual location of the rain or snow.

Thunderstorms

1-26. Although thunderstorms are local and usually last only a short time, they can impede mountain operations. Interior ranges with continental climates are more conducive to thunderstorms than coastal ranges with maritime climates. I n alpine zones, driving snow and sudden wind squalls often accompany thunderstorms. Ridges and peaks become focal points for lightning strikes, and the occurrence of lightning is greater in the summer than the winter. Although statistics do not show l ightning to be a major mountaineering hazard, it should not be ignored and soldiers should take normal precautions, such as avoiding summits and ridges, water, and contact with metal objects.

Traveling Storms

1-27. Storms resulting from widespread atmospheric disturbances involve strong winds and heavy predpitation and are the most severe weather condition that occurs in the mountains. If soldiers encounter a traveling storm in alpine zones during winter, they should expect low temperatures, high winds, and blinding snow. These conditions may last several days longer than in the lowlands. Specific conditions vary depending on the path of the storm. However, when colder weather moves in, dearing at high elevations is usually slow.

1-28. The effects of fog in mountains are much the same as in other terrain. H owever, because of the topography, fog occurs more frequently in the mountains The high inddence of fog makes it a significant planning consideration as it restricts visibility and observation complicating reconnaissance and surveillance. However, fog may help fadlitate covert operations such as infiltration. Routes in areas with a high occurrence of fog may need to be marked and charted to fadlitate passage.

SECTION II - EFFECTS ON PERSONNEL

1-29. The mountain environment is complex and unforgiving of errors. Soldiers conducting operations anywhere, even under the best conditions, become cold, thirsty, tired, and energy-depleted. In the mountains however, they may become paralyzed by cold and thirst and incapacitated due to utter exhaustion. Conditions such as high elevations, rough terrain, and extremely unpredictable weather require leaders and soldiers who have a keen understanding of environmental threats and what to do about them.

1-30. A variety of individual soldier charaderistics and environmental conditions influence the type, prevalence, and severity of mountai n i l ln esses and injuries (see Figure 1-6 on page 1-10). Due to combinations of these characteristics and conditions, soldiers often succumb to more than one illness or injury at a time, increasing the danger to life and limb. Three of the most common, cumulative, and subtle fad ors affeding soldier ability under these variable conditions are nutrition (to indude water intake), decreased oxygen due to high altitude, and cold. Preventive measures, early recognition, and rapid treatment help minimize nonbattle casualties due to these conditions (see Appendix A for detailed information on mountain-specific illnesses and injuries).

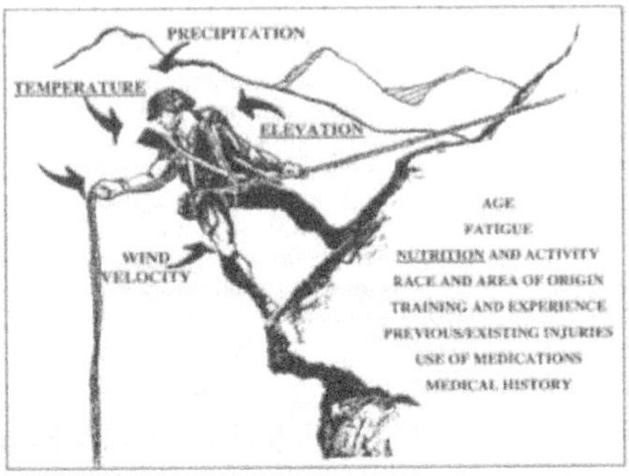

Figure 1-6. Environmental and Soldier Conditions Influencing Mountain Injuries and Illnesses

NUTRITION

1-31. Poor nutrition contributes to illness or injury, decreased performance, poor morale, and susceptibility to cold injuries, and can severely affect military operations. Influences at high altitudes that can affect nutrition include a dulled taste sensation (making food undesirable), nausea, and lack of energy or motivation to prepare or eat meals.

1-32, Caloric requirements increase in the mountains due to both the altitude and the cold A diet high in fat and carbohydrates is important in helping the body fight the effects of these conditions. Fats provide long-term, slew caloric release, but are often unpalatable to soldiers operating at higher altitudes. Snacking on high-carbohydrate foods is often the best way to maintain the calories necessary to function.

1-33. Products that can seriously impact soldier performance in mountain operations include:

- Tobacco. Tobacco smoke interferes with oxygen delivery by reducing the blood's oxygen-carrying capacity. Tobacco smoke in dose, confined spaces increases the amounts of carbon monoxide. The irritant effect of tobacco smoke may produce a narrowing of airways, interfering with optimal air movement. Smoking can effectively raise the"physiological altitude" as much as several hundred meters.
- Alcohol. Alcohol impairs judgement and perception, depresses respiration, causes dehydration, and increases susceptibility to cold injury
- Caffeine. Caffeine may improve physical and mental performance, but it also causes increased urination (leading to dehydration) and, therefore, should be consumed in moderation.

1-34. Significant body water is lost at higher elevations from rapid breathing, perspiration, and urination. Depending upon level of exertion, each soldier should consume about four to eight quarts of water or other decaffeinated fluids per day in low mountains and may need ten quarts or more per day in high mountains. Thirst is not a good indicator of the amount of water lost.

and in cold climates sweat, normally an indicator of loss of fluid, goes unnoticed. Sweat evaporates so rapidly or is absorbed so thoroughly by clothing layers that it is not readily apparent. When soldiersbecomelhirsty.they are already dehydrated. Loss of body water also plays a major role in causing altitude sickness and cold injury. Forced drinking in the absence of thirst, monitoring the deepness of the yellow hue in the urine, and watching for behavioral symptoms common to altitude sickness are important factors for commanders to consider In assessing the water balance of soldiers operating in the mountains.

1-35, In the mountains, as elsewhere, refilling each soldier's water containers as often as possible is mandatory. No matter how pure and clean mountain water may appear, water from natural sources should always be purified or chemically sterilized to prevent parasitical illnesses (giardiasis). Commanders should consider requiring the increased use of individual packages of powdered drink mixes, fruit, and juices to help encourage the required fluid intake.

ALTITUDE

1-36. As soldiers ascend in altitude, the proportion of oxygen in the air decreases. Without proper acclimatization, this decrease in oxygen saturation can cause altitude sickness and reduced physical and mental performance [see Figure 1-7). Soldiers cannot maintain the same physical performance at high altitudethat they can at low altitude, regardless of their fitness level.

Altitude	Meters	Feet	Effects
Low	Sea Level – 1,500	Sea Level – 5,000	None.
Moderate	1,500 – 2,400	5,000 – 8,000	Mild, temporary altitude sickness may occur
High	2,400 – 4,200	8,000 – 14,000	Altitude sickness and decreased performance is increasingly common
Very High	4,200 – 5,400	14,000 – 18,000	Altitude sickness and decreased performance is the rule
Extreme	5,400 – Higher	18,000 - Higher	With acclimatization, soldiers can function for short periods of time

Figure 1-7. Effects of Altitude

1-37. The mental effects most noticeable at high altitudes indude decreased perception, memory, judgement, and attention. Exposure to ait it udes of over 3,000 meters (10,000 feet) may also result in changes in senses, mood, and personality. Within hours of ascent, many soldiers may experience euphoria, joy, and excitement that are likely to be accompanied by errors i n judgement, leading to mistakes and accidents. After a period of about 6 to 12 hours, euphoria decreases, often changing to varying degrees of depression. Soldiers may become irritable or may appear listless. U sing the buddy system during this early exposure helps to identify soldiers who may be more severely affected. High morale and esprit instilled before deployment and reinforced frequently help to mini mize the i mpact of negative mood changes.

1-38. Trie physical effect most noticeable at high altitudes includes vision. Vision is generally the sense most affected by altitude exposure and can potentially affect military operations at higher elevations. Night vision is significantly reduced, affecting soldiers at approximately 2.400 meters (8,000 feet) or higher. Some effects occur early and are temporary, while others may persist after acclimatization or even for a period of time after descent. To compensate for loss of functional abilities, commanders should make use of tactics, techniques, and procedures that trade speed for increased accuracy. By allowing extra time lo accomplish tasks, commanders can minimize errors and injuries.

HYPOXIA-RELATED ILLNESSES AND EFFECTS

1-3g Hypoxia, a deficiency of oxygen reaching the tissues of the body, has been the cause of many mountain illnesses, injuries, and deaths. It affects everyone, but some soldiers are more vulnerable than others. A soldier may be affected at one time but not at another. Altitude hypoxia is a killer, but it seldom strikes alone. The combination of improper nutrition, hypoxia, and cold is much more dangerous than any of them alone. The three most significant altitude-related illnesses and their symptoms, which are essentially a series of illnesses associated with oxygen deprivation, are:

- Acute Mountain Sickness (AMS). Headache, nausea, vomiting, fatigue, irritability, and dizziness.

  High Altitude Pulmonary Edema (HAPE), Coughing, noisy breathing, wheezing, gurgling in the airway, difficulty breathing, and pink frothy sputum (saliva). Ultimately coma and death will occur without treatment,
- High Altitude Cerebral Edema (HACE). HACE is the most severe illness associated with high altitudes. Its symptoms often resemble AMS {severe headache, nausea, vomit i ng), often with more dram at ic signals such as a swaying of the upper body, especially when walking, and an increasingly deteriorating mental status. Early mental symptoms may include confusion, disorientation, vivid hallucinations, and drowsiness. Soldiers may appear to be withdrawn or demonstrate behavior generally associated with fatigue or anxiety. Like HAPE, coma or death wi 11 occur without treatment.

OTHER MOUNTAIN-RELATED ILLNESSES

1-40. Other illnesses and effects related to the mountain environment and higher elevations are:

- Subacute mountain sickness. Subacute mountain sickness occurs in some soldiers during prolonged deployments (weeks/months) to elevations above 3,600 meters (12,000 feet). Symptoms include sleep disturbance, loss of appetite, weight loss, and fatigue. This condition reflects a failureto acclimatize adequately.
- Carbon monoxide poisoning. Carbon monoxide poisoning is caused by the inefficient fuel combustion resulting from the low oxygen content of air and higher usage of stoves, combustion heaters, and engines in enclosed, poorly ventilated spaces.

Sleep disturbances. High altitude has significant harmful effects on sleep. The most prominent effects are frequent periods cf apnea (temporary suspension of respiration) and fragmented sleep Sleep disturbances may last for weeks at elevations less than 5.400 meters (18.000 feet) and may never stop at higher elevations. These effects have even been reported as low as 1,500 meters (5,000 feet).

Poor wound healing. Poor wound healing resulting from lowered immune functions may occur at higher elevations. Injuries resulting from burns, cuts, or other sources may require descent for effective treatment and healing.

ACCLIMATIZATION

1-41. Altitude acclimatization involves physiological changes that permit the body to adapt to the effects of low oxygen saturation in the air. It allows soldiers to achieve the maximum physical work performance possible for the altitude to which they are acclimatized. Once acquired, acclimatization is maintained as long as the soldier remains at that altitude, but is lost upon returning to lower elevations. Acclimatization to one altitude does not prevent altitude illnesses from occurring if ascent to higher altitudes is too rapid.

1-42. Getting used to living and working at higher altitudes requires acclimatization. Figure 1-8 shows the four factors that affect acclimatization in mountainous terrain. These factors are similar to those a scuba diver must consider, and the consequences of an error can be just as severe. In particular. high altitude dimbing must be carefully paced and staged in the same way that divers must pace and stage their ascent tothe surface.

- **Altitude**
- **Rate of Ascent**
- **Duration of Stay**
- **Level of Exertion**

Figure 1-8. Factors Affecting Acclimatization

1-43. For most soldiers at high to very high altitudes. 70 to 80 percent of the respiratory component of acclimatization occurs in 7 to 10 days, 80 to 90 percent of overall acdimatization is generally accomplished by 21 to 30 days, and maximum acdimatization may take several months to years. However, some soldiers may acdimatize more rapidly than others, and a few soldiers may not acclimatize at all. There is no absolute way to identify soldiers who cannot acdimatize, except by their experience during previous altitude exposures.

1-44. Commanders must be aware that highly fit, motivated individuals may go too high too fast and become victims of AMS. HAPE, or HACE. Slow and easy dimbing. limited activity, and long rest periods are critical to altitude acclimatization. Leaves that involve soldiers descending to lower altitudes and then returning should be limited. Acdimalization may be accomplished by either a staged or graded ascent. A combination of the two is the safest and most effective method for prevention of high altitude illnesses.

- Staged Ascent. A staged ascent requires soldiers to ascend to a moderate altitude and remain there for 3 days or more to acdimatize before ascending higher (the longer theduration, the more effedive and thorough the aoclimatization to that altitude). When possible, soldiers should make several slops for staging during ascent to allow a greater degree of acclimatization.

Graded Ascent. A graded ascent limits the daily altitude gain to allow partial acclimatization. The altitude at which soldiers sleep is the critical element in this regard. Having soldiers spend two nights at 2,700 meters (9,000 feet) and limiting the sleeping altitude to no more than 300 meters per day (1,000 feet) above the previous night's sleeping altitude will significantly reducethe incidence of altitude sickness,

1-45. In situations where there is insufficient time for a staged or graded ascent, commanders may consider using the drug acetazolamide to help accelerate acdimatization; however, commanders must ensure soldiers are acclimatized before they are committed to combat. When used appropriately, it will prevent symptoms of AMS in nearly all soldiers and reduce symptoms in most others. It has also been found to improve sleep quality at high altitudes. However, commanders should consult physidans trained in high-altitude or wilderness medidne concerning doses, side effects, and screening of individuals who may be allergic. Asa non-pharmacological method, high carbohydrate diets (whole grains, vegetables, peas and beans, potatoes, fruits, honey, and refined sugar) are effedive in aiding acdimatization.

COLD

1-46. After illnesses related to not being acdimatized. cold injuries, both freezing and nonfreezing, are generally the greatest threat. Temperature and humidity decrease with increasing altitude. Reviewing cold weather injury provontion, training in sholter construction, dressing in layers, and using the buddy system are critical and may preclude large numbers of debilitating injuries. Figure 1-9 lists the cold and snow injuries most common to mountain operations. See FM 3-97.11 and FM 4-25.11 for information regarding causes, symptoms, treatment, and prevention.

- **Frostbite (freezing)**
- **Hypothermia (nonfreezing)**
- **Trench/Immersion Foot (nonfreezing)**
- **Snow Blindness**

Figure 1-9. Common Cold Weather Injuries

1-47. Altitude sickness and cold injuries can occur simultaneously, with signs and symptoms being confused with each other. Coughing, stumbling individuals should be immediately evacuated to medical support at lower levels to determine their medical condition. Likewise, soldiers in extreme pain from cold injurieswhodo not respond to normal pain medications, require evacuation. Without constant vigilance, cold injuries may significantly limit the number of deployable troops and drastically reduce combat power. However, with command emphasis and proper equipment, dothing, and training, all cold-weather injuries are preventable.

SECTION III - EFFECTS ON EQUIPMENT

- Operator/Maintenance Personnel
- Line-of-Sight
- Range
- Thermal Contrast
- Ballistics and Trajectory
- Target Detection and Acquisition
- First Round Hit Capability
- Camouflage and Concealment/Noise
- Mobility
- Wear and Maintenance
- Aerodynamics and Lift
- Functioning and Reliability
- Positioning/Site Selection

Figure 1-10. Weapons and Equipment Factors Affected by the Environment

1-48. No manual can cover the effects of terrain and weather on every weapon and item of equipment within the Army inventory. Although not all-encompassing, the list at Figure 1-10 contains factors that commanders should take into account when considering the effect the mountainous environment may have on their weapons and equipment. Of these, the most important factor is the combined effects of the environment on the soldier and his subsequent ability to operate and maintain his weapons and equipment. Increasingly sophisticated equipment requires soldiers that are mentally atert and physically capable. Failure to consider this important factor often results in severe injury. lowered weapons and equipment performance, and mission failure. The information provided within this manual, combined with the information found in weapon-specific field manuals (FMs) and technical manuals (TMs), provides the information necessary to know how to modify tactics, techniques, and procedurestowin on themountain battlefield

GENERAL EFFECTS

1-49. In a mountainous environment, the speed and occurrence of wind generally increase with elevation, and the effects of wind increase with range (depending on the speed and direction). Due to these factors, soldiers must be taught the effects of wind on ballistics and how to compensate for them. In cold weather, firing weapons often creates ice fog trails. These ice fog trails obscure vision and, at thesametime, allow the enemy to more easily discern the location of primary positions and the overall structure of a unit's defense. This situation increases the importance of alternate and supplementary firing positions,

1-50. Range estimation in mountainous terrain is difficult. Depending upon the type of terrain in the mountains, soldiers may either over- or underestimate range. Soldiers observing over smooth terrain, such as sand, water, or snow, generally underestimate ranges. This results in attempting to engage targets beyond the maximum effective ranges of their weapon systems. Looking downhill, targets appear to be farther away and looking uphill, they appear to be closer. This illusion, combined with the effects of gravity, causes the soldier shooting downhill to fire high, while it has the opposite effect on soldiers shooting uphill.

1-51. Higher elevations generally afford increased observation but low-hanging clouds and fog may decrease visibility, and the rugged nature of mountain terrain may produce significant dead space at mid-ranges. These effects mean that more observation posts are necessary to cover a given frontage in mountainous terrain than in non-mountainous terrain. They also require the routine designation of supplementary firing positions for direct fire weapons. Rugged terrain also makes ammunition resupply more difficult and increases the need to enforce strict fire control and discipline. Finally, the rugged environment creates compartmented areas that may preclude mutual support and reduce supporting distances.

SMALL ARMS

1-52 In rocky mountainous terrain, the effectiveness of small arms fire increases by the splintering and ricocheting when a bullet strikesa rock. M203 and M K-19 grenade launchers are useful for covering close-in dead space in mountainous terrain. Hand grenades are also effective Although it may seem intuitive, soldiers must still be cautioned against throwing grenades uphill where they are likely to roll back before detonation. Grenades (as well as other explosive munitions) lose much of their effectiveness when detonated under snow, and soldiers should be warned that hand grenades may freeze to wet gloves.

1-53. As elevation increases. air pressure and air density decrease. At higher elevations, a round is more efficient and strikes a target higher, due to reduced drag. This effect does not significantly influence the marksmanship performance of most soldiers, however, designated marksmen and snipers should re-zero their weapons after ascending to higher elevations. (See FM 3-25.9 and FM 3-23,10 for further information on ballistics and weather effects on small arms.)

MACHINE GUNS

1-54. Machine guns provide long-range fire when visibility is good. However, grazing fire can rarely be achieved in mountains because of the radical changes in elevation. When grazing fire can be obtained, the ranges are normally short. More often, plunging fire is the result (see Figure 1-11 and FM 3-21.7). In mountainous terrain, situations that prevent indirect fire support from protecting advancing forces may arise. When

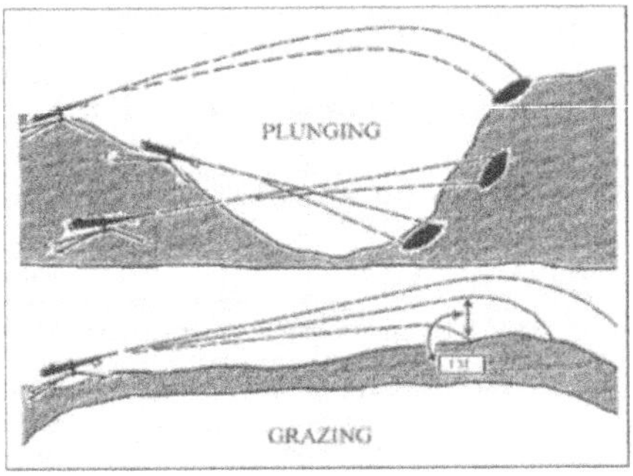

Figure 1-11. Classes of Fire with Respect to the Ground

these occur, the effects of machine-guns and other direct fire weapons must be concentrated to provide adequate supporting fires for maneuvering elements. Again, supplementary positions should be routinely prepared to cover different avenues of approach and dead space.

ANTITANK WEAPONS

1-55. The AT4 is a lightweight antitank weapon ideally suited for the mountainous environment and for direct fire against enemy weapon em placemen ts. Anti-tank guided missiles (ATGMs), such as the Javelin and the tube-launched, optically tracked, wire-guided, heavy antitank missile system (TOW), tend to hinder dismounted operations because of their bulk and weight. In very restrictive mountainous terrain, the lack of armored avenues of approach and suitable targets may limit their utility. If an armored or mechanized threat is present, TOWs are best used in long-range, antiarmor ambushes, while the shorter-range Javelin, with its fire-and-forget technology, is best used from restrictive terrain nearer the kill zone. However, their guidance systems may operate stiffly and sluggishly in extreme cold weather.

SECTION IV - RECONNAISSANCE AND SURVEILLANCE

RECONNAISSANCE

1-56. During operations in a mountainous environment, reconnaissance is as applicable to the maneuver of armies and corps as it is to tactical operations. Limited routes, adverse terrain, and rapidly changing weather significantly increase the importance of reconnaissance operations to focus fires and maneuver. Failure to conduct effective reconnaissance will result in units being asked to achieve the impossible or in missed opportunities for decisive action.

1-57. As in all environments, reconnaissance operations in a mountainous area must be layered and complementary in order to overcome enemy attempts to deny critical information to the friendly commander. In order to gather critical and timely information required by the commander, the activities of reconnaissance assets must be closely coordinated. Strategic reconnaissance platforms set the stage by identifying key terrain, as well as the general disposition and composition of enemy forces. Operational level commanders compare the information provided by strategic assets with their own requirements and employ reconnaissance assets to fill in the gaps that have not been answered by strategic systems and achieve the level of detail they require.

1-53. At the beginning of a campaign in a mountainous environment, reconnaissance requirements will be answered by aerial or overhead platforms, such as satellites, joint surveillance, target attack radar systems (JSTARSs), U2 aircraft, and unmanned aerial vehicles (UAVs). In a mountain AO, it may often be necessary to commit ground reconnaissance assets in support of strategic and operational information requirements. Conversely, strategic and operational reconnaissance systems may be employed to identify or confirm the feasibility of employing ground reconnaissance assets. Special reconnaissance (SR) and long-range surveillance (LRS) teams may be inserted to gather information thai cannot be collected by overhead systems, or to verify data that has already been collected. In this instance, satellite imagery is used to analyze a specific area for insertion for the team. The potential hide positions for the teams are identified using imagery and. terrain and weather permitting, verified by UAVs. See FM 3-100.55 for detailed information on combined arms reconnaissance.

1-59. In harsh mountain terrain, ground reconnaissance operations are often conducted dismounted. Commanders must assess the slower rate of ground reconnaissance elements to determine its impact on theentire reconnaissance and collection process. They must develop plans that account for this slower rate and initiate reconnaissance as early as possible to provide additional time for movement. Commanders may also need to allocate more forces, including combat forces, to conduct reconnaissance, reconnaissance in force missions, or limited objective attacks to gain needed intelligence. Based upon mission, enemy, terrain and weather, troops and support available, time available, civil considerations (METT-TC), commanders may need to prioritize collection assets, accept risk, and continue with less information from their initial reconnaissance efforts. In these cases, they must use formations and schemes of maneuver that provide maximum security and flexibility, to include robust security formations, and allow for the development of the situation once in contact.

1-60. Although reconnaissance patrols should normally use the heights to observe the enemy, it may be necessary to send small reconnaissance teams into valleys or along the low ground to gain suitable vantage points or physically examine routes that will be used by mechanized or motorized forces. In mountainous environments, reconnaissance elements are often tasked to determine:

- The enemy's primary and alternate lines of communication.
- Locations and directions from which the enemy can attack or counterattack.
- Heights that allow the enemy to observe the various sectors of terrain.
- Suitable observation posts for forward observers.
- Portions of the route that provide covert movement.
- Level of mountaineering skill required to negotiate routes (dismounted mobility classification) and sections of the route that require mountaineering installations.
- Suitability of routes for sustained combat service support (CSS) operations.
- Trails, routes, and bridges that can support or can be improved by engineers in order to move mechanized elements into areas previously thought to be impassable.
- Bypass routes
- Potential airborne and air assault drop^pick-up zones and aircraft landing areas.

RECONNAISSANCE IN FORCE

1-61. The compart merited geography and inherent mobility restrictions of mountainous terrain pose significant risk for reconnaissance in force operations. Since the terrain normally allows enemy units to defend along a much broader front with fewer forces, a reconnaissance in force may be conducted as a series of smaller attacks to determine the enemy situation at selected points. Commanders should carefully consider mobility restrictions that may affect plans for withdrawal or exploitation. Commanders should also position small reconnaissance elements or employ surveillance systems throughout the threat area of operations to gauge the enemy's reaction to friendly reconnaissance in force operations and alert the force to possible enemy counterattacks. In the mountains, the risk of having at least a portion of the force cut off and isolated is extremely high. Mobile reserves and preplanned fires must be available to reduce the risk, decrease the vulnerability of the force, and exploit any success as it develops.

ENGINEER RECONNAISSANCE

1-62. Engineer reconnaissance assumes greater significance in a mountainous environment in order to ensure supporting engineers are properly task organized with specialized equipment for quickly overcoming natural and reinforcing obstacles. Engineer reconnaissance teams assess the resources required for clearing obstacles on precipitous slopes, constructing crossing sites at fast-moving streams and rivers, improving and repairing roads, erecting fortifications, and establishing barriers during the conduct of defensive operations. Since the restrictive terrain promotes the widespread employment of point obstacles, engineer elements should be integrated into all mountain reconnaissance operations,

1-63. In some regions, maps may be unsuitable for tactical planning due to inaccuracies, limited detail, and inadequate coverage. En these areas, engineer reconnaissance should precede, but not delay operalions. Because rugged mountain terrain makes ground reconnaissance time-consuming and dangerous, a combination of ground and aerial or overhead platforms should be used for the engineer reconnaissance effort. Data on the terrain, vegetation, and soil composition, combined with aerial photographs and multispectral imagery, allows engineer terrain intelligence teams to provide detailed information that may be unavailable from other sources.

AERIAL AND OVERHEAD RECONNAISSANCE

1-64. During all but the most adverse weather conditions, aerial or overhead reconnaissance may be the best means to gather information and cover large areas that are difficult for ground units to traverse or observe. Airborne standoff intelligence collection devices, such as side-looking radar, provide excellent terrain and target isolation imagery. Missions must be planned to ensure that critical areas are not masked by terrain or other environmental conditions. Additionally, aerial or overhead photographs may compensate for inadequate maps and provide the level of detai l needed to plan operations. I nfrared imagery and camouflage detection film can be used to determine precise locations of enemy positions, even at night. Furthermore, AH-64 and

OH-58D helicopters can provide commanders with critical day or night video reconnaissance, utilizing television or lorward-lcoking infrared.

1-65. Terrain may significantly impact the employment of overhead reconnaissance platforms using radar systems to detect manmade objects. These systems may find themselves adversely impacted by the masking effect that occurs when the mountain terrain blocks the radar beam. Thus, the radar coverage may not extend across the reverse slope of a steep ridge or a valley floor. Attempts to reposition the overhead platform to a point where it can "see" the masked area may merely result in masking occurring elsewhere. This limitation does not preclude using such systems; however, the commander should employ manned or unmanned aerial reconnaissance when available, in conjunction with overhead reconnaissance platforms in order ta minimize these occurrences. The subsequent use of ground reconnaissance assets to verify the data that can be gathered by overhead and electro-optical platforms wi 11 ensure that commanders do not fall prey to del i berate enemy deception efforts that capitalize on the limited capabilities of some types of overhead platforms in this environment.

SURVEILLANCE

1-66. I n the mountains, surveillance of vulnerable flanks and gaps between units is accomplished primarily through well-positioned observation posts (OPs). These OPs are normally inserted by helicopter and manned by small elements equipped with sensors, enhanced electro-optical devices, and appropriate communications. Commanders must develop adequate plans that address not only their insertion, but their continued support and ultimate extraction. The considerations of METT-TC may dictate that commanders provide more personnel and assets than other types of terrain to adequately conduct surveillance missions. Commanders must also ensure that surveillance operations are fully integrated with reconnaissance efforts in order to provide a3dequate coverage of the AO,

1-67. Long-range surveillance units (LRSUs) and snipers trained in mountain operations also contribute to surveillance missions and benefit from the restrictive terrain and excellenl line-of-sight. Overhead platforms and air cavalry may also be used for surveillance missions of limited duration. However, weather may impede air operations, decrease visibility for both air and ground elements, and reduce the ability of ground surveillance elements to remain hidden for prolonged periods without adequate logistical support. As with overhead reconnaissance, terrain may mask overhead surveillance platforms.

Chapter 2

Command and Control

In the mountains, major axes of advance are limited to accessible valleys and often separated by restrictive terrain. The compart merited nature of the terrain makes it difficult to switch the effort from one axis to another or to offer mutual support between axes. The battle to control the major lines of communications of Level I develops on the ridges and heights of Level II. In turn, the occupation of the dominating heights in Level II may leave a force assailable from the restrictive terrain of Level III. Each operational terrain level influences the application of tactics, techniques, and procedures n ecessar y for su ccessf u l operations.

CONTENTS

In mountainous terrain, it is usually difficult to oonduct a coordinated battle. Engagements tend to be isolated, march columns of even small elements extremely long, and mutual support difficult to accomplish. Command and control of all available assets is best achieved if command posts are well forward. However, the mountainous environment decreases the commander's mobility. Therefore, commanders must be able to develop a clear vision of how the battle will unfold, correctly anticipate the decisive points on the battlefield, and position themselves at these critical points.

The success of a unit conducting mountain operations depends on how well leaders control their units. Control is limited largely to a well-thought-out plan and thorough preparation. Boundaries require careful planning in mountain operations. Heights overlooking valleys should be included in the boundaries of units capable of exerting the most influence over them. These boundaries may be difficult to determine initially and may require subsequent adjustment.

During execution, leaders must be able to control direction and speed of movement, maintain proper intervals, and rapidly start, stop, or shift fire. In the mountains, soldiers focus mainly on negotiating difficult terrain. Leaders, however, must ensure that their soldiers remain alert for, understand, and follow signals and orders Although in most instances audio, visual, wire, physical signals, and messengers are used to maintain control, operations may be controlled by time as a secondary means. However, realistic timetables must be based on thorough reconnaissance and sound practical knowledge of the mountain battlefield.

Commanders must devote careful consideration to the substantial effect the mountain environment may have on systems that affect their ability to collect, process, store, and disseminate information. Computers, communications, and other sophisticated electronic equipment are usually susceptible to jars, shocks, and rough handling associated with the rugged mountain environment. They are also extremely sensitive to the severecold often associated with higher elevations. Increased precipitation and moisture may damage electronic components, and heavy amounts of rain and snow, combined with strong surface winds, may generate background electronic interference that can reduce the efficiency of intercept/direction finding antennas and ground surveillance radars. Localized storms with low sustained cloud cover reduce the effectiveness of most imagery intelligence (IMINT) platforms, to include unmanned aerial vehicles (UAVs). The collective effect of mountain weather and terrain diminishes a commander's ability to achieve shared situational understanding among his subordinates. However, increased use of human intelligence (HUMI NT), clear orders and intents, and leaders capable of exercising initiative, allow commanders to dominate the harsh environment of a mountain area of operations.

As in any environment, mountain operations pose both tactical and accident risks. However, since most units do not routinely train for or operate in the mountains, the level of uncertainty, ambiguity, and friction is often higher than in less rugged environments. Commanders must be able to identify and assess hazards that may be encountered in executing their missions, develop and implement control measures to eliminate unnecessary risk, and continuously supervise and assess to ensure measures are properly executed and remain appropriate as the situation changes. Although risk decisions are the commanders' business, staffs, subordinate leaders, and individual soldiers must also understand the risk management process and must continuously look for hazards at their level or within their area of expertise. Any risks identified (with recommended risk reduction measures) must be quickly elevated to the chain of command (see **FM** 3-100.14).

SECTION I - ASSESSMENT OF THE SITUATION

2-1, Although higher-elevation terrain is not always key, the structure of a mountain area of operations (AO) often forms a stairway of key terrain features Identification and control of dominant terrain at each operational terrain level form the basis for successful mountain maneuver. Key terrain features at higher elevations often take on added significance due to their inaccessibility and ease of defense. Tomainlain freedom of maneuver, commanders must apply combat power so that the terrain at Levels II and III can be exploited in the conduct of operations. Successful appl i cat ion of this concept requires commanders to think, plan, and maneuver vertical l y as well as horizontal ly.

2-2. Mountain operations usually focus on lines of communication, choke points, and dominating heights. Maneuver generally attempts to avoid strengths, envelop the enemy, and limit his ability to effectively use the high ground. Major difficulties are establishing boundaries, establishing and maintaining communications, providing logistics, and evacuating wounded. Throughout the plan, prepare, and execute cycle, commanders must continuously assess the vertical impact on the mission, enemy, terrain and weather, troops and support available, time available, civil considerations (METT-TC),

HISTORICAL PERSPECTIVE

Importance of Controlling Key Terrain:

Mustafa Kemal at Gallipoli (April 1915)

On 25 April 1915, the Allies launched their Gallipoli campaign. However, LTC Mustafa Kemal's understanding of the decisive importance of the hilly terrain, his grasp of the enemy's overall intent, and his own resolute leadership preserved the Ottoman defenses. His troops seized the initiative from superior forces and pushed the Allied invasion force back to its bridgehead. The result was nine months of trench warfare, followed by the Allies' withdrawal from Gallipoli.

German Fifth Army Commander General von Sanders expected a major Allied landing in the north, at Bulair. The British, however, were conducting a feint there; two ANZAC divisions were already landing in the south at Ari Burnu (now known as "ANZAC cove") as the main effort. The landing beaches here were hemmed by precipitous cliffs culminating in the high ground of the Sari Bair ridge, a fact of great importance to the defense. Only one Ottoman infantry company was guarding the area. Although prewar plans had established contingencies for using 19th ID, Kemal, the division commander, had received no word from his superiors regarding the developing scenario. Nevertheless, understanding that a major Allied landing could easily split the peninsula, he decided that time was critical and set off for Ari Burnu without waiting for his senior commander's approval. In his march toward Ari Burnu that morning, he recognized that the hilly terrain in general and the Sari Bair ridge in particular were of vital strategic importance: if the enemy captured this high ground they would be in an excellent position to cut the peninsula in half.

Kemal now engaged the enemy. He impressed upon his men the importance of controlling the hilltops at all costs, issuing his famous order: "I am not ordering you to attack. I am ordering you to die. In the time it takes us to die, other forces and commanders can come and take our place." Despite being outnumbered three-to-one, the Turkish counterattack stabilized their position and prevented the Allies from capturing the Sari Bair ridge. Nightfall brought about a lull in the fighting. There was some sniping and a few local encounters on 26 April, and on 27 April Kemal finally received major reinforcements. The front stabilized and the opposing armies settled down into trench warfare. On 16 January 1926, the Allies admitted defeat and withdrew.

The 19th ID's counterattack, which prevented the ANZAC from establishing themselves on the Sari Bair ridge, may well have decided the outcome of the entire Gallipoli campaign. Despite his lack of situational knowledge, Kemal instinctively understood the enemy's intent and, recognizing the importance of controlling the hilltops and ridgelines, was committed to concentrating his combat power to seize and hold this key terrain.

Compiled from "The Rock of Gallipoli," *Studies of Battle Command*, George W. Gawrych

MISSION

2-3. Mission analysis must include the spatial and vertical characteristics of the AO. Although defeating the enemy continues to be the basic objective of tactical operations, the task of controlling specific operational terrain levels will be paramount. At brigade level and below, major tactical objectives are normally translated into tasks pertaining to seizing, retaining, or controlling specific dominating heights at either Level II or Level III. Therefore, it is imperative to identify the tasks and assets necessary to access each operational terrain level.

2-4. At any operational terrain level, defending and delaying are easier at defiles, while attacking is more difficult. Due to the compartmented terrain, units usually execute offensive missions by conducting several simultaneous

smaller-scale attacks, utilizingthefull height, width, and depth of their area of operations. Consequentiy, commanders must always consider the impact of decentralization on security.

2-5. One method of maintaining freedom of action is to seize or hotd key terrain. In the mountains, key terrain is frequently identified as terrain that is higher than that held by the enemy. Seizing this terrain often depends on long and difficult envelopments or turning movements. Therefore, the specified and implied tasks associated with mobility and sustainment, as well as command and control, must be considered in terms of their vertical difficulty.

ENEMY

2-6. An enemy will normally position forces in depth and height along likely avenues of approach. Mountain terrain facilitates wide dispersal, allowing relatively smait units to hold dominant terrain in a connected system of strong points. To prevent bypassing and envelopment attempts, the enemy may adopt a many-tiered, perimeter defense. Aside from the relative size of forces, the type of enemy units and their equipment must be compared with those of friendly forces, to include a comparison of the suitability of forces, tactics, and training. When considering the enemy's ability to operate in mountainous terrain, commanders should consider how well the enemy can accomplish the tasks and actions listed in Figure **2-1.** Again, in analyzing both enemy and friendly factors during mountain operations, the vertical, as well as the horizontal, perspective should be fully integrated into all aspects of the assessment.

- **Utilize the environment to his advantage**
- **Conduct air operations**
- **Conduct decentralized operations**
- **Utilize the terrain in Levels II and III**
- **Employ obstacles or barriers to restrict maneuverability**
- **Conduct limited-visibility operations**
- **Sustain his maneuver elements**

Figure 2-1. Factors Affecting Assessment of the Enemy Situation

TERRAIN AND WEATHER

2-7. As in all military operations, terrain analysis involves observation and fields of fire, cover and concealment, obstacles, key terrain, and avenues of approach (OCOKA). Terrain often influences the conduct of operations more in the mountains than on flatter terrain. The mountains form the nonlinear and vertical structure of the battlefield, and the influences of geography and climate dictate the extent to which commanders modify tactics. Examples of these difficulties are often encountered in the concentration of forces, as well as in the maintenance of command and control.

2-8. In themountains, as elsewhere, surprise is easier to achieve for the force that knows the terrain better and has the skills and equipment necessary to achieve greater mobility. The appropriate use of vertical terrain improves the element of surpriseiftheterrain has been analyzed properly to determine the best means to counter the enemy's reactions. Once the commander decides on a preliminary course of action, he should immediately initiate a detailed terrain reconnaissance,

2-9. In a mountainous environment, the terrain normally favors the defender and necessitates the conduct of limited visibility operations. Highly trained units can achieve significant tactical gains and decisive victories by exploiting limited visibility. However, limited visibility operations in mountainous terrain require precise planning, careful daylight reconnaissance, exceptionally good command and control, and a high degree of training. Imaginative and bold limited visibility operations can minimize the advantage of terrain for the defender and shift the balance of combat power to the side that can best cope with or exploit limited visibility.

OBSERVATION AND FIELDSOF FIRE

2-10. Although mountainous terrain generally permits excellent long-range observation and fields of fire, steep slopes and rugged terrain affect a soldier's ability to accurately estimate range and frequently cause large areas to be hidden from observation. The existence of sharp relief and dead space facilitates covert approaches, making surveillance difficult despite such long-range observation. Four factors that influence what can be seen and hit in mountainous terrain are listed in Figure 2-2.

1. The ability to observe and identify targets in conditions of bright sunlight
2. The ability to estimate range in clear air
3. The ability to apply wind corrections
4. The ability to shoot accurately up and down vertical slopes

Figure 2-2. Factors Affecting Observation and Fields of Fire

COVER AND CONCEALMENT

2-11.The identification and proper use of the cover and concealment provided by mountainous terrain are fundamental to all aspects of mountain operations. The ridge systems found in Level II may provide covert approaches through many areas that are hidden from observation by the vegetation and relief. The difficulties a force encounters in finding available cover and concealment along ridges are fewer than those on the peaks, especially above the timberline. Uncovered portions of an approach leave a force exposed to observation and fire for long periods. The enemy can easily detect movement in this region, leaving commanders with three primary options to improve cover and concealment:

1, Identify and exploit avenues of approach the enemy would consider unlikely, due to their difficult ascent or descent.

2, Negotiate routes during periods of limited visibility.

3, Provide overwhelming route security.

OBSTACLES

2-12, Obvious natural obstacles indude deep defiles, diffs, rivers, landslides, avalanches, crevices, and scree slopes, as well as the physical terrain of the mountains themselves. Obstades vary in their effect on different forces Commanders must evaluate the terrain from both the enemy and friendly force perspective. They must look spedfreal ly at the degree to which obstades restrid operations, and at the ability of each force to exploit the tactical opportunities that exist when obstades are employed. M an-made obstades used in conjundion with restrictive terrain are extremely effedive in the mountains; however, their construction is very costly in terms of time, materiel, transportation assets, and labor. Commanders must know the location, extent, and strength of obstades so that they can be incorporated into their scheme of maneuver.

KEY TERRAIN

2-13. Key terrain generally increases in importance with an increase in elevation and a decrease in accessibility. In the mountains, however, terrain that is higher than that held by the opposing force is often key, but only if the force is capable of fighting there. A well-prepared force capable of maneuver in rugged terrain can gain an even greater advantage over an ill-prepared enemy at higher elevation levels.

2-14. The vast majority of operations in the mountains requires that the commander designate dedsive terrain in his concept of operations to communicate its importance to his staff and subordinate commanders. In operations over mountainouster rain, the analysis of key and ded si ve terrain is based on the identification of these features at each of the three operational terrain levels. There are few truly impassable areas in the mountains. The commander must recognize that what may be key terrain to one force may be an obstacle to another force. He must also recognize that properly trained combatants can use high obstruding terrain as a means to achieve dedsive vidories with comparatively small-sized combat elements.

AVENUES OF APPROACH

2-15. I n mountainous terrain, there are few easily accessible avenues of approach, and they usually run along valleys, defiles, or the crests and spurs of ridges. This type of geography allows the defender to economize in difficult terrain and to concentrate on dangerous avenues of approach A typical offensive tactic is to conduct a coordinated assault with the main effort along accessible avenues of approach, and supporting efforts by one or more maneuver elements on difficult and unexpected avenues of approach. Normally, high rates of advance and heavy concentration of forces are difficult or impossible to achieve along mountainous avenues of approach. Relief features may create large areas of dead space that facilitate covert movement. Units may use difficult and unlikely avenues of approach to achieve surprise: however, these are extremely high-risk operations and are prone to failure unless forces are well trained and experienced in mountaineering techniques. In mountainous terrain, the analysis of avenues of approach should be based on a thorough reconnaissance and evaluated in terms of the factors listed in Figure 2-3 on page 2-8.

WEATHER

2-16. As discussed in Chapter 1, weather and visibility conditions in the mountainous regions of the world may create unprecedented advantages and disadvantages for combatants. To fight effectively, commanders must acquire accurate weather information about their AO. Terrain has a dominant effect on local climate and weather patterns in the mountains. Mountainous areas are subject to frequent and rapid changes of weather, including fog, strong winds, extreme heat and cold, and heavy rain or snow. Thus, many forecasts that describe weather over large areas of terrain are inherently inaccurate. Commanders must be able to develop local, terrain-based forecasts by combining available forecasts with field observations (local temperature, wind, precipitation, cloud patterns, barometric pressure, and surrounding terrain). Forecasting mountain weather from the field improves accuracy and enhances the ability to exploit opportunities offered by the weather, while minimizing its adverse effects (see Appendix B).

- **Ability to achieve surprise**
- **Vulnerability to attack from surrounding heights**
- **Ability to provide mutual support to forces on other avenues of approach**
- **Effect on rates of advance**
- **Effect on command and control**
- **Potential to accommodate deception operations**
- **Ability to support necessary CS and CSS operations**
- **Access to secure rest and halt sites**
- **Potential to fix the enemy and reduce the possibility of retreat**

Figure 2-3. Factors Affecting Analysis of Avenues of Approach

TROOPS AND SUPPORT AVAILABLE

2-17. Commanders must assess the operational and tactical implications of the restrictive environment on mobility, protection, firepower, and logistics. The complex task of arranging activities in time, space, and purpose requires commanders to fully understand the impact of elevation, weather, and visibility on the capabilities of his subordinate elements and relative combat power. Mountainous terrain and weather can greatly enhance the relative combat power of defending forces and, conversely, it can drastically reduce those of the attacking forces. For example, an infantry battalion may be inadequate to defeat a defending infantry company in the mountains. Instead, an infantry battalion may only be capable of defeating a well-positioned infantry platoon. However, commanders must carefully consider each unique situation and weigh all tangible and intangible aspects of combat power (maneuver, firepower, leadership, protection, and information) when comparing strengths and determining the forces necessary to accomplish the mission.

2-18. Commanders must also assess the proper mix of heavy and light forces that capitalizes on the unique strengths that each type of force can bring to mountain operations while minimizing their limitations. While generally complicating command and control, an appropriate mix allows commanders more flexibility in the synchronization of their operations. Additionally, the difficulty providing combat support and combat service support for mountain operations must be evaluated to determine if the proportion of support troops to combat troops is sufficient.

2-19. Prior to and throughout an operation, commanders must continually assess the effect that the rugged mountain environment and sustained combat operations has on the ability of their soldiers to accomplish the mission.

Commanders may need to slow the pace of their operation, transition to the defense for short periods, or rotate units to ensure that their soldiers are physically capable of striking effectively at decisive times and locations. Too often, commanders consider only the operational readiness (OR) rate of equipment and logistics levels when determining their overall ability to continue offensive actions. Failure to consider this intangible human aspect may result in increased loss of lives and mission failure.

2-20. Vertical operations are an integral part of mountain operations and are one means to improve the success of decisive engagements. Commanders must review the state of training of their units to ensure they are adequately prepared to maneuver and fight at various elevations. Increased requirements for aviation support require aviation units to be capable of operating in the specific mountain environment. Units must also have sufficient numbers of pathfinders and trained air assault personnel to select and mark landing zones (LZs) and prepare sling loads.

TIME AVAILABLE

2-21. In the mountains, proper timing is fundamental to creating opportunities to fight the enemy on favorable terms. Restrictive terrain, weather, the accumulation of chance errors, unexpected difficulties, and the confusion of battle increase the time necessary to assemble, deploy, move, converge, and mass combat power, effectively decreasing the amount of time available to plan and prepare. To optimize the time available, commanders must continuously evaluate the impact of reduced mobility caused by the weather and terrain. At times, commanders may need to conduct a tactical pause to facilitate the concentration of combat power at a decisive point. However, they must consider time with respect to the enemy as time available is always related to the enemy's ability to execute his own plan, prepare, and execute cycle. Figure 2-4 summarizes the time considerations that are different from or greater than those encountered on flatter terrain.

- **Adaptability of plans to the terrain and varying weather**
- **Increased time needed to conduct reconnaissance, execute movements, and synchronize events on the battlefield**
- **Significant variance in the number of hours of visibility with season and elevation**

Figure 2-4. Factors Affecting Time Available

CIVIL CONSIDERATIONS

2-22. Generally, civilian population centers will be located at the lower elevations of Level I close to sources of water and along major lines of communications. Refugees and displaced civilians may increase congestion on the already limited road and trail networks normally found in mountainous environments, further complicating maneuver and sustaining operations.

2-23. Commanders must also consider the impact of operations on the often-limited civilian resources available in the mountains. The wisdom of using local resources to lighten in-theater supply requirements must be balanced against the impact on civilians and their local economy. While the purchase of goods and services from the local economy is generally welcomed, it may serve to inflate prices and make it impossible for local civilians to purchase their own scarce and needed supplies.

2-24. In mountainous regions, commanders often encounter a populace of diverse political and ethnic orientation that may support, oppose, or be ambivalent to US operations or the presence of US forces. Depending on friendly force objectives, commanders may conduct public relations, civil affairs, humanitarian assistance, and psychological operations (PSYOP) to influence perceptions and attitudes of neutral or uncommitted parties. Even if commanders cfioose not to commit resources to enlist civilian sympathy and support, they must still adjust their operations to minimize damage and loss of life to innocent civilians.

SECTION II-LEADERSHIP

2-25. To help ease their anxiety in combat, soldiers must have confidence in their leaders. This confidence may diminish rapidly unless leaders demonstrate the ability to lead over formidable terrain and under the most difficult weather conditions. Superficial knowledge of mountain warfare and ignorance or underestimation of mountain hazards and environmental effects may result in mission failureand the unnecessary loss of soldiers' lives.

2-26. Effective leadership in mountain operations combines sound judgment with a thorough understanding of the characteristics of the mountain environment. Commanders must first develop flexible and adaptable leadership throughout the chain of command. They must then be able to understand and exploit the operational and tactical implications of the mountain environment, as well as its effects on personnel, equipment, and weapons. The keys to meeting this challenge are proper training and operational experience in the mountains. To fight effectively, leaders creatively exploit the opportunities offered by the mountain environment while minimizing the adverse effectsit can have on their operations.

2-27. Leadership rapidly becomes the primary element of combat power on the mountain battlefield. Commanders must recognize the distinctive effects created by decentralization of command, develop a depth of leadership that forms the vital link to unity of effort, and organize and direct operations that require minimum intervention. While specific situations require different leadership styles and techniques, the nature of mountain warfare generally necessitates that commanders embrace the philosophy of command and control known as mission command (see FM 6-0). This type of command and control requires subordinates to make decisions rapidly within the framework of the commander's concept and intent. Commanders must be able to accept some measure of uncertainty, delegate, and trust and encourage subordinate leaders at all levels to use initiative and act alone to achieve the desired results, particularly when the situation changes and they lose contact with higher headquarters.

SECTION III - COMMUNICATIONS

2-28. The communications means available to support operations in mountainous regions are the same as those to support operations in other regions of the world. However, rapid and reliable communications are especially difficult to achieve and maintain in mountainous areas. The mountainous environment requires electronic equipment that is light, rugged, portable and able to exploit the advantages of higher terrain. The combined effects of irregular terrain patterns, magnetic and ionospheric disturbances, cold, ice, and dampness on communications equipment increase operating, maintenance, and supply problems and require precise planning and extensive coordination.

COMBAT NET RADIO

SINGLE-CHANNEL GROUND AND AIRBORNE RADIO SYSTEMS (SINCGARS)

2-29. The Single-channel Ground and Airborne Radio System (SINCGARS) family of frequency modulation (FM) radios is good for the control of battalion and smaller-sized units operating in a mountainous environment (see FM 6-02.32 and FM 6-02.18). If available, hands-free radios, such as helmet-mounted radios, are an excellent means of communication for small unil tactics and close-in distances, particularly while negotiating rugged terrain. In colder environments, shortened battery life greatly reduces the reliability of manpacked systems that rely on constant voltage input to maintain maximum accuracy.

2-30. Since even a small unit may be spread over a large area, retransmission sites may be needed to maintain communications and increase range. These sites require extensive preparation and support to ensure the survival of personnel and the continued maintenance of equipment. Retransmission systems are often placed on the highest accessible terrain to afford them the best line-of-sight; however, through simple analysis, these locations are often predictable and make them more vulnerable to enemy interdiction. The importance and difficulty of maintaining adequate communications in mountainous terrain requires commanders to devote additional resources for the protection of these limited assets and operators skilled in the proper use of cover and concealment, noise and light discipline, and other operations security (OPSFC) measures.

2-31. Physical range [imitations, difficulties in establishing line-of-sight paths due to intervening terrain, and limited retransmission capabilities often make it difficult to establish a brigade and larger-sized radio net. However, commanders can, if within range, enter subordinate nets and establish a temporary net for various contingencies. In the mountains or if the mobile subscriber equipment network is not yet fully developed, commanders should consider the increased need for the improved high frequency radio (IHFR) family of amplitude modulation (AM) radios and single-channel tactical satellite communications terminals for extended distances.

SATELLITE COMMUNICATIONS (SATCOM)

2-32. Satellite communications (SATCOM) terminals are light, small, portable ground terminals that are able to communicate in spite of rugged terrain. During operations in mountainous areas having little or no infrastructure to support command and control, satellite communications become the primary means of communications. Single channel SATCOM are currently transmitted over the ultrahigh frequency (UHF) band and readily support forces operating in the mountains, while providing worldwide tactical communications, in-theater communications, combat net radio (CNR) range extension, and linkage between elements of long-range surveillance units (LRSUs) and Army special operations foroes (ARSOF). SATCOM can network with multiple users, communicate while enroute, penetrate foliage while on the ground, and has several other advantages making it an ideal system for mountain communications (see Figure 2-5). However, limitations include restricted access, low-rate data communications, and lack of antijam capability Commanders should review FM 6-02.11 for further information on the employment of SATCOM.

- Greater freedom from siting restrictions
- Extended range, capacity, and coverage
- Mobility and rapid employment
- Extremely high circuit reliability

Figure 2-5. SATCOM Advantages

COMMAND AND CONTROL (C^2) AIRCRAFT

2-33 Using C^2 aircraft can assist the commander in overcoming ground mobility restrictions and may improve communications that would otherwise limit his ability to direct the battle. In the mountains, terrain masking, while making flight routing more difficult, may provide the degree of protection needed to allow an increased use of aircraft. To avoid radar or visual acquisition and to survive, C^2 aircraft must use the same terrain flight techniques employed by other tactical aviation units. This flight method often degrades FM communications and reinforces the requirement for radio relay or retransmission sites

ANTENNAS AND GROUNDS

2-34. Directional antennas, both bidirectional and unidirectional, may be needed to increase range and maintain radio communications. Although easy to fabricate, directional antennas are less flexible and more time-consuming to set up. Positioning of all antennas is also crucial in the mountains because moving an antenna even a small distance can significantly affect reception

2-35. Antenna icing, a common occurrence at high elevations, significantly degrades communications. Ice may also make it difficult to extend or lower antennas, and the weight of ice buildup, combined with increased brittleness, may cause them to break. Antennas should have extra guy wires, supports, and anchor stakes to strengthen them to withstand heavy ice and wind loading. All large horizontal antennas should be equipped with a system of counterweights arranged to slacken before wire or poles break from the excess pressures of ice or wind. Soldiers may be able to remove wet snow and sleet that freezes to antennas by jarring their supports, or by attaching a hose to the exhaust pipe of a vehicle and directing the hot air on the ice until it melts. However, soldiers must exercise great care to ensure that the antenna is not damaged in their attempts to dislodge the ice.

2-36. Ground rods and guy wires are often difficult to drive into rocky and frozen earth. Mountain pitons are excellent anchors for antenna guys in this type of soil. In extreme cold, ropes can be frozen to the ground and guys tied to these anchor ropes. Adequate grounding is also difficult to obtain on frozen or rocky surfaces due to high electrical resistance. Where it is possible to install a grounding rod, it should be driven into the earth as deep as possible or through the ice on frozen lakes or rivers. Grounding in rocky soil may be improved by adding salt solutions to improve electrical flow.

MOBILE SUBSCRIBER EQUIPMENT

2-37. Like FM. mobile subscriber equipment [MSE) requires a line-of-sight transmission path and a tactical satellite or several relay sites to overcome mountainous terrain and maintain MSE connectivity (FM 6-02.55 contains in-depth information concerning the deployment and employment of MSE).

WIRE AND FIELD PHONES

2-38. Wire is normally one of the most reliable means of communication Unfortunately, in rugged mountains and particularly during the winter months, wire is more difficult and time consuming to install, maintain, and protect. Wire may be dispensed in mountain areas by tracked or wheeled vehicle, foot, skis, snowshoes, or oversnow vehicles. As in any environment, units must periodically patrol their wire lines to ensure that they have remained camouflaged and that the enemy has not tapped into them.

2-39. Snow-covered cables and wire can cause the loss of many man-days in recovering or maintaining circuits. This can be avoided by pulling the cable from under the snow after each snowfall and letting it rest just below the surface of the snow. Trees or poles can be used to support wire. Allowances must be made for drifting snow when determining the height above ground at which to support the lines. However, when crossing roads, it is preferable to run the wire through culverts and under bridges rather than bury or raise wire overhead. In addition to ease, this technique reduces maintenance requirements associated with vehicles severing lines, particularly with higher volumes of traffic on limited road networks. If long-distance wire communications are required, the integration of radio relay systems must be considered.

2-40. Great care must be taken in handling wire and cables in extreme cold weather. Condensation and ice on connectors make connecting cables difficult and can degrade the signal path. When rubber jackets become hard, the cables must be protected from stretching and bending to prevent short circuits caused by breaks in the covering. Therefore, all tactical cable and wire should be stored in heated areas or warmed prior to installation. TC 24-20 provides more detailed information on the installation and maintenance of wire and cable.

2-41. Field phones are used in a stationary position, such as a mountain patrol base or an ambush site, although leaders must consider the weight and difficulties encountered in laying and maintaining wire in these sites of limited duration. The batteries that are used to operate field telephones and switchboards are subject to the same temperature limitations as those used to power tactical radiosets.

2-42. When used with a hands-free phone, commercially available rope with a communication wire in it is ideally suited for mountain operations. This system is lightweight and easy to manage, and provides an added measure of security during limited visibility operations. In addition to the standard uses, since it functions as both a rope and a wire, it can be used to control movement on all types of installations, and it can serve as a primary means of communication for climbing teams.

AUDIO, VISUAL, AND PHYSICAL SIGNALS

2-43. Leaders can use simple audio signals, such as voice or whistles, to locally alert and warn. Sound travels farther in mountain air. Although this effect may increase the possibility of enemy detection, interrupting terrain, wind conditions, and echoes can restrict voice and whistle commands to certain directionsand uses.

2-44. Like audio signals, visual signals such as pyrotechnics and mirrors have limited use dueto enemy detection, but may work for routine and emergency traffic at the right time and place. Blowing sand or snow, haze, fog, and other atmospheric conditions may periodically affect range and reliability.

2-45. Units should use hand and arm signals instead of the radio or voice when ever possi bl e, especi al I y wh en cl ose tothe enemy. Luminous tape on t he camouflage band, luminous marks on a compass, or flashlights may be used as signals at night over short distances. Infrared sources and receiving equipment, such as night vision goggles, aiming lights, and infrared filters for flashlights, can be used to send and receive signals at night However, an enemy outfitted with similar equipment can also detect active devices.

2-46. A tug system is a common method of signaling between members of a roped climbing team. However, tug systems are often unreliable when climbers are moving on a rope or when the distance is so great that the friction of the rope on the rock absorbs the signals. Separate tug lines can be installed in Static positions by tying a String, cord, or wire from one position to the next. Soldiers can pass signals quietly and quickly between positions by pulling on thetug line in a prearranged code.

MESSENGER

2-47. Although slew, communication by messenger is frequently the only means available to units operating in the mountains. Messengers should be trained climbers, resourceful, familiar with mountain peculiarities, and able to carry their own existence load. During the winter, advanced skiing skills may also be required. Messengers should always be dispatched in pairs. Air messenger service should be scheduled between units and integrated with the aerial resupply missions. Vehicles may also be employed to maintain messenger communications when conditions of time, terrain, and distance permit.

SECTION IV - TRAINING

2-48. Because US forces do not routinely train in a mountain environment, they must make extensive preparations to ensure individual and unit effectiveness. Ultimate success in the mountains depends largely on developing cohesive, combat-ready teams consisting of well-trained soldiers. To be successful, commanders must understand the stratification of mountain warfare, recognize the unique aspects of leadership required, and implement training programs that prepare soldiers for the rigors of mountain fighting.

2-49. In the mountains, commanders face the challenge of maintaining their units' combat effectiveness and efficiency. To meet this challenge, commanders conduct training that provides soldiers with the mountaineering skills necessary to apply combat power in a rugged mountain environment, and they develop leaders capable of applying doctrine to the distinct characteristics of mountain warfare.

- Mountaineering skills
- Air assault and air movement operations
- Deception
- Stealth and infiltration
- Limited visibility operations
- Patrolling
- Reconnaissance
- Communications
- CS and CSS operations

Figure 2-6. Training Areas of Emphasis

2-50. The ability to apply doctrine and tactics in mountainous environments is not as easy to develop as technical proficiency. Training, study, and garrison experimentation may provide the basis for competence. However, only through experience gained by practical application in the mountains will leaders become skilled in mountain warfare. Proficiency in the areas listed in Figure 2-6 wil l provide commanders with a degree of flexibility in the application of doctrine to a mountain area of operations.

2-51. The best combat and combat support plans cannot ensure victory unless commanders concentrate on developing a leadership climate that is derived from the human dimension of mountain warfare. The complexities of mountain combat make it extremely important to establish training programs that modify the traditional application of tactics so that units can reach their full potential. Training must simulate the tempo, scope, and uncertainty of mountain combat to create the versatility required to capitalize on the harsh environment as a force multiplier,

2-52. Competent units operate effectively in mountains and focus on the battle. Unprepared units, however, may become distracted by the environment and end up expending as much effort fighting the environment as they do fighting the enemy. Soldiers cannot be fully effective unless they have the proper clothing and equipment, and are trained to protect themselves against theeffectsof terrain and frequent and sudden changes in weather.

INITIAL TRAINING ASSESSMENT

2-53. In addition to the questions applicable to every mission, commanders must consider the following when preparing for operations in a mountainous environment:

- What kind of mountains will the unit be operating in?
  - What elevations will the unit be operating at?
  - What are the climatic and terrain conditions of the AO?
    - » Are at least two years of accurate weather reports available (see Appendix B)?
- When must the unit be ready to move?
- What training resources are needed and available?

D Are local training areas and ranges available?

  - If not. what alternative arrangements can be made?
  - What available training areas most closely resemble the AO?

D What special equipment does the unit require?

D What training assistance is available?

  - Does the unit have former mountain warfare instructors, military mountaineers, or others with experience in a mountainous environment?
  - Are instructors available from outside the unit?

P What special maintenance is required for weapons and equipment?

- What is the level of physical fitness?

0 What additional combat, combat support, and combat service support units are necessary to accomplish the operational missions?

  - Can specific units be identified for possiblecoordinated training?

D Will allied and multinational troops participate?

2-54, As commanders get answers to these and other questions, they must develop training programs to bring their units to a level where they will be fully capable of operating successfully in mountainous conditions. To do this, they must establish priorities for training. The training requirements listed in Figure 2-7 are only a guide. Commanders should add, delete, and modify the tasks as necessary, depending on the specific AO, the state of readiness of their units when they begin preparations for mountain operations, and the time and facilities available (see FM 7-10).

PHYSICAL CONDITIONING

2-55. Soldiers who have lived and trained mostly at lower elevations tend to develop a sense of insecurity and fear about higher elevations - many are simply afraid of heights in general. With this in mind, leaders must plan training that accustoms soldiers to the effects of the mountain environment. Physical conditioning must be strictly enforced, since "new musde" strain associated with balance and prolonged ascents/descents quickly exhausts even the most physically fit soldiers. Even breathing becomes strenuous, given the

TRAINING REQUIREMENTS	ALL	STAFF AND LEADERS	TEAMS AND CREW MEMEBERS	SPECIALISTS
Physical Conditioning and Acclimatization	✔	✔	✔	✔
Mountain Illnesses and Injuries	✔	✔	✔	✔
Mountain Living and Survival	✔	✔	✔	✔
Mountain Navigation Techniques	✔	✔	✔	✔
Mounted and Dismounted March Planning	✔	✔	✔	✔
Communications Techniques	✔	✔	✔	✔
Weapons/Equipment Training	✔	✔	✔	✔
Additional Maintenance Requirements	✔	✔	✔	✔
Camouflage and Concealment	✔	✔	✔	✔
Obstacles		✔	✔	
Above-ground Fortifications	✔	✔	✔	✔
Level 1 Mountaineering	✔	✔	✔	✔
Level 2 and 3 Mountaineering				✔
Driver and Pilot Training				✔
Air Assault/Air Movement Operations	✔	✔	✔	✔
NBC Operations	✔	✔	✔	✔

Figure **2-7.** Mountain Preparatory Training

thinner atmosphere at higher altitudes. Therefore, training must emphasize exercises designed to strengthen leg muscles and build cardiovascular (aerobic) endurance (see FM 3-25.20). Frequent marches and dimbswith normal equipment loads enhance conditioning and familiarize soldiers with mountain walking techniques,

MOUNTAIN LIVING

2-56. Successful mountain living requires that personnel adjust to special conditions, particularly terrain and weather. To develop confidence, soldiers should train in conditions that closely resemble those they will face. Lengthy exercises test support facilities and expose soldiers to the isolation common to mountain operations. The mountain area of operations can be harsh, and training should develop soldiers who possess the necessary field craft and psychological edge to operate effectively under mountainous conditions. Although FM 4-25.10 and FM 3-25.76 do not specifically address mountain environments, much of their information applies. Regardless of the level of technical mountaineering training required, all soldiers deploying to a mountainous region should be trained in the areas listed in Figure *2-8* on page 2-**18.**

2-57. Navigation in the mountains is made more difficult because of inaccurate mapping, magnetic attraction that affects compass accuracy, and the irregular pace of the soldiers It is easy to mistake large terrain features that are very far away for features that are much closer The increased necessity for limited-visibility operations restricts the use of terrain techniques as the primary means of determining and maintaining direction. Individuals must train to use a variety of equipment, such as a compass, an altimeter, global positioning system devices, and maps, as well as learn techniques pertaining to terrestrial navigation, terrain association, dead reckoning, resecfioning, and artillery marking (seeFM 3-25.26).

- Temperature extremes and clothing requirements
- Bivouac techniques and shelter construction
- Elevation and rarified air effects
- Hygiene, sanitation, and health hazards
- Locating and purifying water
- Food-gathering techniques

Figure 2-8. Mountain Living Training

MD EQUIPMENT

2-58. Nearly every weapon or piece of equipment familiar to the soldier is affected to some degree by the mountain environment. In addition to honing skills, training must focus on the specific operational area and ways to overcome anticipated environmental impacts when using weapons and equipment.

2-59. Individual marksmanship training must emphasize the effect of wind and include practical training in wind measurement techniques and adjusted aiming points (holdoff). Practical training in range estimation techniques, combined with using laser range finders, M19 binoculars, target reference points, and range cards, helps lo overcome difficulties in range estimation.

2-60. In the conduct of their preparations, commanders should strive to increase the number of qualified snipers within their units, as they are ideal in the mountains and can be used to adversely affect enemy mobility by delivering long range precision rifle fire on selected targets. They can inflict casualties, slow enemy movement, lower morale, and add confusion to enemy operations A single sniper team in well-concealed positions, such as mountain passes, can severely impede enemy movement (see FM 3-21.20 and FM 3-91.2 for further information on sniper employment).

;E AND CONCEALMENT

2-61. The basic principles of camouflage and concealment also apply in mountain operations (see FM 3-24 3). However, certain elements must be adjusted for snow. With snow on the ground, standard camouflage nets and paint patterns are unsuitable. In areas where snow cover is above 15 percent of the background, winter camouflage nets should take the place of standard nets and temporary white paint should be used over the green portions of vehicles. In terrain with more than 85 percent snow cover, the vehicles and equipment should be solid white. However, with less than 15 percent snow cover, standard patterns should be maintained.

2-62. Snow provides excellent conditions for threat thermal and ultraviolet sensor detection. To counter these types of sensors, soldiers must be trained to utilize the terrain to mask themselves and their weapons and equipment from enemy detection. The mountainous terrain often limits the access routes to and from selected positions. Commanders must take appropriate measures to conceal vehicle tracks and limit movement times to periods of limited visibility. Snow presents a significant problem, making movement discipline an absolute requirement. When moving, leaders should be trained to follow the shadows along windswept drift lines as much as possible. Drivers should learn to avoid sharp turns, which are easily recognizable in the snow, and follow existing track marks where possible.

FORTIFICATIONS

2-63. Fighting and protective positions in the mountains do not differ significantly from other environments, except in areas of snow and rock (see FM 3-34.112 for more information on common survivability positions and FM 3-97.11 for positions created in snow). Digging positions in rocky ground is difficult and often impossible. If demolitions, pneumatic drills, and jackhammers are available, positions may be blasted or drilled in the rock to afford some degree of protection. More often, it will be necessary to build above-ground positions by stacking boulders, stones, and gabions. If possible, existing rock formations should be used as structural wall components.

2-64. If above-ground positions are to be used, considerable care should be taken to avoid siting them in view of any likely enemy avenues of approach. Even a two-man position is difficult to conceal if it is above the timberline. Camouflage nets and the use of background rocks are necessary to break up theoutlineof theposition and hide straight edges.

2-65. Positions should be built of the largest rocks available, wedged securely together. Extreme care should be taken that the walls are stable and not leaning or sloping downhill. An unstable wall is more of a liability than an asset, as the first Impact may cause it to collapse onto the defenders. Rocks and gabions should be stacked to systematically overlap each joint or seam to help ensure stable construction. Larger rocks or stones can be used to help bond layers of rock beneath. If possible, a layer of sandbags should be placed on the top of and around the inside of the wall. Substantial overhead cover is normally required in rocky areas. The effects of artillery bursts within and above a protective position are greatly enhanced by rock and gravel displacement or avalanche Figure 2-9 on page 2-20 shows simple examples of the right and wrong way to build these positions.

MILITARY MOUNTAINEERING

2-66. The skills required for movement are often difficult to learn and usually very perishable. Commanders must understand the application and mechanics of technical mountaineering systems needed for mobility and movement of soldiers and equipment. In the mountains, a unit may be ineffective unless it has the prerequisite technical training. However, some mountains may feature terrain that is relatively benign, requiring minimal specialized techniques. Other areas will mandate the need for more advanced mountaineering skills. One key to quickly determining the type and extent of training required is to analyze and classify the level of individual movement required according to the dismounted mobility classification table introduced in Chapter 1. Once commanders have determined the specific level and tasks required, TC 90-6-1 will provide them with detailed information on specific mountaineering techniques and equipment (described below).

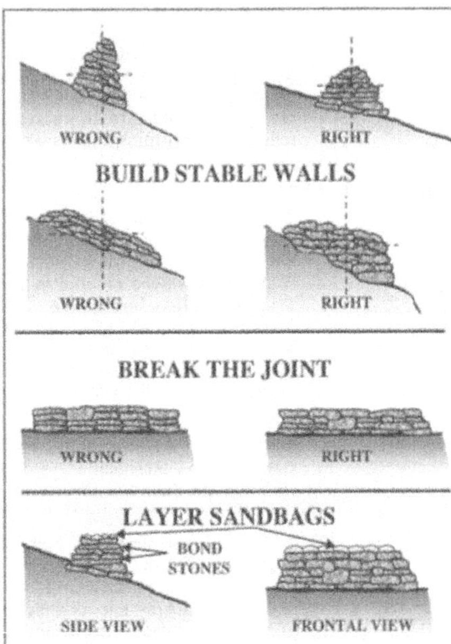

2-67. Military mountaineering training provides tactical mobility in mountainous terrain that would otherwise be inaccessible. Soldiers with specialized training who are skilled in using mountain climbing equipment and techniques can overcome the difficulties of obstructing terrain. Highly motivated soldiers who are in superior physical condition should be selected for more advanced military mountaineering training (Levels 2 and 3) conducted at appropriate facilities. Soldiers who have completed advanced mountaineering training should be used as trainers, guides, and lead climbers during collective training. They may also serve as supervisors of installation teams (see Chapter 4) and evacuation teams (see Chapter 5). Properly used, these soldiers can drastically improve mobility and have a positive impact disproportionate to their numbers. Units anticipating mountain operations should strive to achieve approximately ten percent of their force with advanced mountaineering skills.

LEVEL 1: BASIC MOUNTAINEER

2-68 The basic mountaineer, a graduate of a basic mountaineering course, should be trained in the fundamental travel and climbing skills necessary to move safely and efficiently in mountainous terrain. These soldiers should be comfortable functioning in this environment and, under the supervision of qualified mountain leaders or assault climbers, can assist in the rigging and use of all basic rope installations. On technically difficult terrain, the basic mountaineer should be capable of performing duties as the "follower" or "second" on a roped dimbing team, and should be well trained in using all basic rope systems. These soldiers may provide limited assistance to soldiers unskilled in mountaineering techniques. Particularly adept soldiers may be selected as members of special purpose teams led and supervised by advanced mountaineers. Figure 2-10 lists the minimum knowledge and skills required of basicmounlaineers

☐ Characteristics of the mountain environment (summer and winter) ☐ Mountaineering safety ☐ Use, care, and packing of individual cold weather clothing and equipment ☐ Care and use of basic mountaineering equipment ☐ Mountain bivouac techniques ☐ Mountain communications ☐ Mountain travel and walking techniques ☐ Hazard recognition and route selection ☐ Mountain navigation ☐ Basic medical evacuation	☐ Rope management and knots ☐ Natural anchors ☐ Familiarization with artificial anchors ☐ Belay and rappel techniques ☐ Use of fixed ropes (lines) ☐ Rock climbing fundamentals ☐ Rope bridges and lowering systems ☐ Individual movement on snow and ice ☐ Mountain stream crossings (to include water survival techniques) ☐ First aid for mountain illnesses and injuries

Figure 2-10. Level 1: Basic Mountaineer Tasks

2-59. In a unit training program, level 1 qualified soldiers should be identified and prepared to serve as assistant instructors to train unqualified soldiers in basic mountaineering skills. **All** high-risk training, however, must be conducted under the supervision of qualified level 2 or 3 personnel.

LEVEL 2: ASSAULT CLIMBER

2-70, Assault climbers are responsible for the rigging, inspection, use, and operation of all basic rope systems. They are trained in additional rope management skills, knot tying, belay and rappel techniques, as well as using specialized mountaineering equipment. Assault dimbers are capable of rigging complex, multipoint anchors and high-angle raising/lowering systems. Level 2 qualification is required to supervise all high-risk training associated with Level 1. At *a* minimum, assault climbers should possess the additional knowledge and skillsshown in Figure2-11 on page 2-22.

LEVEL 3: MOUNTAIN LEADER

2-71, Mountain leaders possess all the skills of the assault climber and have extensive practical experience in a variety of mountain environments in both winter and summer conditions. Level 3 mountaineers should have well-developed hazard evaluation and safe route finding skills over nil types of mountainous terrain. Mountain leaders are best qualified to advise commanders on all aspects of mountain operations, particularly the preparation and leadership required to move units over technically difficuft, hazardous, or

- Use specialized mountaineering equipment
- Perform multipitch climbing:
  - Free climbing and aid climbing
  - Leading on class 4 and 5 terrain
- Conduct multipitch rappelling
- Establish and operate hauling systems
- Establish fixed ropes with intermediate anchors
- Movement on moderate angle snow and ice
- Establish evacuation systems and perform high angle rescue
- Perform avalanche hazard evaluation and rescue techniques
- Familiarization with movement on glaciers

Figure 2-11. Level 2: Assault Climber Tasks

exposed terrain. The mountain leader is the highest level of qualification and is the principle trainer for conducting mountain operations. I nstructor experience at a military mountaineering training center or as a member of a special operations forces (SOF) mountain team is critical to acquiring Level 3 qualification. Figure 2-12 outlines the additional knowledge and skills expected of mountain leaders. Depending on the specific AO, mountain leaders may need additional skills such as snowshoeing and all-terrain skiing

Figure 2-12. Level 3: Mountain Leader Tasks

DRIVER TRAINING

2-72. Driving in mountains is extremely difficult. To be successful, drivers must know their equipment's limitations and capabilities Training should center on practical exercises in mountainous terrain that gradually introduce drivers to more complex terrain and weather conditions. The exact nature of the mountainous terrain determines the training (see Figure 2-13).

- Identification and recognition of potential dangers
- Movement along steep grades combined with:
  - Narrow roads and sharp curves
  - Loose rock and gravel
  - Ice and snow (to include using tire chains for wheeled vehicles)
  - Towed loads
- Increased cold weather maintenance requirements

Figure 2-13, Driver Training

ARMY AVIATION

2-73. The mountainous environment, particularly its severe and rapidly changing weather, affects aircraft performance capabilities, accelerates crew fatigue, and influences basic flight techniques. These techniques can be acquired only through a specific training program for the particular type of mountainous terrain. Additionally, limited visibility operations in the mountains are extremely hazardous and require extensive training for those aviation units involved. Common problems associated with mountain operations become much more complex at night, even when using night vision devices. Few Army aviation units regularly train for mountain operations, so it is critical to alert them as soon as possible to fadlitate the required training to ensure safe and successful mission execution.

RECONNAISSANCE AND SURVEILLANCE

2-74. Training in reconnaissance and surveillance should focus on trafficability (route, mobility, and bridge classification), potential drop zones or landing areas, likely defensive positions, and potential infiltration routes. Infiltration and exfiltration are relatively easy in mountainous terrain and constitute a significant threat tothe maneuver elements and their support units.

TEAM DEVELOPMENT

2-75. The decentralized nature of mountain combat and the need for the exercise of a mission command philosophy of command and control involve assigning missions to independenlty operating small teams that may be isotated from their higher headquarters. The disruptive influences of the environment and sustained physical stress further increase the perception of isolation (see FM 4-02.22). The most important factor that sustains a soldier in combat is the powerful psychological support that he receives from his primary group, such as a buddy team, squad, or platoon. He is less likely to feet the stress of loneliness under the isolated conditions of mountain warfare if his primary group maintains its integrity.

2-76, The soldier's ability to survive and operate in the mountains is the basis for the self-confidence needed to feel accepted by the team. Leaders must develop small-unit cohesion down to the buddy team. Each soldier must have a buddy to share both responsibilities and rewards. The leader must not simply assign two soldiers as a buddy team, but pair soldiers whose skills and attributes complement each other. Each soldier can then learn his buddy's specialized skills adding depth to the unit if one soldier becomes disabled. Soldiers work with their buddies, as well as function as part of the larger squad team. The combined strengths of buddies enhance both unit effectiveness and combat power. FM 6-22 has more information on team development.

Chapter 3

Firepower and Protection of the Force

Employing fire support systems, which are an integral part of maneuver, is included in this chapter. This arrangement, however, does not suggest any change in the close doctrinal relationship between fires and maneuver during mountain operations.

SECTION I – FIREPOWER

FIELD ARTILLERY

3-1. The basic tactical principles for artillery remain valid in mountains, subject to the limitations imposed by terrain and weather.

MOVEMENT AND POSITIONING

3-2, Rugged terrain and reduced mobility increase the reliance on field artillery fire support. However, the employment and positioning of field artillery systems may be severely impacted by the extreme difficulty of ground mobility in mountainous terrain Self-propelled artillery is often limited to traveling on the existing road and trail networks and positioning in their immediate vicinity. Towed field artillery is usually more maneuverable; it can be brought into position with the aid of trucks, tractors, and fixed or rotary-winged aircraft. Therefore, gun crews should be

CONTENTS

proficient in equipment-rigging techniques and air assault procedures, and possess ample sling-load equipment. Field artillery emplaced by helicopter normally requires continued airlift for subsequent displacement and ammunition resupply, and often necessitates substantial engineer support,

3-3. Light field artillery may require forward displacement of gun sections by helicopter to provide forward troops the necessary support. Medium field artillery may give the longer range required, but may be limited by high-terrain crest clearance. Normally, field artillery is employed far enough to the rear to take advantage of increased angles of fall. Flat areas, such as dry riverbeds, villages and towns, and farmland, can usually accommodate firing units, however, these positions present particular problems in the mountains for the following reasons:

- Dry riverbeds are hazardous because of the danger of flash flooding.
- Towns and villages usually have adequate flat areas such as parks, schoolyards, and playing fields but they are relatively scarce and often targeted by the enemy.
- Farmland is often difficult to negotiate from spring to fall. In the winter, if the ground is frozen, farmland may provide good firing positions; however, frozen ground may cause difficulty emplacing spades, base plates, and trails.

3-4, Good artillery positions, selected for cover, flash defilade, and accessibility to road nets and landing zones (LZs). are difficult to find, and their relative scarcity makes it easier for the enemy to target probable locations. In some instances, it may be necessary to by-pass the best position for one less suitable to reduce the enemy's counterfire effects. Commanders must ensure that positions on dominant terrain provide adequate defilade. Positions on commanding terrain are preferable to low ground positions because there is-

- A reduction in the number of missions requiring high-angle fire.
- A reduced amount of dead space in the target area.
- Less exposure to small arms fire from surrounding heights.
- Less chance of being struck by rockslides or avalanches.

3-5. Seme weapons may be moved forward to provide long-range interdiction fires or. in extreme cases, direct fires to engage a road-bound enemy in mountain passes or along valley floors. Because of rugged terrain, higher angles of fire, and reduced ranges, it is generally neoessary to displace artillery more frequently than on level terrain to provide continuous support. In the mountains, commanders must often employ field artillery in a decentralized manner because of the limited space for gun positions.

ACQUISITION AND OBSERVATION

3-6. Because of high angle fire requirements, radar can be effective against enemy indirect fire systems. In many instances, terrain masking and diminished line-of-sight may degrade its effectiveness. Sites should be selected on promi nent terrain to obtain the lowest possible screening crest. However, it is often difficult to obtain a low and consistent screening crest in mountainous terrain. Too low a screening crest drives the search beam into the ground. Too high a screening crest allows the enemy to fire under the beam and avoid detection. When positioning weapons locating radars, commanders should also consider thefollowing:

- Although time-consuming, visibility diagrams are extremely useful in determining the probability of acquiring targets within the sectors of search of the radar.
- To limit search areas, radars should focus on terrain that can be occupied by artillery and mortars.
- Accurate survey control is essential because of the extreme elevation variations in mountainous terrain. Helicopters may be useful in performing survey by use of the Position Azimuth Determining System (PADS) I f possible, digital radar maps may be used to minimize the time required for height correction of the weapon system. Digital maps allow the Firefinder systems to initially locate weapon systems to within 250 meters. This allows the radar operator to make only two to three visual elevation adjustments to accurately locate the weapon system.
- Impact predict is computed at the radar's elevation, therefore, excessive errors in the impact predict can be expected.
- Firefinder radars in the same area must not face one another and radiate at the same time. This causes interference and emissions burnout, resulting in equipment failure. If radars need to face one another to accomplish the mission, commanders must coordinate to ensure that they do not radiate at the same time.
- Computing track volume may become a critical task in determining a radar's effectiveness for a proposed position (see FM 3-09,12 for computation s).
- Units will use more shelling reports (SHELREPs) to determine enemy firing locations.

3-7. The majority of afl field artillery fires in mountains will be observed, especially dose support and defensive fires. Unobserved fires are frequently unreliable because of poor maps and rapidly changing meteorological conditions that cause registration corrections for high angle fire to be valid for only short periods of time.

3-8. Generally, field artillery observation posts should be emplaced on the highest available ground to increase observation. Low clouds or fog may require moving them to preplanned emplacements at lower elevations. Observers must be prepared to perform assault climbing to reach the most advantageous observation site. Commanders may use aerial observers or unmanned aerial vehicles (UAVs) to detect long-range targets and complement forward observers by adjusting fires beyond terrain masks, in deep defilade, and on reverse slopes. However, in extremely high mountains aerial observers may be confined to valleys and lower altitudes due to altitude limitations on different types of aircraft,

3-9. Laser weapons demand increased emphasis on observation techniques Laser target ranging and designation systems help to overcome difficulties in range estimation by providing accurate directional distance and vertical angle information for use in locating enemy targets. However, when positioning with a laser designator, an observer should consider line-of-sight with the target, as well as cloud height. Cloud ceilings that are too low will not allow laser guided munitions enough time to lock on and maneuver to the target.

TARGETING

3-10. Because of the decentralized nature of mountain operations, targets warranting massed fires may present themselves less often than in open terrain. However, narrow defiles used as routes of supply, advance, or withdrawal by the enemy are potentially high payoff targets for interdiction fires or large massed fires. Large masses of snow or rocks above enemy positions and along main supply routes are also good targets, because they can be converted into highly destructive rockslides and avalanches that may deny the enemy the use of roads and trails, and may destroy dements in ddilade. In the mountains, suppression of enemy air defenses takes on added importance because of the increased dependence on all types of aircraft. Commanders and thdr staffs should carefully review FM 3-60. A clear understanding of the targeting methodology combined with the knowledge of the capabilities and limitations of target acquisition and attack systems in a mountain environment is crucial to the synchronization of all available combat power.

3-11. To provide accurate and timely delivery of artillery fires in mountainous terrain, commanders must take into account the foil owing:

- High angles of elevation and increased time of flight for rounds to impact.
- Targets on reverse slopes, which are more difficult to engage than targets on flat ground or rising slopes, requiring more ammunition for the same oover age.
- I ncr eased amounts of dead space that cannot be hit by artillery fires.
- I ntervening crests that require detailed map analysis.

[] When the five requirements for accurate predicted fire (target location and size, firing unit location, weapons and ammunition information, meteorological information, and computational procedures) are not achievable, registration on numerous checkpoints becomes essential because of the large variance in elevation (see FM 3-09.40 for more detailed information).

MUNITIONS

3-12 Terrain and weather also affect the use of field artillery munitions Considerations for munitions employment in the mountains are discussed below.

- Impact fuze, high explosives (HE) shells and dual-purpose improved conventional munitions (DPI CM s) are very effective on rocky ground, scattering stones and splintering rocks, which themselves become missiles. However, deep snow reduoes their bursting radius, making them approximately 40 percent less effective. The rugged nature of the terrain may afford added protection for defending forces; therefore, large quantities of H E may be required to achieve the desired effects against enemy defensive positions.

- Variable time (VT) or time fuzes should be used in deep snow conditions and are particularly effective against troops on reverse slopes. There are some older fuzes that may prematurely detonate when fired during heavy precipitation (M557 and M572 impact fuzes and M564 and M548 time fuzes).

- Smoke, DPICM, and illuminating fires are hard to adjust and maintain due to swirling, variable winds and steep mountain slopes. Smoke (a base-ejecting round) may not dispense properly if the canisters become buried in deep snow. In forested mountains, DPI CM s may get hung up in the trees. These types of munitions are generally more effective along valley floors.

- Using the artillery family of scatterable mines (FASCAM) and Copperhead is enhanced when fired into narrow defiles, valleys, and roads. FASCAM may lose their effectiveness on steep terrain and in deep sncw. Melting and shifting sncw may cause the anti-handling devices to detonate prematurely the munitions, however, very little settling normally occurs at temperatures lower than 5 degrees Fahrenheit. Remote antiarmor mine system (RAAMS) and area denial artillery munitions (ADAM) must come to rest and stabilize within 30 seconds of impact or the submunitions will not arm. and very uneven terrain may keep theADAM trip wires from deploying properly.

MORTARS

3-13. Mortars are essential during mountain operations. Their high angle of fire and high rate of fire is suited to supporting dispersed forces. They can deliver fires on reverse slopes, into dead space, and over intermediate crests, and, like field artillery, rock fragments caused by the impact of mortar rounds may cause additional casualties or damage.

3-14. The 60mm mortar is an ideal supporting weapon for mountain combat because of its portability, ease of concealment, and lightweight ammunition. The 81mm mortar provides longer range and delivers more explosives than the 60mm mortar. However, it is heavier and fewer rounds (usually no more than two per soldier) can be man-packed. The 120mm mortar may be more desirable in some situations, since they can fire either white phosphorous (WP) or HE at greater ranges than tighter mortars and have a significantly better illumination capability. However, because of the weight of these mortars and their ammunition, it may be necessary to transport fewer of them into mountainous terrain and use the remaining gun crews as ammunition bearers, or position them close to a trail network in a valley or at lower elevations. The second technique may be satisfactory if the movement of the unit can be covered and sufficient firing positions exist.

AIR SUPPORT

3-15, Air interdiction and dose air support operations can be particularly effective in mountains, since enemy mobility, like ours, Is restricted by terrain. Airborne forward air controllers and close air support pilots can be used as valuable sources of information and can find and designate targets that may be masked from direct ground observation. Vehides and personnel are particularly vulnerable to effective air attack when moving along narrow mountain roads. Predsion-guided munitions, such as laser-guided bombs, can quickly destroy bridges and tunnels and, under proper conditions, cause landslides and avalanches to dose routes or collapse on both stationary and advandng enemy foroes. Moreover, air-delivered mines and long-delay bombs can be employed to seriously impede the enemy's ability to make critical route repairs. Predsion-guided munitions, as well as fuel air explosives, can also destroy or neutralize well-protected point targets, such as cave entrances and enemy forces in defilade.

3-16. Low ceilings, fog, and storms common to mountain regions may degrade air support operations. Although, global positioning system (GPS) capable aircraft and air delivered weapons can negate many of the previous limitations caused by weather. Terrain canalizes low altitude air avenues of approach, limiting ingress and egress routes and available attack options, and increasing aircraft vulnerability to enemy air defense systems. Potential targets can hide in the crevices of diffs and the niches of mountain slopes, and on gorge floors. Hence, pilots may be able to detect the enemy only at short distances, requiring them to swing around for a second run on the target and giving the enemy more time to disperse and seek better cover. Additionally, accuracy may be degraded due to the need for pilots to divert more of their attention to flying whi le simultaneously executing their attack.

ELECTRONIC WARFARE

3-17. The ability to use electromagnetic energy to deceive the enemy, locate his units and facilities, intercept his communications, and disrupt his command, control, and target acquisition systems remains as important in the mountains as elsewhere. Theeffectsof terrain and weather on electronic warfare (EW) systems are often *a* result of the effects on the components of those systems (particularly soldiers, communications, and aviation). Although a number of the effects are discussed in more detail elsewhere in this manual (and in applicable FMs and TMs), for ease some of the more common degrading effects of the mountainous environment on the components of electronic warfare systems are described in Figure 3-1 on page 3-8.

SECTION II - PROTECTION OF THE FORCE

AIR DEFENSE ARTILLERY

3-18. The severe mountain environment requires some modification of air defense employment techniques. Suitable positions are scarce and access roads are limited. In some instances, supporting air defense weapons may not be able to deploy to the most desirable locations. Consequently, the man-portable air defense systems (MANPADS) may be the only air defense weapon capable of providing dose-in protection to maneuver elements.

3-19. Mountain terrain tends to degrade the electronic target acquisition capabilities of air defense systems. This degradation makes it more difficult for the air defense planner to locate and select position to provide adequate coverage for the force, and increases the importance of combined arms for air defense (CAFAD) and passive air defense measures (see FM 3-01.8). Individual and crew-served weapons can mass their fires against air threats. The massed use of guns in local air defense causes enemy air to increase their standoff range for surveillance and weapons delivery, and increase altitude in transiting to and from targets. These reactions may make the enemy air more vulnerable to air defense artillery (ADA).

3-20. Enemy aircraft will probably use defiles and valleys in mountainous terrain for low-altitude approaches to take advantage of terrain masking of radar. Congested roads and trails, and their junctions, may become lucrative targets for enemy air strikes. Enemy pilots may avoid early detection by using terrain-dearance or terrain-following techniques to approach a target. Rugged mountain terrain degrades air defense detection, but. at the same time, mountain ridges and peaks tend to canalize enemy aircraft. Detailed terrain analysis, coupled with predictive analysis to identify probable enemy air avenues of approach, aids in effective site selection.

EW COMPONENT	ENVIRONMENTAL FACTORS							REMARKS
	CLOUDS	FOG	RAIN	SNOW	WIND	TEMP	TERRAIN	
Soldiers[1]	✔	✔	✔	✔	✔	✔	✔	■ Clouds, fog, precipitation, and terrain affect visibility and observation. ■ Precipitation, temperature, and the rugged terrain affect soldier performance and ability to operate systems.
Electronics and-wire/cables[2]		✔	✔	✔		✔	✔	■ Extreme cold, combined with rugged terrain, increases fragility and breakage. ■ Precipitation and humidity affect electronic components.
Antennas[2]			✔	✔	✔		✔	■ Strong winds damage or prevent erection. ■ Precipitation and cold create ice, causing breakage (increased load and wind resistance) and reduce effectiveness. ■ Terrain affects masking and line-of-sight restrictions.
Aircraft[3]	✔	✔	✔	✔	✔	✔	✔	■ Clouds, fog, and precipitation degrade visibility and may prevent aircraft from flying under visual flight rules (VFR), precluding missions requiring aircraft landing at unimproved mountain LZs. ■ Cold and precipitation lead to icing, which impedes lift. ■ Compartmented terrain affects flight routes and target acquisition.
Vehicles			✔	✔		✔	✔	■ Rain, snow, and rugged terrain decrease mobility.
Radars/ Sensors		✔	✔	✔	✔		✔	■ Wind increases background noise, reducing efficiency. ■ Terrain affects masking and line-of-sight restrictions. ■ Fog and precipitation decrease infrared and electro-optical systems effectiveness.
Batteries						✔		■ Terrain reduces effectiveness and battery life – some systems may not even work under reduced power.

1 See Chapter 1 (Effects on Personnel)
2 See Chapter 2 (Communications)
3 See Chapter 4 (Helicopters) and the Previous Section (Air Support)

Figure 3-1. Effects of the Mountainous Environment on EW Systems

3-21 Movement to and occupation of positions in mountainous terrain require additional time. Planners must consider slope (pitch and roll), site preparation, and access route improvement prior to movement. Bradley Stinger fighting vehicle (BSFV) units often are unable to accompany small, lightly equipped maneuver elements, and may be restricted to supporting elements in more accessible areas of the battlefield. Avenger fire units can be sting-loaded by heavy lift aircraft and MANPADS airlifted into otherwise inaccessible positions. However, equipment emplaced by helicopters is resupplied and repositioned by the same means. When moving dismounted, MAIMPAO teams are limited to one missile per soldier, unless olher members of the unit are tasked to carry additional missiles.

3-22. Because of terrain masking of radars and the difficulty in establishing line-of-sigbt communications with lhe Sentinel or light and special division interim sensor (LSDIS) radar, early warning for short-range air defense (SHORAD) systems may be limited. Soldiers must maintain continuous visual observation, particularly along likely low-level air avenues of approach. Therefore, when possible. Sentinel or LSDIS radars should be emplaced on the highest accessible terrain that provides the best air picture for target detection and early warning, not necessarily peaks and summits.

ENGINEER OPERATIONS

3-23. Engineer combat support requirements increase in mountainous terrain because of the lack of adequate cover, the requirement for construction of field fortifications and obstades, and the need to breech or reduce enemy obstades. With such an enormous multitude of tasks, effective command and control of engineer assets is essential for the optimal utilization of these relatively scarce resources (see also the discussion of engineer augmentation and employment in themobility section of Chapter 4),

3-24, Digging fighting positions and creating temporary fortifications above the timberline is generally difficult because of thin soil with underlying bedrock. As described in Chapter 2, boulders and loose rocks may be used to build hasty, aboveground fortifications, Wei l-assembled positions constructed in rock are strong and offer good protection, but they require considerable time and equipment to prepare,

3-25. Engineers assist maneuver units with light equipment and tools carried in or brought into position by ground vehicles or helicopters. Bulldozers, armored combat earthmovers (ACEs). and small emplacement excavators (SEEs) can be used in some situations to help prepare positions for command bunkers and crew-served weapons. They can also be used to prepare positions off existing roads for tanks, artillery, and air defense weapons. Conventional equipment and tools are often inadequate in rocky terrain, and extensive use of demolitions may be required- In the mountains, a greater number of engineer assets will be devoted to maintaining mobility and maneuver and unit commanders should assume that available engineer support will be limited to assist them with their survivability efforts. To enhance survivability and mobility a minimum of two soldiers per maneuver platoon should be capable of using standard demolitions.

NBC PROTECTION

3-26. Terrain and weather dictate a requirement for a high degree of nuclear, biological, and chemical (NBC) defense preparedness in mountainous areas. Due to limited mobility, viable tactical positions, and limited communication abilities, friendly units must be self-sufficient in protecting themselves against NBC weapon system effects.

3-27. Wearing mission-oriented protective posture (MOPP) gear at high elevations, when possibly combined with altitude sickness, increased dehydration, and increased physical exertion, degrades performance and increases the likelihood of heat casualties. Commanders should make every effort to keep soldiers out of MOPP gear until intelligence indicators reveal that an NBC attack is imminent or it is confirmed that a hazard actually exists (see FM 3-11.4 for a discussion on vulnerability analysis). When precautions must be taken against hazards, commanders must make decisions early and allow extra time for tactical tasks. Commanders should also refer to TC 3-10 for greater detail on tactics, techniques, and procedures necessary to operate under NBC conditions.

NUCLEAR

3-28 A mountainous environment can amplify or reduce the effects of and distort the normal circular pattern associated with nuclear blasts. The irregular patterns reduce the accuracy of collateral damage prediction, damage estimation, and vulnerability analysis.

3-29. Air blast effects are amplified on the burst side of mountains (see Figure 3-2). Mountain walls reflect blast waves that can reinforce each other, as well as the shock front. Therefore, it is passible that both overpressure and dynamic pressure, and their duration will increase. An added danger is the creation of rockslides or avalanches A small yield nudear weapon detonated 30 kilometers or more from the friendly positions may still cause rockslides and avalanches, and easily close narrow roads and canalized passes. On the other hand, there may be little or no blast effects on the side of the mountain away from the burst.

3-30. Hills and mountains block thermal radiation, and trees and other foliage reduce it. Low douds, fog, and falling rain or snow can absorb or scatter up to 90 percent of a burst's thermal energy. During colder weather, the heavy dothing worn by soldiers in the mountains provides additional protection. However, the reflection from snow and the thin atmosphere of higher elevations may increase the effects of thermal radiation. Snow and ice melted by thermal radiation can result in flash flooding.

3-31. Frozen and rocky ground may make it difficult to construct shelters for protection from the effects of nuclear weapons. However, natural shelters such as caves, ravines, and diffs provide some protedion From nudear effects and contamination, in some instances, improvised shelters built of snow, ice, or rocks may be the only protedion avai lable. The dear mountain air extends the range of casualty-produdng thermal effeds. Within this range, however, the soldiers' added dothing reduoes casualties from these effects.

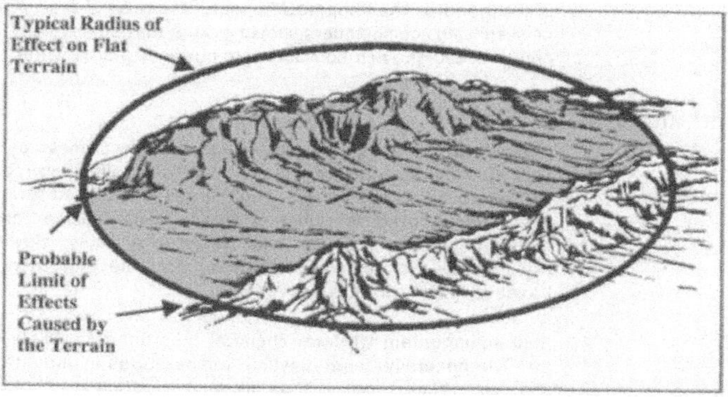

Figure 3-2. Effects of Mountains on Radiation and Blast

3-32. In mountainous regions, the deposit of radiological contamination is very erratic in speed and direction because of variable winds. Hot spots may occur far from the point of detonation, and low-intensity areas may occur very near it. Limited mobility makes radiological surveys on the ground difficult, and the difficulty of maintaining a constant flight altitude makes air surveys highly inaccurate. Additionally, melting snow contributes to the residual radiation pattern. After a nudear detonation, streams should be checked for radiation contamination before using them for drinking or bathing. As with the other effeds, the pattern of initial and induced nudear radiation may be modified by topography and the height of the burst,

BIOLOGICAL

3-33. Most biological pathogens and some toxins are killed or destroyed by the ultraviolet rays in sunlight. Above the timberline, there is little protedion from the sun; thus, the effectiveness of a biological attack may be reduced Downwind coverage may be greater because of the frequent occurrence of high winds over mountain peaks and ridges. Additionally, inversion conditions favor the downwind travel of biological agents through mountain valleys. Typically, winds flow down terrain slopes and valleys at night and up valieys and sunny slopes during the day. The effeds of mountainous terrain and rapidly changing wind conditions on the ability to predid and provide surveys of contamination for biological agents are similar to that for nudear radiation.

3-34 Temperatures and humidity also affect the survivability of biological agents. Generally, cool temperatures favor survival, and higher humidity increases the effectiveness of the agents. Extreme cold weather and snow deposited over a biologically contaminated area can lengthen the effective period of the hazard by allowing the agent to remain alive but dormant until it is disturbed or the temperature rises, if the use of biological agents is known or suspected, commanders should ensure that soldiers pay added attention to personal hygiene and consume only purified/treated water.

CHEMICAL

3-35. Wind and terrain can also cause the effectiveness of chemical agents to vary considerably. Depending on conditions, effects can be significantly enhanced or almost ineffective. High winds and rugged terrain cause chemical agent clouds to act in a manner similar to radioactive fallout. Inversions in mountain valleys may afso effectively cap an area, slowing the dissipation rate. Because of terrain and winds, accurate prediction of the downwind travel of toxic agent clouds is difficult

3-36. In mountain warfare, chemical munitions are likely to be delivered by air. The generally cooler daytime temperatures in mountainous terrain slow the evaporation process, thus allowing a potential contamination hazard to remain active longer Midday temperatures favor using persistent or blister-type agents, since nonpersistent agents dissipate too rapidly to cause any effect and unsupervised personnel are more likely to remove protective clothing for oomfort

3-37. The actions to protect against chemical agents in the mountains are not significantly different than from the requirements in less mountainous terrain. However, in extreme cold weather, survey and monitoring is often limited to the individual team mission, the FOX system may be limited to roads and trails, and the detection of vapor hazards is limited when the temperature falls below 32 degrees Fahrenheit. Decontamination may be more difficult due to freezing conditions, and the virulency period of contamination hazard for persistent agents may increase.

SMOKE AND OBSCURANTS

3-33. Smoke operations in mountainous areas are characterized by difficulties encountered due to terrain and wind Inadequate roads enhance the military value of existing roads, mountain valleys, and passes and add importance to the high ground that dominates the other terrain. Planners can use smoke and flame systems to deny the enemy observation of friendly positions, supply routes, and entrenchments, and degrade their ability to cross through tight, high passes and engage friendly forces with direct and indirect fires.

3-39. Thermally induced slope winds that occur throughout the day and night increase the difficulty of establishing and maintaining smoke operations, except in large and medium sized valleys. Wind currents, eddies, and turbulence in mountainous terrain must be continuously studied and observed, and their skillful exploitation may greatly enhance smoke operations rather than deter them. Smoke screens may be of limited use, due to enemy aerial observation, to include UAVs. and observation by enemy forces located on high ground. Smoke units may be required to operate for extended periods with limited resupply unless petroleum, oils, and lubricants (POL) supplies are emplaced in hide positions with easy access.

Chapter 4
Maneuver

The mountain environment requires the modification of tactics, techniques, and procedures. Mountains limit mobility and the use of large forces, and restrict the full use of sophisticated weapons and equipment. These limitations enable a well-trained and determined enemy to have a military effect disproportionate to his numbers and equipment. As such, mountain campaigns are normally characterized by a series of separately fought battles for the oontrol of dominating ridges and heights that overlook roads, trails, and other potential avenues of approach. Operations generally focus on smaller-unit tactics of squad, platoon, company, and battalion size. Because access to positions is normally difficult, adjacent units often cannot provide mutual support and reserves cannot rapidly deploy. Attacks in extremely rugged

CONTENTS

terrain are often dismounted, with airborne and air assaults employed to seize high ground or key terrain and to encircle or block the enemy's retreat. While the mountainous terrain is usually thought to offer the greatest advantage to the defender, the attacker can often gain success with smaller forces by effectively using deception, bold surprise actions, and key terrain.

Although mountains often increase the need to employ light forces, commanders should not be misled into believing that this environment is the sole domain of dismounted units. On the contrary, the integrated use of mounted and dismounted forces in a mountainous environment, as elsewhere, increases a commander's capabilities while reducing his limitations. However, the employment of mixed forces must be based on sound mission, enemy, terrain and weather, troops and support available, time available, civil considerations (METT-TC) analysis of the specific mountain area of operations (AO), The infantry, armor, and combined arms series of field manuals, at both battalion and brigade level, provide the capabilities and limitations for each type force, planning and safety considerations, as well as, various concepts for employment. In all cases, commanders should assign complementary missions to each type force that capitalizes on their strengths and reduces their weaknesses, and takes into consideration the differential in operations tempo. Working together on the mountain battlefield, armored and dismounted forces can offset each other's weaknesses and provide much greater lethality than any one alone.

SECTION I - MOVEMENT AND MOBILITY

4-1. To move decisively in all directions without losing momentum in a mountain area of operations requires meticulous planning and careful preparation. In a mountainous environment, numerous conditions exist that affect mobility. The force that can maintain its momentum and agility under these conditions has the best chance of winning. Reduced mobility is a primary limitation to be considered during all phases of planning for mountain operations. Rugged terrain, the time of year, the weather, and the enemy have a decisive influence on movement in the mountains. Commanders must ensure that they have sufficient time and space to deploy their forces for battle by maintaining constant security and selecting proper routes and movement techniques. Additionally, they must closely manage limited off-road areas. Tactical operations centers, artillery units, aid stations, air defense artillery, battalion trains, and other supporting units will compete for limited space in restrictive mountainous terrain.

4-2. At any elevation level, movement is generally considered to be either movement across or along terrain compartments. When moving across terrain compartments from one ridge to another, elements should use bounding overwatch. Lead elements should secure the high ground and provide overwatching fires as the rest of the element crosses the low ground. When moving along a terrain compartment, forces should move on the high ground without silhouetting themselves or, at a minimum, place an element there to securetheir flanks

4-3. Maneuver forces should move by stealth and exploit the cover and concealment of terrain. Using rough, unlikely routes and movement during limited visibility helps avoid enemy detection. All movements must exploit known weaknesses in enemy detection capabilities. Whenever possible, movement should be planned to coincide with other operations that divert the enemy's attention,

4-4. Because of the narrow routes sometimes encountered, especially in the higher elevations, formations may be compressed to columns or files. To reduoe vulnerability to the enemy, forces should move separated from each other on multiple and unlikely routes. When moving dismounted along unlikely routes, special teams construct fixed ropes, hauling systems, traverse systems, and other mountaineering installations to provide access to higher elevation levels and increased mobility.

4-5. The danger of surprise attack is most acute in terrain that makes deployment from the march impossible. Even with well-thought-out movement plans, maneuver elements must take both active and passive security measures at all times. Restrictive terrain facilitates templating and determining the movement of forces, making the actions of an armored force more predictable. Elements may avoid detection by using planned fires to destroy known enemy sensors and observation posts or by pfadng fires to divert the enemy's attention away from an exposed area through which the element must move. However, the placement of fires in a particular area or along a route may compromise operational security.

4-6. When the danger of rockslides or avalanches exists, the distance between elements should be increased as much as four to six times more than required on flat terrain. The more the conditions vary for each unit, the more thorough the planning must be, especially if units must reach the objective simultaneously. Often, a reserve of time must be programmed if units move on multiple routes, over unfamiliar terrain, or during limited visibility, or if they face an uncertain enemy situation.

MOUNTED MOVEMENT

WHEELED AND TRACKED VEHICLES

4-7. Generally, the mountain terrain above the valley floor severely limits movement of wheeled vehides and is too restricted for tracked vehides. Trafficable terrain tends to run along features with steep slopes on either side, making mounted movement vulnerable to vehicular ambushes and attack aircraft. Recovery vehides must always accompany mounted forces in mountainous terrain to rapidly remove disabled vehides from the limited and narrow trail network.

4-8. Tanks and other armored vehides, such as infantry fighting vehides (IFVs), are generally limited to movement in valleys and existing trail networks at lower elevations. Even at these levels, the trails may require extensive engineer work to allow tracked vehides to pass over them. Tanks,

Bradley fighting vehicles (BFVs), and cavalry fighting vehicles (CFVs) can support by fire if accessible firing positions are available; however, it will rarely be possible for them to accompany dismounted infantry in the assault. In such cases, commanders may seek to use their increased firepower to isolate the objective for the dismounted assault. If employed above Level I, armored vehicles are forced to fight in smaller numbers, yet a single tank at a critical pant may have a decisive effect. Although antitank weapons employed from higher elevations can easily penetrate the top of armored vehides, in many situations, the inability to elevate the weapon system's main gun suffidently to return fire may further increase its vulnerability.

4-9. Low atmospheric pressure considerably increases the evaporation of water in storage batteries and vehide cooling systems, and impairs cylinder breathing. Consequently, vehicles expend more fuel and lubricant, and engine power is reduced by four to six percent for every 1,000-meter (3.300-foot) increase in elevation above sea level. This translates to a fuel and oil increase of approximately 30 to40 percent or more.

4-10. Figure 4-1 contains questions that are part of any mounted movement plan. Limited road networks and restricted off-road mobility significantly increase their importance. In the mountains, failure to address these questions in detail may seriously jeopardize the overall mission.

HELICOPTERS

4-11, Utility and cargo helicopters are key to the rapid movement of soldiers and equipment in the mountains. However, any operation that depends primarily on continuous aviation support to succeed is extremely risky. H rgh eievat ions and rapidly changing and severe weather common to mountainous regions is very restrictive to aviation operations and makes availability of aviation support very unpredictable. At high altitudes, weather that appears to be stable to the ground observer may significantly affect helicopters- The effects of fog, frontal systems, winds, and storms are readily discernible. Additionally, higher altitudes may restrict aircraft lift capabilities and decrease aircraft allowable gross weight in mission profile. Aircraft idng is common at high altitude and may occur suddenly. De-ice/anti-ice capabilities exist for rotor blades, however, idng may still decrease lift and. in severe cases, prevent flight altogether. Therefore, commanders must become intimately familiar with the conditions that may limit the full effectiveness of Army aviation when operating in a mountain environment (see FM 3-04.203).

- ☐ **How fast can the march be conducted?**
- ☐ **Will there be other traffic on the route?**
- ☐ **Are there potential areas that offer covered, off-road positions?**
- ☐ **Are there any locations along the route that could be used for resupply?**
- ☐ **Are there alternate routes?**

Figure 4-1. Mounted Movement Planning

4-12. Additionally, commanders must consider the effect of altitude on soldiers when planning air assault operations (see Chapter 1). If possible, commanders should use soldiers acdimatized at or above the elevation level planned for the air assault. Depending on the situation, it may be better to have troops walk in rather than fly them to the necessary elevation level.

4-13. Rugged, mountainous terrain complicates flight route selection and places an additional navigational load and strain on the entire crew, as they have little margin for error. Direct routes can seldom be flown without exposing aircraft to an unacceptable risk of detection and destruction by the enemy. Tactical flight routes follow valley corrida's, where it is possible to obtain cover and concealment while maintaining the highest possible terrain flight altitude. Terrain flight in the mountains may preclude using dosed formations. Multi-helicopter operations are normally flown in "loose" or "staggered trail" format ions with increased spadng between aircraft.

4-14. Terrain suitable for multiple helicopter landing zones (LZs) in mountainous regions is limited. Level areas that are suitable for mountain LZs frequently require little preparation beyond the dearance of loose material, since the ground is usually firm enough to support helicopters. Conversely, if LZs must be developed, dearing may be difficult due to the rocky ground. Stand-off space from rock wall faces must be deared and a level landing surface must be created. Demolitions may be required to dear large rocks but care must be used to prevent rockslides or avalanches started by the explosive shock. During the winter, snow must be packed to prevent whiteouts. Similarly, sandy or dusty LZs should be dampened with water to prevent brownouts.

4-15. When only single aircraft landing zones are available, in-flight spadng between helicopters must be significantly increased. Although helicopter LZs should be located on the windward side of ridges or peaks to take advantage of the more stablewinds, conceal ment from enemy observation and the mission are the most i m port ant factors in site selection in forward areas. When it is impossible for helicopters to land, personnel may rappel and light equipment may be sling-loaded into a LZ or. in some situations, lowered by rope while the helicopter hovers However, this may increase turnaround time and aircraft vulnerability Since available landing sites are often limited, the enemy can be expeded to target all likely locations. Personnel should secure terrain that dominates a landing site before using it. They must extensively suppress enemy air defense weapons during air assault or supply operations

4-16, Attack helicopters can be well suited for a mountain environment; however, commanders must be continuously mindful of weather and elevation effects on their employment. They can be the commander's most mobile maneuver forces in mountain warfare, enabling him to concentrate combat power quickly and exploit enemy weaknesses. During stable weather conditions, attack helicopters equipped with a variety of ordnance can rapidly engage targets beyond the range of other weapons or those masked by intervening crests. As discussed earlier, higher altitudes and icing conditions affect lift and subsequently armament loads. Ice can also prevent attack helicopters from firing their weapons altogether.

4-17, Employment time and fuel consumption increases because of the few direct routes. Terrain compartments provide excellent terrain masking and radar and visual acquisition avoidance, and allow for rapid movements to the flanks and rear of an isolated enemy force. However, these same compartments may limit aircraft maneuverability and necessitate smaller flight formations, which, in turn, may affect target engagement techniques. The compartmented terrain, combined with extended distances, may require engagement without the support of other combined arms. If terrain precludes placement of fuel and arming points in the forward area, turnaround time increases and on-station lime decreases. Since ground-to-air communication is often degraded by intervening terrain, in-flight operational control over extended ranges may be difficult.

4-18. Enemy motorized and mechanized forces may be slowed and canalized as they move up steep grades, down narrow valleys, and along mountain trails. These types of conditions allow attack helicopters to engage slower moving targets that have little room to maneuver or hide. However, these same conditions also make it difficult for pilots to select positions that allow line-of-sight to the target, sufficient tracking distance, acceptable standoff range, and adequate cover and concealment. Positions located high on a ridgeline may support successful target acquisition, tracking, and standoff, but create dangerous silhouettes and look-down angles (the angle from the aircraft to the target) that exceed aircraft weapon constraints. Lower positions, possibly in draws or saddles, may provide concealment to the flanks and an extensive backdrop to help conceal positions, but can decrease a pilot's ability to locate and track largets. Intervisibility lines may mask targets, and extreme terrain relief within the aircraft's optics field of view may Inhibit tracking.

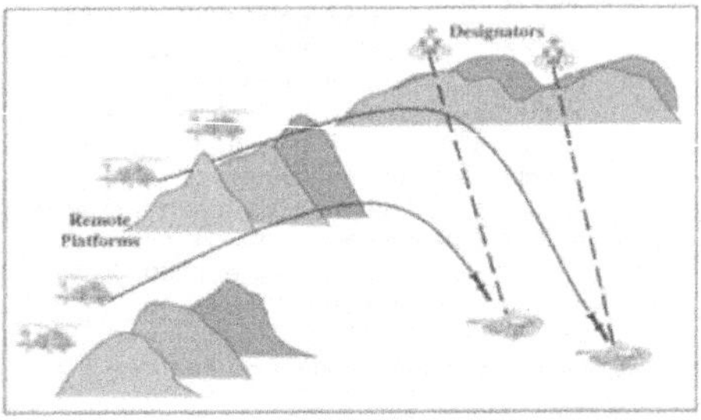

Figure 4-2. Remote Hellfire Engagements

4-19. Remote Hellfire engagements avoid most of these problems (see Figure 4-2), but they may increase the time of flight of the missile. A remote engagement limits the number of aircraft exposed for tracking and lasing targets. When engaging with remote fires, a designating team is plaoed in a position overwatching the enemy's route of march. The remaining aircraft position themselves in covered and concealed battle positions oriented on the engagement area. If possible, these battle positions should be located below the reverse military crest or the counterslope of the ridgeline paralleling the engagement area. Aircraft in these battle positions will act as remote platforms by providing missiles to the designators without unmasking. This tactic prevents the enemy from achieving lineof sight on the firing aircraft.

4-20. The AH-64D Longbow Apache will provide an even greater killing capability for the mountain commander. It is able to detect, classify, prioritize, and engage targets with Longbow H el If ire missiles without visually acquiring the target. The commander of the AH-64D Longbow Apache utilizes the its fire control radar and mast-mounted site to target all of the vehicles in the enemy's march formation. A data transfer handover is then executed to all other AH-64D Longbow Apaches within the company. After the data transfer has been completed, the company can then engage the enemy march column without exposing the remote platforms. When employing the AH-64D Longbow Apache with radar guided Heflfire missiles, engagement times are greatly reduced and aircraft exposure to the enemy is limited. Regardless of the type of helicopter, a thorough terrain analysis and early involvement of aviation operational planners are key to successful application of Army aviation assets in mountainous terrain.

DISMOUNTED MOVEMENT

4-21. Dismounted movement is often extremely slow and arduous, and may require the skills of technical mountaineering teams to secure the advance. For example, movement in Level 11 may dictate that elements secure the high ground in Level II I. As with any type of movement, proper movement techniques and formations and conslant security to avoid unplanned enemy contact are some of the keys to successful dismounted movement.

4-22. Foot marches in the mountains are measured in time rather than distance. When making a map reconnaissance, map distance plus one-third is a good estimate of actual ground distance. One hour is added for each 300 meters of ascent or 600 meters of descent to the time required for marching a map distance. Figure 4-3 on page 4-8 shows dismounted movement calculations for an estimated 16-kilometer march on flat and mountainous terrain. Although not included in this example, commanders must also consider acclimatization, fatigue, soldiers' loads, limited visibility, and other factors that affect movement times (see FM 3-25.18 for additional factors influencing dismounted march rates).

4-23. Commanders cannot permit straggling or deviations from the selected route. Every aspect of march discipline must be rigorously enforced to keep a column closed with the knowledge that the interval between individuals depends on terrain and visibility.

4-24. To prevent an accordion effect, soldiers must allow enough distance between themselves todimb without causingthe following individual to change pace. In mountainous terrain, a slow, steady pace is preferred to more rapid movement with frequent halts. Commanders must incorporate scheduled rest halts into their movement plans based on distance, availability of covered and concealed positions, and other fadors described above. If possible, commanders should not con dud rest halts during steep ascents or descents. At the start of a march, soldiers should dress lightly so that they begin slightly chilly. However, a short halt should be taken to adjust dothing and equipment after the first 15 minutes of movement. In addition, soldiers must put on special mountaineering equipment before reaching steep terrain.

Flat Terrain	Mountain Terrain
Total 16 kilometers / 4 kilometer per hour = 4 hours	Normal Time 16 kilometers / 4 kilometer per hour = 4 hours
	Ascent 600 meters / 300 meters per hour = 2 hours
	Descent 600 meters / 600 meters per hour = 1 hour
	Total 4 + 2 + 1 = 7 hours

Figure 4-3. Example Dismounted Movement Calculations

4-25. In glacial areas, the principal dangers and obstacles to movement are crevices, snow and ice avalanches. Exposure to the hazards of glaciated mountainsis increased at company-level and above, and movement should be limited to separate platoon and lower levels. When moving on glaciers, an advance element should be used- This group identifies the best routes of advance, marks the trail, and provides directions and distances to follow-on units. A marked trail is especially important during inclement weather and low visibility, and provides a route for retrograde. Commanders must carefully weigh the advantages of a marked route against the possibility of ambush and the loss of surprise.

MOBILITY

4-26. During mountain operations where limited mobility exists, it is critical that units maintain security and control of available road/transportation networks. This includes securing key bridges, fords, crossing sites, intersections, and other vulnerable choke points. These locations must be protected against enemy air. obstacle, and ground threats. However, commanders must carefully balance their available combat power between protecting their freedom of mobility/maneuver and allocating forces to critical close combat operations. Effective risk analysis and decisions are essential. Route clearance operations, patrols, traffic control points (TCPs), and other security operations aid commanders in securing routes. During offensive operations, commanders may need to commit forces to seize key terrain and routes that afford their forces greater mobility and tactical options against the enemy.

4-27. Engineer support in front of convoys and combat formations is often necessary to dear and reduce obstades, such as washouts, craters, mines, landslides, and avalanches, as well as, snow and ios in colder regions. Redudng obstades is more difficult in mountainous areas because of reduced maneuver space, lack of heavy equipment, and an increased competition for engineer support. Minefields should normally be breached, since bypassing properly sited obstades is often impossible. In the mountains, using mechanical mine plows and rollers is frequently impossible due to the lack of roads and trails, and removal of mines by hand or through demolitions is often required. Commanders must exerdse extreme caution when employi ng demolitions in the vidnity of snow and rock covered slopes because they can cause dangerous rockslides, avalanches, and secondary fragmentation FM 3-34 2 has information on breaching operations and synchronization required.

4-2S. Creating new road systems in mountainous regions is usually impractical because of the large amount of rock excavation required. Therefore, roadwork is generally limited to the existing roads and trails often requiring extensive construction, improvement, maintenance, and repair towithstand the increased military traffic and severe weather conditions. In certain mountainous areas, materials may be difficult to obtain locally and impossible to make full use of conventional heavy engineer equipment for road and bridge construction or repair. In such cases, large numbers of engineers are required and units must rely heavily on hand labor, light equipment, and demolitions.

4-29. Secondary roads and trails should be steadily improved to accommodate trucks and infantry fighting vehides, and. eventually, heavier vehicles. Their seledion depends on necessity and the speed with which the routes can be put into service. Abnormal gradients on roads may be necessary to ensure that construction keeps pace with lactical operations. Sidehill cuts are the rule, and the same contour line is followed to avoid excessive fills or bridging. Turnouts should be installed approximately every 500 meters to reduce traffic congestion on single-lane roads or trails. Drainage requirements must be considered in detail because of the effeds of abnormally steep slopes, damaging thaws, and heavy rains.

4-30. Stream and river crossing operations are difficult and must usually be accomplished by expedient means. Bridging operations in mountainous terrain are normally limited to spanning short gaps and reinfordng existing bridges by using prefabricated materials and fixed spans from floati ng bridge equipment. However, standard design or improvised suspension bridges may still be needed for longer spans. Because exist i ng bridges may have low vehide load classifications, standard fixed tactical bridges and bridging materials should be on hand to quickly ranforce or repl ace them. I n extremely rough terrai n, cableways and tramways may be construded lo move l ighl loads and personnel across gorges, and up and down steep slopes.

COUNTERMOBI LITY

4-31. Obstades become more important because of the compartmented terrain and already limited road and trail networks. It is easy to create effective obstructions in mountains by cratering roads, fully or partially destroying bridges, or inducing rockslides and avalanches. Units can use antitank minefields effectively to canalize the enemy, deny terrain, or support defensive positions. Commanders should remember that dearing or redudng these same obstades may be extremely difficult and a hindrance to future operations. Using reserve and situational obstades. lanes and gaps, and plans to rapidly reduce friendly obstades must be an integral part of all defensive operations. Commanders must also consider the enemy's ability to create similar obstades and minefields when developing courses of action that hinge on speed of movement or a particular avenue of approach.

4-32. Reinforcing obstacles can be used effectively with the natural ruggedness of mountains to deny the enemy terrain and to delay and impede his movement. As in all environments, the engineer and maneuver force commander must site obstacles based on terrain and the availability of weapon systems.

4-33. Antitank mines are laid along the comparatively narrow approaches suitable for mounted attacks. Flash floods and excessive runoff may dislodge mines from their original location; however, they normally remain armed. Family of scatterable mines (FASCAM), particularly artillery-delivered and helicopter-delivered mines, increases the flexibility of the maneuver unit commander, reduces the engineer effort, and is a valuable resource in protecting rear areas from enemy envelopment and breakthroughs. Using FASCAM should be weighed against thetime in delivery, displacement of the artillery, and the additional logistics burden that may be involved

ENGINEER AUGMENTATION AND EMPLOYMENT

4-34. Because mountain terrain requires small-unit decentralized operations, an engineer platoon or company should be allocated to each maneuver battalion, light or heavy. Allocation in this manner may leave division and brigaderear areas short of engineer support.

4-35. An additional corps engineer battalion (wheeled) and an engineer light equipment company may be needed to augment an infantry division. Platoons from the engineer light equipment company may be tasked to assist divisional platoons with the engi neer effort in each maneuver battalion area. The corps combat engineer battalion (wheeled) provides heavy equipment and dump trucksrequired to support road improvement and maintenance in division and brigade rear areas. Also, this corps combat engineer battalion (wheeled) can accomplish such tasks as constructing or reinforcing bridges. To operate efficiently, additional items, such as compressors, jackhammers. power drills, chain saws, and bulldozers, may be necessary, as well as large amounts of explosives and obstacle materials.

SPECIAL PURPOSE TEAMS

4-36. On steep, exposed, or technically difficult terrain, soldiers with advanced mountaineering skills may be required to maintain or improve mobility. Advanced climbers may deploy ahead of maneuver forces during limited visibility or inclement weather to erect aids that will help maneuver elements move in difficult terrain. They may also be committed to lead forces or to operate independently to strike the enemy suddenly over unlikely routes and to occupy certain key heights that can be defended easily because of their position. The specific employment of special purpose teams is based on the mission, tasks, and requirements of the commander.

4-37. Commanders must analyze operational terrain levels and identify the mobility requirements necessary to obtain and maintain freedom of both tactical maneuver and operational movement (see Figure 4-4). It is critical that special purpose teams are properly organized before a mission begins. Once movement is underway, unplanned deviations have little chance of success. Bypassing obstacles in mountainous terrain is almost always difficult or impossible. In many instances, the best available bypass will channel friendly forces into enemy kill zones or ambushes.

4-38. To enable a force to move personnel, equipment, and supplies on the mountain battlefield with limited delays due to terrain, visibility, or obstacles, commanders should organize soldiers with advanced mountaineering training as guides, lead climbing teams, installation teams, and evacuation teams (evacuation teams are covered in the combat health support portion of Chapter 5).

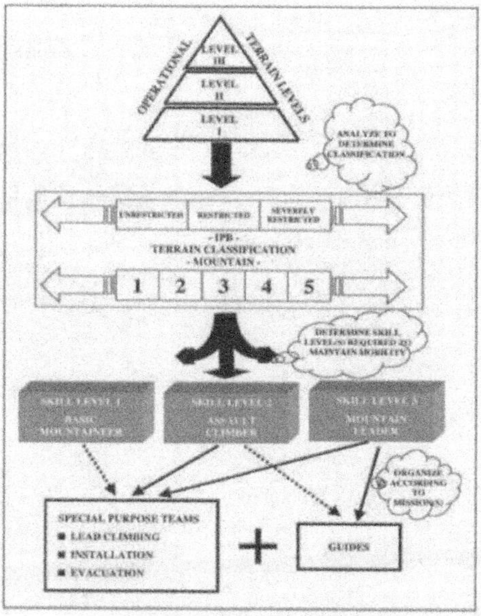

Figure 4-4. Relationship of Terrain to Skills and Special Purpose Teams

HISTORICAL PERSPECTIVE

The 10th Mountain Division and Riva Ridge (February 1945)

After attempts to capture Bologna in Italy's Po Valley during the winter of 1944-1945 failed, Allied commanders focused on the possibility of wide sweeping movements aimed at encircling Bologna and all the German armies in the region. The plan called for using the 10th Mountain Division to capture a series of mountain peaks and ridges dominating a 10-mile section of Highway 64, one of two main routes leading from Pistoia through the northern Apennines to Bologna. This would provide the Allies with a better position before starting the spring offensive, since seizure of this terrain would result not only in the Germans' being unable to protect their vital lines of communications, but also in the Americans' being able to observe German activity almost all the way to the Po Valley, approximately 40 miles away. The Division's baptism-by-fire was to be in a region dominated by two ridges whose highest peaks rose 3,000 to 5,000 feet. One of these ridges, Pizzo di Campiano-Monte Mancinello, became known as Riva Ridge (see Figure 4-5).

These heights appeared to be impregnable, as it was doubtful that any force large enough to overwhelm the Germans could be massed unobserved for an assault. The Germans had all the advantages of the commanding heights, and there was little cover for troops crossing the barren, snow-covered ground. It was clear that Riva Ridge would have to be cleared before the decisive attack could advance up Mount Belvedere and along the ridge towards Monte della Torraccia. Therefore, the plan was for the mountain troops to climb the 1500-foot cliff and surprise the Germans, who would not be expecting the attack up the face of the cliff.

Figure 4-5. Riva Ridge

The 86th Mountain Infantry Regiment's 1st Battalion and one company from its 2nd Battalion were assigned to negotiate the cliff and capture Riva Ridge, thereby setting the conditions for other assault elements to achieve their objectives. Select teams of climbers assembled their ropes, pitons, and other gear. The teams climbed in the dead of night, hammering pitons into the rock, attaching links to them, and fastening ropes to the links. These served as fixed lines to assist those who followed in their ascent of the vertical face of the ridge. The lead climbing teams reached the top around midnight, and battalion units began their ascent in force. By 0400 on 20 February, all three of 1st Battalion's companies and the company from the 2nd Battalion had reached their individual objectives on top of the ridge without being seen. They attacked the defending Germans and completely surprised them. German daylight counter-

> attacks were repulsed and the division's left flank was secured on Riva Ridge opening the way for the rest of the division to accomplish its mission.
>
> The capture of Riva Ridge was the only significant action in which the division put to use its specialized stateside training. Nevertheless, no one would deny that this single mountain operation justified all the demanding training the 10th Mountain Division had undergone.
>
> Adapted from *See Naples and Die*, Robert B. Ellis.

GUIDES

4-39. Mountain guides are useful for any operation, particularly *on* terrain above Class 3, which requires accurate judgment and extensive technical knowledge. They act primarily as advisors to unit commanders and assist with planning when technical mountaineering problems affect the tactical scheme of maneuver. They must be experienced in all aspects of mountaineering (usually a Level 3 mountaineer) and capable of ensuring that the force is never beyond the margin of its operational capabilities while operating on mountainous terrain. Mountain guides must possess the skills, knowledge, and experience necessary to develop a true perspective of the obstades and conditions that must be overcome.

4-40. Mountain guides should perform the following functions for the commander:

1. Conduct a terrain analysis of the assigned area.
2. Select the best march routes
3. Identify danger areas, obstades, and hazards.
4. Estimate the effects of elevation on force capabilities.
5. Determine the technical means necessary to employ the force.
6. Estimate the time of unit movements.
7. Develop a movement sketch indicating the azimuth, time, and mountaineering requirements
8. Lead units, usually company-level and above, on difficult terrain.
9. Supervise al l aspects of mountaineering safety.

LEAD CLIMBING TEAMS

4-41. As with all spedal purpose teams, lead dimbing teams have highly skilled soldiers qualified in advanced mountaineering techniques (normally Level 2 mountaineers). To a lesser degree, lead dimbing teams perform many of the same functions as guides. They also accompany a unit over unprepared routes and assist with actually conducting its mission by maintaining and improving mobility. In some instances, lead dimbing teams may operate independently of other forces to accomplish spedfic missions.

4-42. These teams are capable of climbing at night with the aid of night-vision goggies and can conduct operations over any type of terrain. Lead climbing teams should be utilized to install fixed ropes to assist personnel over exposed terrain. They make the most difficult climbs and act as rope leaders. Members of the lead climbing teams must be extremely proficient in the technical aspects of mountaineering, since they select the specific routes to be climbed. This responsibility emphasizes the importance of accurate judgment, since a single mistake could jeopardize the success of the unit's mission. Commanders should consider assigning lead dimbing teams to, or developing an organic capability within:

- Ground reconnaissance elements.
- Forward observer parties.
- Air defense sections.
- Communications sections.
- Security elements.
- Assault elements.
- Sniper sections.

4-43. The number of lead dimbing teams required is dependent on the mission and difficulty of the elevation level or dismounted mobility classification. If more than one point of attack is to be used or a more mobile patrol is needed, it may be necessary to employ additional lead climbing teams. Lead dimbing teams may also assist in the attack of very steep objectives by negotiating the most unlikely avenues of approach. They can be dispatched ahead of the attacking elements to secure the advance at night and during periods of inclement weather. The commander's imagination and their limited availability are the only restralnls toward using lead dimbing teams.

4-44. In predpilous terrain, lead dimbing learns can be used alone to conduct small reconnaissance patrols or to form the nudeus of larger patrols. They may also be used to expedite movement of flank security elements over difficult terrain and during poor visibility. At least one team should be attached to each element, depending on the size of flank security and the operational terrain level in which operations take place. When the AO dictates a rate of march on the flanks that is slower than that of the main body, flank security elements should be located adjacent to the advance echelon. Lead dimbing teams are detached as necessary to reconnoiter and hold dominant terrain features on the flanks of the line of march. As the trail element passes and the security position is no longer required, the lead dimbing team's flank outposts joi n the mai n body or proceed forward to another security location.

INSTALLATION TEAMS

4-45. Installation team members are qualified in the construction and maintenance of technical mountaineering systems, referred to as installations, which fadlitate unit movement. These teams deploy throughout the AO. in any weather or visibility conditions, to erect mountaineering installations that overcome obstacles to the movement of friendly forces and supplies.

4-46. In most situations, installation teams should consist of six qualified members, enough to build most installations. Installation teams deploy early and prepare the AO for safe, rapid movement by constructing various types of mountaineering installations (see Figure 4-6). Following construction of an installation, theteam, or part of it, remains on site to monitor the system, assist with the control of forces across it, and make adjustments or repairs during its use. After passage of the unit, the installation team may then disassemble the system and deploy to another area as needed.

- Fixed ropes
- Pulley systems
- Hauling lines
- Rope ladders
- Rope bridges
- Handrails
- Traverse systems

Figure 4-6. Types of Installations

4-47. Although installation teams can assist, commanders remain responsible for establishing a traffic control system and the rate of negotiation to expedite tactical movement and prevent unnecessary massing of personnel and equipment on either side of t he obstacle- Plans for the traffic control system includearrangements for staging and holding areas, as well as movement. A prearranged order of movement, dissemi nated to all elements of the force, facilitates rapid movement. The traffic control plan also includes instruct ions for dispersing forces on the far side once they negotiate the obstacle. The successful negotiation of systems depends on how quickly a force can consolidate on the far side and continue its mission.

4-48. Commanders must carefully consider the proportion of installation team assets allocated to maneuver with those assigned to sustain essential logistics activities. When considering a unit's scheme of maneuver, the number and types of installations depend on METT-TC, emphasizing factors listed in Figure4-7.

- Number of routes required by the maneuver force
- Size of the maneuver force
- Amount and type of maneuver force equipment
- Number of installations required to extend logistics support
- Weather and visibility conditions
- Number of installation teams available

Figure 4-7. Factors Influencing Mountain Installations

4-49. Once operations stabilize, installation teams can direct and assist engineers in the construction of fixed alpine paths. Fixed alpine paths consist of permanent or semi-permanent mountain aids that assist troops in traversing rugged mountain terrain and facilitate the movement of equipment and supplies to and from forward areas.

SECTION II - OFFENSIVE OPERATIONS

4-50. Offensive operations in the mountains vary depending on the degree of restrictions dictated by mountains of different heights and character, but are normally planned and conducted utilizing the movement to contact and the attack. Exploitation and pursuit are conducted, but less frequently than in other environments.

4-51. Mountain operations will most likely be fought to gain control of key or decisive terrain. The goal is to seize objectives that are important for continuing the battle, such as lines of communication, passes, ridges, and choke points. Every attempt must be made to fight from the heights down. Frontal attacks against defended heights have little chance of success and attacks are usually made along the flanks and to the rear of the enemy. Consequently, envelopment becomes the preferred form of maneuver.

4-52. The missions assigned to units operating in a mountainous region remain the same as those in lowland regions. Commanders should place an increased emphasison;

- Limited-visibility operations,
- Mobility and survivability.
- Securing friendly lines of communications while severing those of the enemy.
- Reconnaissance and security,
- Centralized planning and decentralized small-unit actions.
- Environmental factors favoring operations of short duration and violent execution.

PLANNING CONSIDERATIONS

4-53. When conducting offensive operations in the mountains, surprise is often a dominant consideration, rather than a supporting one. Units achieve surprise by achieving superior situational understanding, especially of the terrain, and by using that knowledge to do the unexpected. Friendly forces should conduct bold and imaginative operations to exploit enemy weaknesses or inability to operate in a mountainous environment. A well-trained force can achieve surprise by infiltrating and attacking the enemy's rear or attacking during periods of limited visibility, such as night, rain, or snow. They can use helicopters and their technical mountaineering skills to conduct decisive operations anywhere in the AO. The effects of surprise can be dramatically increased if commanders select objectives in restrictive terrain that decrease the enemy's mobility and ability to react effectively.

4-54, The mountainous terrain increases the threat to concentrated formations. Usually, it is difficult to coordinate all forces by time and location so that they can rapidly support each other and achieve massed effects. The compartmented terrain separates adjacent units, precluding mutual support, and may adversely affect supporting distances. Therefore, it is critical to anticipate the concentration of forces and fires before the battle begins to achieve effective synchronization.

4-55. Commanders usually select decisive points for the attack based, in part, on their ability to seize and control key terrain. Each of these objectives often necessitates the seizure of one or more intermediate objectives. The compartmented terrain and resulting dispersion make it difficult for commanders to visualize the entire AO and complicates command and control, while the terrain often affords the defender excellent observation and decreases the attacker's ability to concentrate forces undetected.

4-56. Once a battle is joined, helicopters are the only rapid means for massing forces on terrain higher than Level I. When weather conditions permit, attack helicopters and tactical air assets are essential because they can move and strike over a large AO in a short time *Therefore,* the suppression of enemy air defense may become a high priority during offensive operations.

4-57. Commanders must personally acquaint themselves with the terrain to the fullest extent possible They combine this knowledge with other factors of METT-TC to develop simple, yet precise plans and orders. As previously discussed, compartmented terrain tends to divide the battle into many isolated engagements that are difficult to control. A complex and inflexible plan wilt not withstand the changing situations associated with this type of decentralized combat. An uncomplicated plan with a well-thought-out intent that is clearly communicated ensures understanding at the lowest level and allows subordinates to exploit battlefield opportunities even if communications fail.

PREPARATION

4-58. The length of the preparatory phase is typically longer in a mountainous environment. An offensive action against an enemy defensive position must be based on thorough reconnaissance and orderly preparation. The primary factor in determining the technique to be used in destroying the enemy position is the strength of the enemy's defenses. The stronger the defenses, the more deliberate the attack must be. In mountainous terrain, evaluation and exploitation of the ground are essential. Commanders must prepare plans of attack that take advantage of the weaknesses found in the enemy's defensive system. In the mountains, a larger number of reconnaissance assets and additional time may be needed to determine the strength of enemy positions on the objective and all surrounding heights, and favorable routes to and past the objective.

4-59. Difficult approach routes should be marked and prepared for safe passage. Easily traversed slopes, broad hills, plateaus, and valley floors, as well as mountainous terrain with well-developed road and transportation nets, permit deployment in breadth. High ranges with ridges and crests leading to the objective require organization in depth with extended lines of communications

4-60. In trackless mountain terrain, company-sized teams usually conduct attacks, tfthearea assigned toa battalion permits, companies should approach the objective separately on multiple routes. In restrictive terrain, adequate maneuver space may not always be available and several units may be required to move along the same avenue of approach. It may even be necessary to conduct shaping operations to seize sections of terrain from which the enemy can dominate the forces' movement and approach.

4-61. The preparatory phase may also indude feints and demonstrations to shape Ihe battlefield. In mountainous lerrain, the defending force has a number of advantages that allow it to defeat an attacking force much larger than its own, such as long-range observation, rugged natural cover, concealment, advance siting of weapons, and operations on familiar terrain. Feints and demonstrations mask friendly operations, expose enemy vulnerabilities, disorganize the defense, and allow the attacking force to engage the enemy discriminately. In a mountain area of operations, the benefits achieved from a successful deception effort may well outweigh the difficulties involved in mounting it.

FORMS OF MANEUVER

4-62 The forms of offensive maneuver are common to all environments, to indude mountainous terrain. While frequently used in combination, each form of maneuver attacks the enemy in a different way, and some pose different challenges to the commander attacking in the mountains.

INFILTRATION

4-63. Infiltration is frequently used in the mountains. The difficult terrain and recurring periods of limited visibility allow for undetected movement. Infiltration in a mountainous environment is used to shape the battlefield by attacking enemy positions from the flank or the rear, securing key terrain in support of the decisive operations, or disrupting enemy sustaining operations. Infiltration is normally conducted using one of threetechniques: movement in small groups along one axis, movement in one group, or movement in small groups along several routes at the same time. Regardless of the technique used, units must move in a covert manner to reduce the chance of enemy contact.

4-64 With movement in small groups along one axis, all members of the force use the best route. Small groups are harder to deled and easier to control, and do not compromise the total force if detected. This technique may require an excessive amount of time and an increased number of guides and lead dimbing teams, and does require an assembly area or linkup point prior to oondud of the dedsive action. If the lead group is detected, other groups may be ambushed.

4-65. The enemy can more easily deted movement in one group. If the force is deteded, the overall mission may be endangered. However, this technique has no reassembly problems, since reassembly is not required. Everyone uses the same route, easing navigation and redudng the number of guides and lead dimbing teams required. A large force can fight out of a dangerous situation more easily than a small one This technique minimizes coordination problems with other infiltrating units.

4-66. Movement in small groups along several routes at the same time has several advantages. It avoids putting the total force in danger and is less likely to be seen. Il forces the enemy to react in many locations and makes il harder for him to determine the size of the force or its mission. Groups travelling over severely restrictive terrain may have significant assembly, control, and sustainment difficulties, and may require more guides, lead climbing teams, and installation teams.

ENVELOPMENT AND TURNING MOVEMENT

4-67. The envelopment and its variant, the turning movement, are used extensively in mountain operations. Both forms of maneuver seek to avoid the enemy's strength and attack the enemy at a decisive point or points where the enemy is weakest or unprepared, and both rely on superior agility. In the mountains, the ability to react faster than the enemy may be achieved through greater mountaineering skills, using airborne and air assaults, and, depending on the specific AO, amphibious assaults.

PENETRATION

4-68. Mountainous terrain normally makes penetration extremely dangerous or impossible because of the difficulty in concentrating overwhelming combat power in the area of penetration. Due to mobility restrictions, it is also difficult to develop and maintain the momentum necessary to move quickly through a gap and on toward the objective. The area of penet ration is always vulnerable to flank attack, and this vulnerability significantly increases in mountainous terrain A penetration may be useful when attacking an enemy that is widely dispersed or overextended in his defense. If a penetration must be conducted, flank defensive positions must be eliminated before the initial breach of enemy positions. Successful penetration of a defensive positron in restrictive mountainous terrain requires using limited visibility, stealth, and covered and concealed terrain at selected breach points.

FRONTAL ATTACK

4-69. Frontal attacks in hilly or mountainous areas, even when supported by heavy direct and indirect fires, have a limited chance of success. Mountain terrain adds to the relative combat power of the defender. The frontal attack exposes the attacker to the concentrated fire of the defender while simultaneously limiting the effectiveness of the attacker's own fires. In the mountains, the frontal attack is an appropriate form of maneuver to conduct as a shaping operation designed to fix a force, while the decisive operation uses another form, such asthe envelopment, to defeat the enemy.

MOVEMENT TO CONTACT

4-70. The fundamentals regarding movement to contact also apply in a mountainous environment with the added likelihood of surprise attack and ambush. Limited mobility and dependence on restrictive terrain make it difficult to rapidly deploy from the movement formation. Plans and movement formations and techniques should be based on maintaining fiexibility and providing continuous security.

4-71. During a movement to contact, the advance guard normally advances in column, moving continuously or by bounds, until it makes contact. While requiring less physical exertion, movement along the topographical crest of a ridgeline increases the possibility of enemy observation and should normally be avoided. Given adequate concealment, this exposure may be reduced by moving along the military crest. Ridgelines and crests can often provide a tactical advantage to the force that controls them. Their control may allow rapid movement from one terrain compartment to another and afford excellent observation into lower terrain levels. In all cases, commanders must address the control or clearance of ridgelines that dominate their planned avenues of approach.

4-72. The main body should never be committed to canalizing terrain before forward elements have advanced far enough to ensure that the main body will not become endrded. This is a critical fador when employing mixed heavy and light forces that have sharp differences in operational tempo. Combat service support must be decentralized and readily available to sustain the combat elements. Major terrain compartments may physically separate maneuver units moving as part of a larger force. Continuous reconnaissance to the front and flank security is essential to prevent the enemy from infiltrating the gaps between units.

4-73. As the enemy situation becomes better known, commanders may shorten the distance between elements to decrease reaction time, or they may begin to deploy in preparation for the attack. Lateral movement between adjacent columns is frequently difficult or impossible. However, every attempt should be made to maintain at least visual contad. Commanders must emphasize the use of checkpoint reporting, contact patrols, and phased operations to coordinate and control the movement of the overall force. Control measures should not be so numerous as to impede operations and stifle initiative. Proper control ensures that units and fires are mutually supporting, objectives arecorrectly identified, and unitsare in position to attack.

ATTACK

4-74. Speed, flexibility, and surprise, normally advantages enjoyed by the attacker, are limited by restridive terrain and the defender's increased ability to see and acquire targets at greater distances. These limitations make it difficult for units above the company team level to conduct a hasty attack against prepared positions. In the mountains, commanders usually need more time to coordinate fire support, pick routes to prevent enemy observation and detedion. and select control measures to coordinate and control the operation. Therefore, deliberate attacks requiring a detailed scheme of maneuver and well-developed fire support plan become the norm at battalion-level and above. Since daylight contributes to the defender's ability to see and, thereby, reduces the attacker's chances of success, commanders should seek opportunities to exploit the advantages of limited visibility. Although these conditions slow movement even more and make coordinating forces more difficult, they decrease the enemy's ability to accurately sense what is happening and react effectively.

4-75. In planning and conduding the attack, commanders should recognize that the enemy will generally seek to control the valleys and trail networks, induding adjacent slopes and high ground. Defenses normally be anchored around obstades, and long-range, di rect fire weapons employed in poorly t rafficable terrain, often on slopes and protruding high ground. The enemy will attempt to engage the attacker in the valleys and low ground with flanking fires and artillery, often in a direct fire mode. Commanders must analyze the terrain to determine not only how the enemy will organize his defensive positions, but also how the terrain might contribute to the enemy's ability to counterattack As friendly forces attempt to deploy for the attack, the enemy, using his advance knowledge of the terrain and prepared routes, may maneuver forces to counterattack from the flank or rear.

4-76, All terrain features that can be occupied by even a small enemy force should be secured. I n many instances, overwatch positions may not be readily available within the range capability of organic weapons. Infiltration, technical dimbing, and extensive breaching may be required to position weapons to support the assault. On many occasions artillery support, espedally in high mountains, may not be available. In other instances, commanders may need to identify intermediate objectives for maneuver forces based on the need to ensure that artillery units have suitable, secure firing positions to range the enemy and support the attack. As in all environments, commanders must identify fire support requirements and allocate fires based on the ability to support and available ammunition. Because resupply may be limited and extremely difficult, they may need to place restrictions on the amount of ammunition expended on spedfictargets

4-77. Fire and movement during the assault are extremely difficult. In situations where machine guns can be positioned effectively, a rifle platoon can provide itself with support from a flank or from a height. However, during an assault up a slope, supporting fire cannot come from an overwatch position and must originate from the flanks or through gaps between the assaulting soldiers. Control is difficult to maintain when the assault is in steeply rising terrain. Commanders must pay speda! attention to the dangers of fratricide.

4-78. Fire and movement are easier in an assault over a downward slope Down-slope assaults often have the advantage of good observation, but dead spaces and intervening terrain may reduce the effectiveness of supporting fires. Defensive positions laid out by a skillful enemy on a reverse slope significantly increase the effect of unfavorable down-slope conditions. This type of defense compels the attacking force to position its supporting weapons and observation posts on exposed crests. In this sit u at ion, support dements must be positioned to avoid terrain masking and crest dearancs problems.

4-79. Breaching obstades and preparing bypass routes that allow the assault force to move into the defensive position must bean integral part of the commander's plan. In rugged terrain, man-made obstades that are covered by fire create a particularly dangerous and formidable barrier. Command and control of a covert, in-stride, deliberate, or assault breaches is more difficult than in open terrain, and mobility support is extensive if the obstade cannot be reduced. Assaults in mountainous terrain almost always involve preparing routes that allow the assault force to rapidly move over difficult natural obstades and into the objectives.

4-80. Commanders should maintain a strong reserve, if possible. In the mountains, as elsewhere, commanders can use their reserves to restore the momentum of a stalled attack, defeat enemy counterattacks, and exploit success. Reserves must becarefulty positioned and organized so difficult terrain, limited road networks, or unpredictable weather does not delay their arrival.

Once committed, commanders make every effort to reconstitute another reserve from available units.

4-81. An attack should not be halted on a summit or on a ridgeline objective, which enemy artillery and mortar fire will likely target. Reorganization is generally best conducted well forward of a crest line on the next suitable slope. Commanders must ensure that the enemy is not allowed the opportunity to counterattack to recapture key terrain. Rapid adjustment of positions and coordination with flanking units are essential. Support weapons, especially mortars, should be brought forward as quickly as possible. Helicopters are useful for this purpose and may be used for backhaul of casualties

COUNTERATTACK

4-82. Counterattacks in the mountains must exploit the aspects of terrain that impair enemy momentum and make it difficult for him to mass and maneuver. Obstructing terrain that canalizes movement and restricts mobility significantly increases the potential for counterattacks. In planning a counterattack, the commander must carefully consider the enemy's weaknesses or inability to operate in a mountainous environmen l. A counterattack, even on a very small scale, can have a decisive impact in mountainous terrain.

RAID AND AMBUSH

4-83. The restrictive terrain also affords increased opportunities to conduct raids and ambushes. These operations should take advantage of limited visibility and terrain that the enemy may consider impassable. In steep terrain, movement time increases significantly, and only light equipment can be taken. The force should use special climbing techniques to negotiate the difficult routes during limited visibility. Commanders must carefully consider the routes and methods used for extraction to ensure that the combat force does not become isolated after executing the mission. They can ensure a successful operation by avoiding detection through proper movement techniques and by skillfully using natural cover and concealment. It may be necessary to reposition some indirect fire support assets to cover dead space or use attack helicopters and close air support. The ambush or raid commander must know in advance if supporting fires can not cover his routes to and from the objective.

DEMONSTRATIONS AND FEINTS

4-84 Because maneuver space is usually limited or confined and restricts the number of avenues of approach for heavier forces, deception plays an important part in the mountain battle To mislead the enemy regarding friendly intentions, capabilities, and objectives, commanders should plan systematic measures of deception.

EXPLOITATION AND PURSUIT

4-85. In a mountainous environment, exploitation and pursuit operations must be conducted discriminately and the mountain commander must always prepare for success. A battalion may exploit its own success to a limited extent, but it normally participates in the exploitation as part of a larger force. Air assault and attack helicopter units can be used to augment exploitation and pursuit operations. The exploiting commander must compensate for the ground mobility restrictions imposed by terrain and weather. Speed can best be achieved by isolating enemy positions with the smallest force possible Engineer support should be well forward with the necessary equipment to allow combat troops to maintain momentum and avoid delay by enemy obstacles. The commander must be careful to prevent overextending either the exploiting force or its sustaining logistics. A withdrawing force is capable of establishing numerous strong points and firing positions on heights that allow it to quickly dissipate the combat power of the exploiting force.

MOTTI TACTICS

4-86. Motti tactics are presented here to demonstrate how forces can exploit superior mobility skills and knowledge of the mountainous terrain and environment to defeat the enemy. The Finns developed these tactics during the Finnish-Russian War in 1939-1940. They are characterized by attacks on rear areas, bivouac sites, and command posts.

4-87. The Finnish word "motti" means a pile of logs ready to be sawed into lumbo; in effect, setting the conditions so that a larger force can be defeated in detail. These tactics were most successful in the forested areas of Finland during the arctic winters. During the Finnish-Russian War, the Soviets were neither prepared for, nor trained for, warfare under such conditions. They were almost totally trail-bound, with few ski troops. In the 1980s, the Soviet Union experienced similar difficulties in the mountains of Afghanistan, In both instances, the road and trail-bound nature of their forces and their basic tactics left them vulnerable to motti tactics in mountainous terrain.

4-88. Generally, the force utilizing motti tactics never becomes decisively engaged. It disrupts the enemy's supply lines, denies him warmth and shelter, infiltrates his bivouacs, and destroys his rear areas to the point where he must remain in a high state of alert. These attacks, in combination with the environment, help to destroy the enemy's will to fight Commanders should not only develop a thorough understanding of how to apply these tactics, but also understand the conditions that may leave their own forces vulnerable to its use (see Figure 4-B on page 4-24).

Motti tactics generally follow the sequence of:

1. Locating and fixing the enemy.

2. Isolating the enemy.

3. Attacking to defeat or destroy the enemy

4-89 Reconnaissance is conducted to locate an enemy force moving in or toward an area that will restrict his movements to roads, trails, or linear terrain. Once identified, the force must be fixed so that it presents a linear target along the axis of advance to which it is bound. This is accomplished using obstacles and a series of squad and platoon sized ambushes and raids Obstacles may be natural, such as snow, crevices, deep mud, steep terrain, and water obstacles, or man-made, for example mines, landslides, avalanches, or destroyed bridges.

4-90. The ambushes and raids not only fix the enemy, they also disturb his composure, create an air of uncertainty, and prevent uninterrupted sleep and rest. Friendly units attack the enemy from the high ground. They make maximum use of night vision deviceSr as well as the difficult restrictive terrain. They avoid enemy security and interdict his operations. As a further result of these actions, the enemy is compelled to use more forces on security tasks. Unless the enemy can be easily defeated or destroyed, the attacking force rapidly withdraws after forcing the enemy to deploy. In general, this series of attacks confuses the enemy as to the friendly unit's exact location and intent, and slows his decision-making cyde so that he reads ineffectively to subsequent operations.

FORCES CAN USE MOTTI TACTICS WHEN THEY:

- Have superior technical mobility skills necessary to negotiate Class 4 and 5 terrain
- Are able to operate effectively in a noncontiguous area of operations with limited support, and despite temperature extremes and inclement mountain weather
- Are able to navigate in high mountainous terrain, dense vegetation, darkness, storms, and fog while making good use of available cover and concealment
- Maintain the element of surprise

FORCES ARE VULNERABLE TO MOTTI TACTICS WHEN THEY:

- Operate within noncontiguous areas of operations
- Have limited mobility skills restricting their movements to roads, trails, and Class 1 and 2 terrain
- Have inadequate reconnaissance and security

Figure 4-8. Conditions Affecting the Use of Motti Tactics

4-91. The attacking force then isolates the enemy into smaller groups. Once isolated, the friendly force maneuvers to envelop and attacks to defeat or destroy the isolated elements. As the enemy exhausts himself in an effort to break out, the attacking force may regroup and repeat the sequence. It is imperative that the attacking force seal off the enemy and keep avenues of approach closed, and not ignore the threat to its flanks, which may increase as the attack progresses.

4-92. Overall, motti tactics wear the enemy down to a point where he is vulnerable to more direct attacks or to the point where it is no longer beneficial or feasible to continue operations in the area. Motti tactics employed alone only prove decisive over a long period of time, depending on the enemy's capabilities, strength, and resolve. Based on METT-TC. friendly forces normally must increase the operation's tempo to gain a quick, decisive outcome. Sti 11.

these type tactics may complement other more direct offensive operations in support of the overall pfan.

SECTION III - DEFENSIVE OPERATIONS

4-93. In the mountains, more so than in the lowlands, the strength of the defense depends on its selection and use of key and decisive terrain. Key and decisive terrain provides the defender-and usually denies the attacker-excellent observation and fighting positions. Reinforcing obstades significantly enhance the natural obstacles of rugged mountainous terrain.

4-94. The immediate objective of a mountainous defense is to deny the enemy access to key terrain that helps him conduct further operations Therefore, it is necessary to defend in terrain that restricts and contains the enemy, as well as control the high ground that dominates this terrain. The effects of rapidiy changing weather, visibility, and mountain hazards must be continually assessed. The terrain provides the defender with cover, concealment, and camouflage that can deceive the enemy regarding the strength and dispositions of friendly forces. The advantages of knowing the terrain, having fortified positions, siting weapons in advance, stockpiling supplies, and identifying or preparing lateral trail networks favor the defense. They allow the defender to shift forces on the ground more rapidly than the attacker. Delaying operations are particularly effective in the mountains and can be accomplished by a smaller force. These advantages combine to make the mountai ns an ideal place for defensive operations. Regardless of the scale of defensive operations, key factors in achievi ng success in the mountai ns are having good observation and aggressive reconnaissance, white denying the same to the enemy.

PLANNING CONSIDERATIONS

4-95. Defending commanders must develop flexible pEans for control of fire, maneuver, communications, and logistics. Initially, the attacker has the initiative and decides where and when combat will take place. The defender must be agile enough to maintain control of the heights, strike effectively, and shift his effort quickly without losing momentum and flexibility. Tactical flexibility depends on planning in detail, organizing in depth, and retaining an adequate, mobile reserve

4-96. Although the mountains generally allow observation at greater distances, intervening terrain features and weather often prevenl commanders from seeing the area of operations beyond the area to their immediate front and flanks. Consequently, commanders normally allocate more assets for reconnaissance and security, echeloned in depth and in height, to ensure that they are able to sense all aspects of the AO and gain the time needed to decisively apply combal power.

4-97. Commanders must prevent the enemy from concentrating overwhelming combat power against isolated sections of their defense. The restrictive terrain is one of the primary advantages of the mountain defender, as it interferes with the attacker's synchronization, canalizes his movement, and impedes his ability to maneuver. However, unless oommanders carefully analyze the terrain from both the friendly and enemy viewpoints, to indude the horizontal and vertical perspectives, they leave themselves vulnerable to infiltration and possible attack from the flanks and rear along difficult and unexpected routes.

4-98. In the mountains, commanders usually organize for a perimeter defensetobe prepared to defeat the enemy from any approach, to indude those that may appear impassable. Although preparing for an all-around defense, they should weight a portion of the perimeter to cover the most probable direction of enemy attack. Rocky terrain may make it more difficult to prepare defensive positions and rapidly changing weather may halt preparations altogether. If suffident forces are not available, the commander must economize in some areas and rely moreheavily on prepared positions, to ind ude alternate and supplementary positions, obstades, and well-planned indired fires to cover gaps and dead space.

4-99. The width of an area to be defended depends mainly on the degree to which terrain is an obstade. Terrain that significantly restricts enemy movement tends to lavor a larger AO. Normally, an area should be approximately as deep as it is wide, and may indude the entire length and surrounding heights of a valley. Ridges that run at right angles to the enemy's direction of attack also permit increased width with less depth. A defense in depth is required in valleys that run in the direction of the enemy's attack. In either case, it is essential to have forces deploy on the dominating heights that control approach routes.

4-100. Ideally, reserves should be mobile enough to read to enemy adion in any portion of the perimeter. Less mobile reserves are positioned to block the most dangerous avenues of approach and assigned on-order positions on other critical avenues. Sharply compartmented terrain may require the creation of more than one reserve. Helicopters may be used to deploy reserves, but their use depends on the availability of suitable, secure LZs and favorable weather conditions.

PREPARATION

4-101. The process of preparing the defense must begin with a thorough reconnaissance. Preparations for a mountain defense require more time than in other terrain, and as units arrive they must begin immediate preparation of their defensive positions. In some instances, technical mountaineering skills may be needed to establish effective security and to emplace crew-served weapons properly. However, commanders must weigh the advantages gained from these inaccessible positions against difficulties in repositioning and resupply. Preparations for the defense must alsoindudeinstalling communications, stocking forward supply points with particular attention to Class IV. empladng medical elements, adjusting air defense coverage, and arranging for security of installations in the rear area. Commanders must ensure that time is available to develop alternate routes and positions, rehearse and time movements between positions and along routes, and rehearse counterattacks.

4-102. Commanders must seek every opportunity to recapture the initiative from the attacker and transition to offensive operations. Preparations for a counterattack in the mountains must include caching ammunition, preparing counterattack positions and routes to attack downhill, identifying crew-served weapon positions, and establishing rally points that are usually on the reverse slope.

ORGANIZATION OF THE DEFENSE

4-103. Defensive operations in the mountains derive their strength, balance, and freedom of action from the effective use of terrain. The mobility restrictions found in mountainous areas, combined with the necessity to hold dominating ground, dictates that an area defense be used. Mountain defenses use security forces, continuous reconnaissance and combat patrols, as well as numerous observation posts. The mountain AO is usually organized into a security area, main battle area (MBA), and rear area.

SECURITY OPERATIONS

4-104. While a screening force is often thought to be the most preferable form of security in extremely rugged mountainous terrain, all forms of security operations, to include guard, cover, and area, may be employed effectively in a mountain AO based on the factors of METT-TC with particular emphasis on:

- Forces available for security operations.
- Ability to maintain a mobility advantage.
- Size of the security area and the number of avenues of approach.
- Likelihood of enemy action,
- Size of the expected enemy force.
- Amount of early warning and reaction time needed.

4-105 A screen primarily provides early warning to the protected force and is usually an economy-of-force measure. The compartmented nature of mountainous terrain often serves to create multiple gaps and exposed flanks The rugged terrain may also serve to restrict the movement of not only advancing enemy forces, but also the movement and mobility of larger friendly security forces. In these instances, commanders may choose to use minimum combat power to observe, identify, and report enemy actions at these locations, and engage and destroy enemy reconnaissance within the screening force's capability. The screening force may be able to avoid decisive contact by withdrawing into restrictive terrain that forces the enemy to utilize difficult climbing techniques if he continues the pursuit.

4-106. In the mountains as elsewhere, the screening force should adjust to the enemy advance and continue to screen as far forward as possible, even though elements of the force may have to withdraw. Retention of selected forward positions may allow surveillance and targeting forward of the MBA, upsetting the enemy's coordination. By allowing the enemy to bypass advance positions, the screening force can facilitate counterattack to the front of the forward edge of the battle area (FEBA) by providing observation of. and access to. the flanks and rear of attacking forces.

4-107. If a significant enemy force is expected or a significant amount of time and space is required to provide the required degree of protection, commanders usually resource a guard or cover mission instead of a screen. As long as trie security force can maintain a mobility advantage over the enemy, it can effectively delay and attack the enemy force by using obstacles and the restrictive terrain to its advantage. Although utilizing a greater proportion of his combat power, the appropriate use of a guard or cover force should provide the mountain commander greater depth in his security area and the ability to defeat, repel, or fix lead elements of an enemy ground force before they can engage the main body with direct and indirect fires.

4-108. No matter the type of security used, defending forces must prevent enemy infiltration by carefully positioning observation posts (OPs) and conducting continuous patrols and ambushes. Combat reconnaissance patrols and other intelligence gathering assets observe the enemy as far ahead of friendly positions as possible and report his strength and composition, as well as his route of movement. To accomplish this, reconnaissance patrols may need to rely heavily on technical climbing skills. Ground surveillance radar and unattended ground sensors can be used effectively, but the defender must besureto cover all gaps and dead spaces. The defender must make best use of his time to study the ground and determine ail possible infiltration routes.

MAIN BATTLE AREA

4-109. In rugged mountainous terrain, it may be difficult to maintain mutual support and overlapping observation. Elements should be employed to man observation posts, assist the passage of security forces into the MBA. cover obstacles and avenues of approach by fire, screen gaps between defensive positions, and ambush enemy infiltrators.

4-110. Defensive positions along ridges or domi nating heights should include as much of the forward and reverse slopes as possible to add depth and all-around security. The actual size of unit positions is terrain-dependent. At a minimum, fighting positions and observation posts should be echeloned vertically, as well as in depth. When defending a mountain valley, forces should establish fighting positions that are located on adjacent heights and in depth to permit coveringthevalleywith interlocking fire. These positions must also be anchored to restrictive terrain or adjacent defensive forces to prevent enemy envelopment. In wooded terrain, defensive positions may be organized on theforwardedgeofthe woods, as well as on commanding heights. Obstacles should be widely employed to slow or stop enemy movement throughout.

4-111, Mountain warfare demands that forces conduct an aggressive defense. Defending units must infiltrate enemy units and attack headquarters, supply lines, and rear areas. Smaller patrols and OPs should be deployed well forward to direct artillery fire and attack aircraft on targets of opportunity, and to conduct personnel and antiarmor ambushes. Disruption operations should be conducted to force the enemy to deploy additional assets to protect lines of communication and delay and upset preparations for the attack. In the mountains, enemy forces can frequently be isolated if they are discovered in time and reserves are effectively placed and highly mobile.

4-112. In the defense, the reserves' primary purpose must preserve the commander's flexibility In mountainous terrain, the commander may need to rely on the reserves as his principal means of restoring his defense's integrity or exploiting opportunities through offensive action. Because of the difficulties of movement, small reserves may be located near primary defensive positions, ready for immediate counterattack. This type of small, responsive counterattack may be much more effective than a large-scale, major counterattack. It can catch the enemy exhausted after an uphill assault and before position consolidation. Large, centrally placed counterattack forces are normally unableto intervene in time unless the terrain permits mounted movement, or sufficient helicopter lift assets are committed to the reserve force or made rapidly available.

REAR AREA

4-113. To minimize the vulnerability of sustaining operations and extended lines of communication, command and control, as well as support operations, in the rear area must be dispersed, redundant, and as far from potential enemy approaches as possible. Because of limited space available in the rear area, the commander must be careful in selecting and locating positions for combat service support activities. These positions are likely to be confined to small valleys. Therefore, they are high-priority targets for enemy artillery and air attack or raids by small combat patrols, particularly at night or in bad weather. When possible, combat service support elements must avoid the most obvious positions and occupy atypical sites. However, they should be in the vicinity of a defined road network and an air loading area, even if the network or area is within Level 11 Locating base defenses at Level II elevations may allow more access to supply bases for air resupply during inclement weather, such as the heavy fog often encountered in valleys and at lower elevations

4-114. A perimeter defense is planned for each combat service support unit within the defensive area. Defensive positions should be selected on the dominating high ground. Sensors, OPs. and radars are used to cover avenues of approach and gaps between positions. Rear area forces must routinely conduct patrols and ambushes around the perimeter, especially at night and during other periods of limited visibility. Air defense assets should be located to protect rear area facilities. Tactical combat forces (TCP) must be prepared to respond rapidly to rear area threats and should be prepared to move to any of their objectives by multiple routes. However, units within the rear area must not fall into the trap of relying solely on a TCP for their security. No matter how well-organized or mobile the TCP, rear area units must provide their own well thought-out and active security measures, even at the cost of a reduced ability tosustain the force.

REVERSE SLOPE DEFENSE

4-115. Reverse slope defenses apply particularly well to mountain operations and pursue offensive opportunities through surprise and deceptive operations by defending in a manner for which the enemy is unprepared. This defense seeks to reduce the effects of massed indirect fire from mortar, artillery, and dose air support, and draws the battle into the small arms range of infantry weapons. The overall goal is to make the enemy commit his forces against the forward slope of the defense, causing his forces to attack in an uncoordinated fashion across the exposed topographical crest. Once this type of defense is employed, subsequent use may be of limited value, due to the loss of the key element of surprise.

4-116. All or partsof thedefending force may use reverse slope techniques. In many instances, mountainous terrain favors a defense that employs combined forward and reverse slope positions to permit fires on enemy approaches around and over the crest and on the forward slope of adjacent terrain features. Key factors to this type of defense are mutually supporting covered and concealed positions, numerous natural and man-made obstacles, theability to bring fire from all availableweapons onto the crest, and a strong and mobile counterattack force.

4-117. The reverse slope defense is organized so that the main defensive positions are masked from enemy observation and direct fire by the topographical crest (see Figure 4-9). It extends rearward from the crest only to the maximum effective range of small arms fire. Observation and fires are maintained over the entire forward slope as long as possible to continue to destroy advancing enemy forces and prevent him from effectively massing for a final assault. A successful reverse slope defense is based on denying the topographical crest to the enemy, either by fire or by physical occupation. Although the crest may not be occupied in strength, control of the crest by fire is essential for success. For more detailed discussions of the reverse slope defense, see FM 3-100.40 and FM 3-21.30,

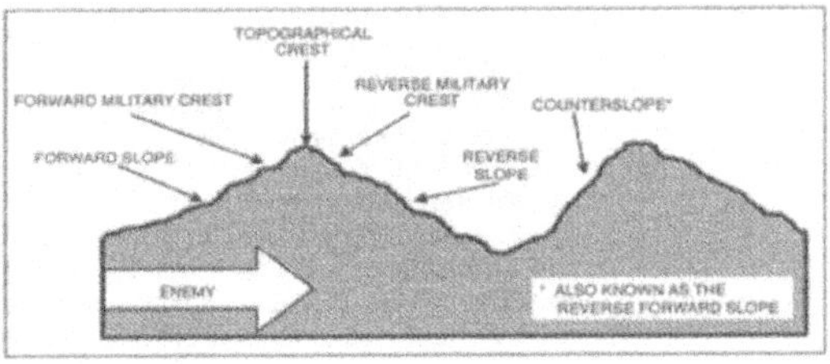

Figure 4-9. Reverse Slope Defense

RETROGRADE OPERATIONS

4-118. Retrograde operations of delay and withdrawal can be conducted in mountainous terrain with fewer assets because of the mobility difficulties of an advancing enemy. Delaying operations are particularly effective in the mountains. Numerous positions may exist where elements as small as a machine-gun or sniper team can significantly delay a large force. When conducting retrograde operations in mountainous terrain, the friendly force must accomplish several tasks.

- it must make maximum use of existing obstacles- However, the addition of relatively few reinforcing obstacles, such as the antitank mining of a route with very steep sides, often increases the value of existing obstacles.
- The force must conduct detailed reconnaissance of routes to rearward positions. Routes of withdrawal are not as numerous in mountainous terrain and often do not intersect as they do on flat terrain. These

factors complicate subsequent link-up operations and necessitate meticulous planning.

- It musl protect the flanks and rear to prevent encirclement, particularly by air assault. There are only a few LZs and they can significantly influence the outcome of a battle. At a minimum, they must be covered by fire.

STAY-BEHIND OPERATIONS

4-119. The compartmented terrain in the mountains lends itself tothe employment of stay-behind forces as a tool for offensive action. Stay-behind operations involve the positioning of friendly elements within operational areas before the enemy advances through the area. Stay-behind forces conceal their location and allow themselves to be bypassed as the enemy advances. (Figure 4-10 outlines the important tasks that stay-behind forces can accomplish for the mountain commander.)

- Attack the enemy throughout the depth of his formations
- Disrupt the cohesion of the enemy offense by interrupting lines of communication and logistics
- Detract from the enemy's main effort by forcing him to allocate combat forces to rear areas
- Provide immediate intelligence
- Call for and control indirect fire and close air support

Figure 4-10. Tasks for Mountain Stay-Behind Forces

4-120. Stay-behind forces may be positioned forward of the FEBΛ, in the MBA, and not participate in the initial fight, or, under certain conditions, in the MBA after the fighting has started. When planning for stay-behind operations in the mountains, commanders must consider the following:

- Stay-behi nd forces should be a combined arms force that indudesengineers.
- Indirect fire support or dose air support should be available throughout their operations.
- Return routes must be well planned and reconnoitered in advance. Exfiltration, regardless of element size, should follow covered and concealed routes, and rally points should be designated forward of and behind the lines of friendly forces. Reentry must be carefully coordinated to prevent fratridde.
- After an attack, stay-behind forces may be unable to reach a hide position to subsequently return to friendly lines by exfiltration. If this occurs, they must be prepared to condud a breakout,
- The tactical situation and logistics supplies that were stockpiled or cached in the AO have an impad on the length of time stay-behind forces remain in enemy territory.

Chapter 5

Logistics and Combat Service Support

Mountainous terrain poses great challenges to combat service support (CSS) forces and complicates sustaining operations. Existing roads and trails are normally few and primitive, and cross-country movement is particularly demanding. Highways usually run along features that have steep slopes on either side, making them vulnerable to disruption and attack. Rivers become major obstades because of rapid currents, broken banks, rocky bottoms, and the lack of bridges. Landslides and avalanches, natural as well as man-made, may also pose serious obstacles to CSS operations. Mountainous areas have wide variations in climate and are subject to frequent and sudden changes of weather that may preclude reliance on continuous aviation support. Together, these conditions compound the obstacle-producing effects of mountainous terrain and create major challenges for the CSS planner. Therefore, the forward distribution of supplies may depend upon the knowledge, skill, and proficiency of CSS personnel in both basic mountaineering and aerial resupply operations.

CONTENTS

HISTORICAL PERSPECTIVE

The Importance of Lines of Communications: The Satukandav Pass (Soviet-Afghan War, November-December 1987)

One of the characteristics of the Soviet-Afghan War (December 1979 - February 1989) was the attempt by both sides to control the other's lines of communications (LOCs). In an effort to deprive the guerrillas of their source of sustainment, the Soviets used various methods to drive the rural population into exile or into cities. For their part, the Mujahideen regularly interdicted supply routes through the establishment of blocking positions and vehicular ambushes. In some regions, they were able to effectively interdict supply routes for weeks, months, and even years at a time. The Soviet main supply route was a double-lane highway network winding through the rugged and inhospitable Hindu Kush Mountains. The continued Soviet presence in Afghanistan depended, in large part, on their ability to keep the roads open. Therefore, much of heavy Soviet combat was a fight for control of this road network, with this control often changing hands during the course of the war.

In the fall of 1987, the Mujahideen had established a series of blocking positions that severely limited the supply of weapons, ammunition, and food to Soviet forces in the Khost district. In response, the Soviets planned and

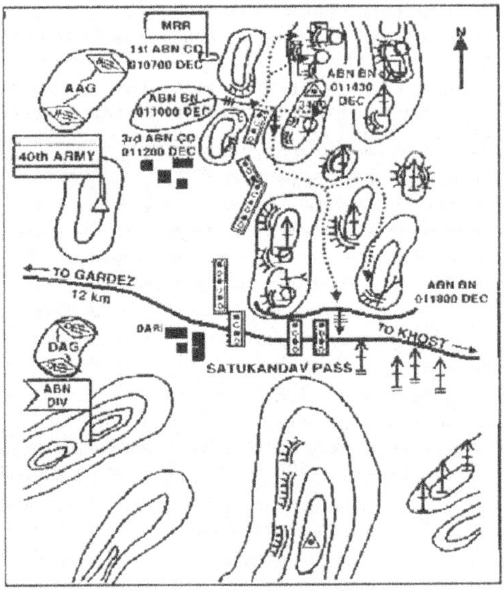

Figure 5-1. Satukandav Pass

conducted Operation Magistral, "main highway," to open the LOCs (see Figure 5-1). The guerrilla forces had established strong positions in the Satukandav Pass, virtually the only way through the mountains between Gardez and Khost. For the operation, the Soviets massed a motorized rifle division, airborne division, separate motorized rifle regiment (MRR), separate airborne regiment, various 40th Army regiments, special forces, and other subordinate units, and regiments from the Afghan Armed Forces. On 28 November, in order to determine the location of Mujahideen positions, particularly air defense systems, the Soviets conducted a ruse in the form of an airborne assault using dummy paratroopers. When the Mujahideen fired at the dummies, Soviet artillery reconnaissance was able to pinpoint enemy strong points and firing positions. The Soviets hit these positions with air strikes and a four-hour artillery barrage. The next day, however, an MRR failed to make its way up the foothills to seize the dominant terrain along the crest, and suffered heavy casualties. The 40th Army Commander, General Gromov, nevertheless, decided to continue to press his advance using the 1st Airborne Battalion and a battalion of Afghan commandos. On 1 December, two airborne companies captured key terrain and used this to support the decisive operation against the dominant peak to the south. This flanking attack took the Mujahideen by surprise and they began to withdraw. While calling in artillery fire on the retreating guerrilla forces, primarily on the reverse slope and along the probable avenue of approach for the commitment of enemy reserves, the Soviet battalion commander used this hard-won, key terrain to support a simultaneous, two-prong attack to the south toward the Satukandav Pass. Now it was the Soviets who were in a position to cut off supplies, especially fresh drinking water, from the Mujahideen. The latter were forced to withdraw, and the two battalions captured the pass. However, while the operation itself was a success, Soviet and Afghan Army forces could keep the road open for only 12 days, after which the Mujahideen once again cut off the supply route to Khost.

Both sides recognized the vital importance of LOCs, and this shift of LOC control was a constant feature throughout the entire duration of the Soviet-Afghan War. The Mujahideen's ability to interdict the LOCs prevented the Soviets from maintaining a larger occupation force there, a key factor in the eventual Soviet defeat.

Compiled from *The Other Side of the Mountain* and *The Bear Went Over the Mountain*.

SECTION I - PLANNING CONSIDERATIONS

5-1. Mountainous areas of operations dictate that commanders foresee needs before demands are placed upon CSS personnel The main logistical differences between mountain operations and operations in other terrain are a result of the problems of transporting and securing material along difficult and extended lines of support Logistics support must emphasize a continuous flow of supplies to specific locations, rather than the build-up of stocks at supply points along the main supply routes. Supply point operations alone are insufficient; the proposed support structure must plan for redundancy in the ability to distribute supplies directly to units operating from predetermined supply routes. If possible, commanders should plan to use multiple supply routes designed to support maneuver elements moving on separate axis.

5-2. Commanders must be concerned not only with the sustainment of current operations but also with the support of future operations. A detailed logistics preparation ol the theater (LPT) to identify the potential lines of communication plays a major part in determining the conduct of CSS operations. A detailed reconnaissance should be conducted to determine:

- The type and maximum number of vehicles that the road network can support in the area. New roads may need to be constructed or improvements made to existing ones to support protracted operations in isolated areas.
- Classification of bridges.
- Suitable sites for drop zones (DZs), loading zones (LZs), and short, tactical airstrips.
- Availability of water sources.
- Availability of local resources, facilities, and service and support activities.

5-3. Because of terrain constraints, it may be necessary to disperse support units over a wider area and ensure that supplies are positioned closer to supported units. Dispersion reduces vulnerability of CSS assets, which also creates problems with command, control, and security CSS units are often high-priority targets, and must ensure adequate protection against ground and air attacks.

5-4, In mountainous terrain, battalion CSS elements are normally echeloned into combat and field trains to increase responsiveness, provide adequate space, and decrease the logistics footprint. Combat trains are routinely located in ravines or valleys on the rear slope of the terrain occupied by the unit. This permits the personnel officer (S1) and logistics officer (S4) to operate in close proximity to the tactical operations center (TOC), and allows them to keep abreast of unit requirements,

5-5. In the mountains, unresolved logistical problems can quickly lead to mission failure. Ground operations may increase fuel consumption rates of individual vehicles by 30 to 40 percent, requiring more frequent resupply operations. The operation of equipment in mountainous terrain has proven that maintenance failures far exceed losses due to combat, and most breakdowns can be attributed to operator training. Air operations are characterized by a significant increase in lift requirements; however, increased elevations decrease aircraft lift capabilities,

5-6. Commanders must carefully consider combat loads in the mountains, based upon a thorough mission analysis. Excess equipment and supplies reduce the efficiency of the individual soldier and seriously impede operations. In steep terrain above 1.500 meters (5,000 feet), soldier loads may need to be reduced by nearly 50 percent. Commanders must develop priorities, accept risk, and require the combat foroe to carry only the bare essentials needed for its own support. Nonessential equipment should be identified, collected, and stored until it is needed. In situations where there are conflicts between the weighl of ammunition and weapons, experience has shown that it is belter lo carry more ammunition and fewer weapons. In the mountains, commanders should strive to achieve the imperatives indicated in Figure 5-2

- **Limit supplies to essentials.**
- **Lighten the individual soldier's combat load.**
- **Improvise methods and supply sources, to include utilizing captured enemy supplies and equipment.**
- **Use aviation assets to increase responsiveness.**
- **Anticipate maintenance requirements.**
- **Develop plans that place realistic demands on the CSS system.**

Figure 5-2. Mountain Supply Imperatives

SECTION II - SUPPLY

5-7. Units operating in mountainous terrain transport supplies by a combination of wheeled vehicles, over snow vehicles, i ndigenous pack animals and personnel (see FM 3-05.27), assisted by Army and Air Force lift assets. These combinations depend on equipment availability, location of combat units, type of terrain, and weather. However, any combination of resupply usually i ndudes combat soldiers man-packing supplies to their positions

5-8. Since combat operations in the mountains are decentralized, CSS operations are correspondingly decentralized. This decentral ization serves to create heavier man-loads, while rough, steep terrain decreases the amount soldiers are able lo carry. Although most soldiers are eventually able to acdimate themselves to higher elevations, their pace and subsequently the overall pace of the entire operation slows down as elevation increases.

5-9. Mountain warfare is highly dependent on accurate logistical planning if supply operations are to function smoothly. To win in any area of operations (AO), commanders normally seek to move and strike as rapidly as possible. Rapidly changing tadical situations may cause long supply lines, resulting in delay or complete disruption of supply operations. To mitigate these risks, situational understanding, rapid dedsions, and continuous coordination between tactical and logistical planners are essential. Stockpiling and caching supplies may also help to decrease the risks to resupply.

5-10. The total tonnage of supplies required by the force may also decrease. For example, while individual vehide petroleum, oils, and lubricants (POL) consumption may increase, overall consumption may decrease because of lower vehide movement. The quantity of supplies needed by the individual soldier normally increases. Soldiers consume more food because of increased energy expenditure, and need many additional items of equipment, such as extra clothing, sleeping bags, climbing equipment, tents, and stoves, all of which must be stored and transported.

SUPPLY ROUTES

5-11. Main supply routes are generally limited to the roads located along major valleys and, through necessity, to the smaller, more restrictive trails that follow or parallel the ridgelines. The limited number of routes increases the volume of traffic and places heavy demands on engineer units to maintain them. In most cases, engineer units require assistance in clearing and developing, as well as in securing, these routes. Travel times for ground transportation assets are significantly increased due the generally poor quality of mountain roads and trails, frequent switchbacks, and steep grades that require lower vehicle speeds. Traffic control assumes increased importance due to the limited number of routes in the mountains, and may require an increased number of military police dedicated to the task of battlefield circulation control. In particular-

- Existing roads should be rapidly analyzed for bottlenecks, deployment areas, passing places, and turnarounds for various vehicles.
- Routes should be classified as one- or two-way, and schedules developed for the use of one-way routes.
- Signs should be placed for both day and night moves on difficult and dangerous routes.
- Whenever possible, separate routes should be designated for vehicular and dismounted movement. Additionally, separate routes should be designated for wheeled and tracked vehicles, particularly if the latter are likely to damage road surfaces.

5-12, The enemy will emphasize destroying logistical units and interdicting supply activities. Enemy units will infiltrate and seize key terrain that dominates supply routes in an effort to disrupt and isolate units from their logistics support. Using mountain trails and roads without securing the high ground on both sides invites ambush Patrols must be continually conducted at irregular intervals to verify the status of roads and prevent enemy infiltration. Patrols must be continuously alert for ambush and they must be skilled at locating and identifying mines. However, a combination of patrols and aerial reconnaissance is the best means of providing route security-Observation posts on dominant terrain along supply routes are also essential for early warning of enemy infiltration into rear areas.

5-13. Most often, units have to use the narrrow ridge trails as alternate supply routes, in some instances as main supply routes, to reduce the volume of traffic on the main supply routes located along valley floors. This involves movements in much more restrictive terrain and exposure to excellent observation and fire by the enemy. Supply columns moving along separate routes face the same problems as combat units; they face the difficulties of being able to provide mutual support due to compartmented terrain, should one column come under attack. Movement of supplies at night may reduce vulnerability to enemy attack, but night marches present other hazards due to the difficult terrain, and require daylight reconnaissance, careful route preparations, and using guides

CLASSES OF SUPPLY

CLASS I: RATIONS AND WATER

5-14. The strenuous activities required during mountain operations increase caloric requirement to 4,500 calories or more per day. Improper or too little food means soldiers will lack the stamina to accomplish the mission. Although combat rations are normally used, unitized group rations (UGRs) should be provided once a day if the situation permits. Individual packages of oatmeal and dehydrated soup mixes should be issued if the UGR cycle cannot be maintained.

5-15. In abrupt ascents to high altitude, soldiers do not have time to acclimate themselves, so their entire circulatory system labors to supply oxygen to the body. In this situation, standard rations are hard to digest and special rations, such as the ration, cold weather (RCW), that allow soldiers to eat light and often should be procured. The totally self-contained operational ration consists of one full day's feeding in a flexible, white-camouflaged meal bag. It contains cooked, freeze-dried, or other low moisture entrees, as well as a variety of items such as oatmeal, a nut-raisin mix, and fruit-cookie bars. The RCW provides sufficient calories (approximately 4.500 kilocalories) to meet the increased energy expenditure during heavy exertion, while limiting sodium and protein content to reduce the risk of dehydration. Because of rapidly changing weather conditions and the difficulty of resupply, each soldier may need to carry two to three days' supply of rations. However, this increases the soldier's load by approximately 10 to 15 pounds.

5-16. Proper water production, resupply, and consumption are essential and a constant challenge during mountain operations. In low mountains, planners should count on at least four quarts of water per soldier per day when static and up to eight quarts per day when active. In high mountains, planner should increase those requirements by about two quarts per soldier. In the mountain environment, medical care often requires an increased water supply and must be considered as part of the original planning and contingency factors.

5-17. Units should always be prepared to use natural water sources to help reduce the logistics burden However, far above the timberline, water is extremely difficult to find. Special measures must be taken to protect it from freezing in cold weather, such as placing canteens in the chest pockets of the extended cold weather clothing system (ECWCS) coat, hanging a two-quart canteen on a strap under the coat, or utilizing a camel-back type, commercially available, canteen under overgarments. Purification and chemical sterilization are always necessary no matter how clean mountain water may appear. Micro-organisms present in mountain water may cause serious illness and rapidly degrade the strength of a unit. If above ground water sources cannot be located or are not reasonably available, drilling for underground sources may become a critical engineer task. Once engineer units access the water, quartermaster units have responsibility for completing the water points and purifying the water.

CLASS II: GENERAL SUPPLIES

5-18. General supplies indude expendable administrative items, individual dothing and equipment, tentage, and other items authorized by common tables of allowance. All units must deploy with enough Class 11 items lo last until routine resupply can be established. Spedal items, such as extended cold weather dothing, gloves, dimbing equipment, extended cold-weather sleep systems, batteries, and one-burner cook stoves, will be in great demand. Due to the rugged nature of the terrain, mountain operations also increase requirements for replacement items of individual dothing and equipment. Combat boots, for example, may be expected to last approximately two weeks in harsh rocky terrain.

CLASS III: FUEL AND PACKAGED PETROLEUM PRODUCTS

5-19. Individual vehicles need much more fuel in mountainous terrain. However, limited road nets and steep slopes reduce the volume of vehide traffic and overall fuel consumption. The heavy reliance on aviation assets for resupply and movement increases aviation fuel requirements. A commander must routinely plan for the emplacement of a forward arming and refueling poinl (FARP) within their AO to support intensive aviation operations. Battalions should establish a fuel point in the field trains using collapsible fuel drums. These drums should be operational as soon as the field trains are established and prepared to receive fuel from the forward supply company as soon as it arrives. When terrain makes refueling operations vulnerable to attack, units should conduct forward refueling using supply point distribution, and dispense fuel to unit vehides using the tailgate technique. The lack of suitable terrain normally increases the percentage of forward refueling done by this method

5-20. At 600 meters (2,000 feet), multi-fuel cook staves operate at about 75 percent effidency. When soldiers refuel cook stoves, they must avoid using automotive fuel. Fuel points must supply units with refined or white gasoline that is spedfically produced for pressurized stoves. Relatively large quantities of this fuel will be used when procuring water and preparing food. Adequate quantities of five-gallon cans, nozzles, and one-quart fuel bottles musi be on-hand before deployment.

CLASS IV: CONSTRUCTION, BARRIER, AND FORTIFICATION MATERIALS

5-21. Soldiers should make maximum use of local materials to reduce Class IV requirements and demands on the transportation system. Gabion-type material is espedally versatile during mountain operations. Gabions are widely used in the mountains for constructing obstades, fighting positions, anchors, mountain installations, traverse platforms, and helicopter LZs; for creating landslides or rockfalls; and for repairing roads. Units should stock adequate quantities of easily transportable sizes of reinforcing mesh and other suitable materials for constructing gabions.

CLASS V: AMMUNITION

5-22. Because of terrain, ammunition resupply is difficult, making strict fire control and discipline an absolute necessity. Ammunition transfer poi nts need to be as far forward as possible without revealing friendly unit locations or pfad ng ammunition stocks at risk of capture or destruction. Di rect delivery to the user may be required using aeri al resupply. Innovation and flexibility are critical, in the mountains, the traditional mixes of tank ammunition may be less effedive. Depending upon the spedfic threat, more rounds may be needed to attack light vehicles and fortified positions and less may be needed to engage tanks. Ammunition consumption for direct fire weapons may be low, however, consumption of indirect fire munitions, such as grenades, mortars, and artillery, may be high because of the dead space common to mountainous terrain. Planners must ensure that increased consumption of indirect fire munitions is induded in computating required supply rates.

CLASS VII AND IX: MAJOR END ITEMSAND REPAIR PARTS

5-23. Rugged terrain and dim ate extremes cause an increase in repair parts consumption. However, overall vehide utilization decreases in many situations Because it is difficult to transport large end items to toward units, the commander must place additional emphasis on preventive maintenance and repair.

CLASS VIII: MEDICAL SUPPLIES

5-24. The medical platoon obtains medical supplies for the battalion from the supporting forward support medical company (FSMC) or similar task organized medical element Medical supply organizations may distribute supply by various means: supply point distribution, unit distribution, or a combination of both. Mountainous terrain necessitates using suppty point distribution to a great extent. Medical supply activities must maximize use of empty evacuation assets moving forward to execute unit distribution of supplies as often as possible. The terrain will severely constrain ground movement operations. Pre-planned unit distribution via air assets is a must for emergent situations, such as mass casualty scenarios. Medical supplies must have a high priority for movement. Distribution of Class V! II via air lines of communications (LOCs) should occur as often as tactically feasible.

SECTION III — TRANSPORTATION AND MAINTENANCE

5-25. Transportation assets For mountain operations are often limited, and their use requires sound planning. Although vehicles are used to move supplies as far forward as possible, they may not be able to reach deployed units. Usi ng smai ler cargo vehides with i mproved cross-country mobi lity and dedicated aircraft is paramount to sustaining units in the mountains. Locally obtained animals, indigenous personnel, or combat soldiers must often move supplies from roads and trails to unit positions. The poor quality of road networks requires increased engineer effort. The rugged mountain terrain aids in infiltration increasing security requirementsalong the route.

5-26. Air resupply should always be considered lo reduce the transportation burden on ground assets. Therefore, support personnel should be well-trained in aerial resupply and sling-load operations. Aerial resupply, either by parachute drop, free drop, or cargo helicopter, may be available for a variety of tactical situations. However, unpredictableweather and air currents, cloud cover, and lack of suitable landing zones make aerial delivery unreliable, and higher elevations decrease overall aircraft lift capabilities. The integrated use of available helicopter transport should be used for forward transport of mail, replacements, returnees, and personnel service support providers, such as chaplains and finance support teams.

- **Tires**
- **Tracks**
- **NVG Batteries**
- **Communications Equipment**
- **Cooling Systems**

Figure 5-3. Key Repair Parts

5-27. Fixing equipment as far forward as possible takes on added importance during mountain operations. In low mountains, equipment recovery and maintenance teams are critical in keeping limited routes clear and returning damaged vehicles to the battle in the shortest possible time. In high mountains, these teams are generally less critical to units operating there because terrain often limits vehicle use. Helicopter repair teams are critical in all mountainous environments due to helicopters flying at or near the maximum limits of their operational capabilities to meet increased needs for helicopter support, 1 n all cases, maintenance turn-around time increases to compensate for fatigue and the other effects of the environment on maintenance personnel. Figure 5-3 lists some of the critical repair parts that often fail or require frequent replacement during mountain operations.

5-28. Drivers well trained in proper maintenance and driving techniques eliminate a great deaf of unnecessary maintenance and reduce maintenance requirements and vulnerable bottlenecks. All soldiers must devote increased attention to applicable FMs and TMs for their weapons and equipment and must conduct preventative maintenance, to include the availability and use of suitable cleaning solvents and lubricants, appropriate for the weather and terrain conditions.

SECTION IV - PERSONNEL SUPPORT

5-29. Personnel support in the mountains is not unlike that provided toother lypesof operations except for the limitations on soldiers and equipment posed by the environment. Key personnel support missions are to provide manning and personnel service support to unit commanders, soldiers, and Army civilians.

5-30. Personnel units normally depend on large quantities of automation equipment to successfully accomplish their mission. Adverse weather and rugged terrain may decrease their reliability and commanders should take extra precautions to protect this equipment. Additionally, the difficulty in establishing and maintaining communications may require an increased reliance on manual strength reporting until communications and automated systems are firmly established.

5-31. Postal services establish the link between soldiers and their family and friends and assist in defeating the isolation caused by the compartmented terrain and the resulting dispersion of units. However, the limited lines of communications in mountainous terrain may adversely affect mail distribution. Inefficient distribution of mail can quickly undermine morale, regardless of the AO. The timely delivery or mail is especially important in countering the shock of entering a new environment. Commanders should consider devoting a high priority to the distribution of mail on arrival in the theater of operations. FM 1-0 describes in detail the critical personnel systems and functions essential to providing manning and personnel service support.

SECTION V - COMBAT HEALTH SUPPORT

PLANNING

5-32. Combat health support (CHS) in the mountains is characterized by–

- Difficulty in accessing casualties in rugged terrain.
- Increased need for technical mountaineering skills for casualty evacuation.
- Longer periods of time needed for casualties to be stabilized.

5-33. When planning CHS, commanders must consider the tactical situation, the nature of the terrain, and speed of movement along the chain of evacuation. Aid stations should be centrally located in relation to supported units. The exact location should be based on the ability to provide shelter from the elements, cover and concealment from the enemy, ease of evacuation, and expected casualty rates.

5-34. The decentralization in the mountain area of operations also forces the decentralization of CHS. Commanders may need to establish casualty collection points, operated by aidmen from the evacuation section, to support isolated elements. These points are designated at intermediate points along the routes of evacuation where casualties may be gathered. Additionally, multiple ambulance exchange points may be required to transfer casualties from one type of transportation to another.

EVACUATION

5-35. Aeromedical evacuation remains the preferred form of casualty evacuation in mountain operations. Aircraft provide the best capability of evacuating casualties from isolated locations and transporting them to treatment centers. However, in many instances, even lightly wounded personnel may not be able to move unassisted over rough terrain and LZs may not be available.

5-36, Medical evacuation teams must complete reconnaissance and install necessary evacuation systems along each evacuation route before the onset of casualties. Litter relay stations may be required at predetermined points to conserve the stamina of litter bearers and aocelerate evacuation. The evacuation plan must include measures to care for wounded soldiers at points along the route of evacuation where delays are possible. The plan must also depict all evacuation routes and provide for proper disposition of medical personnel along the lines of evacuation (see FM **4**-02,2). Evacuating the wounded from mountain combat zones normally requires a larger number of medical personnel and litter bearers than on flat terrain. The number and type of evacuation systems depend on mission, enemy, terrain and weather, troops and support available, time available, civil considerations (METT-TC) and the factors listed in Figure **5-4.**

- **The patient's condition**
- **Anticipated casualty rate**
- **Importance of rapid movement**
- **Number of available evacuation teams**
- **Number of evacuation routes available**
- **Transportation assets and equipment available**
- **Availability of suitable and secure LZs**

Figure 5-4. Factors Affecting Evacuation Systems

MOUTAIN EVACUATION TEAMS

5-37. As the battle to control the heights evolves, combatants attempt to exploit technical aspects of terrain. Consequently, commanders must anticipate the need for evacuation teams, normally Level 2 mountaineers, that have the capability to reach, stabilize, and evacuate casualties in rugged terrain. Ground evacuations are generally classified as steep slope (non-technical) or high angle (technical). The mission of trained mountain evacuation teams is to move casualties over cliffs, obstacles, and other technical terrain that would significantly impede the mobility of standard litter bearers. Using evacuation systems to negotiate obstacles shortens routes and increases the speed of evacuation.

5-38. Because units normally deploy over a wide area and the availability of qualified technical evacuation teams is likely to be limited, all soldiers should be trained to conduct less technical, steep-slope evacuations. Two of the most qualified evacuation teams should be identified in each battalion-sized unit prior to planned operations. They should be designated as the battalion's technical evacuation assets and should undergo more advanced mountaineering training and rigorous evacuation training. These soldiers can also devetop and teach a program of instruction that will increase the proficiency of the company evacuation teams.

5-39. Mountain evacuation teams must install the necessary evacuation systems before casualties occur. They must man the systems, move casualties over the obstacle until the evacuation route is no longer required, disassemble the system, and redeploy as necessary. Depending on the specific terrain, evacuation teams may demand extensive additional training in some of the areas listed in Figure 5-5.

Figure 5-5, Mountain Evacuation Team Tasks

TREATMENT

5-40. Treatment of Ihe wounded in forward areas by medical personnel is extremely difficult in restrictive terrain, since even a single company is often deployed over a wide area. Combat in the mountains demands a greater reliance on self-aid, buddy-aid, and unit combat lifesaver techniques. Emphasis must be placed on lifesaving and life-preserving measures to be performed before medical personnel arrive. Unit combat lifesavers must be identified and trained to perform in the absence of medics. Units operating in mountainous areas should strive to meet or exceed Army standards for the number of combat lifesavers required for their specific unit See FM 4-02.92 for more information on combat lifesavers.

5-41. Soldiers in mountain regions are exposed to many and varied types of illnesses and injuries. Appendix A describes the cause, prevention, symptoms, and treatment of common mountain illnesses and injuries.

Appendix A

Mountain Illnesses and Injuries

Table A-1. Chronic Fatigue and Its Effects

CHRONIC FATIGUE (ENERGY DEPLETION)			
CAUSE	PREVENTION	SYMPTOMS	TREATMENT
Low blood sugar. Sources of energy are depleted. Insufficient caloric intake.	Provide adequate food (type and quantities). Monitor food intake and ensure soldiers eat 4,500 calories or more per day. Eat small, frequent meals rather than large, infrequent meals. Snack lightly and often. Increase amounts of fat in diet.	Difficulty sleeping. Fatigue, irritability, and headache. Difficulty thinking and acting coherently -- impaired judgement. Victims begin to stumble and become clumsy and careless. Energy depletion resembles and aggravates hypothermia. The body does not have enough fuel to maintain proper body temperature. As a result inadequate sources of energy, coupled with cold, create a compound or synergistic effect.	Proper diet and rest. Treat synergistic effects if required.

Table A-2. Dehydration and Its Effects

DEHYDRATION			
CAUSE	**PREVENTION**	**SYMPTOMS**	**TREATMENT**
Loss of too much fluid, salt, and minerals due to poor hydration. <u>Contributing Factors:</u> Water loss occurs through sweating, breathing, and urine output. In cold climates, sweat evaporates so rapidly or is absorbed so thoroughly by clothing layers that it is not readily apparent. In cold weather, drinking is inconvenient. Water is hard to resupply, heavy to carry, and freezes in colder climates. Lack of humidity in the dry mountain air. Diminished thirst sensation induced by hypoxia.	Drink 3 to 4 quarts of water per day when static and up to 8 quarts during increased activity. Adequate rest. Avoid caffeine (coffee, tea, soda) and alcohol, as they compound dehydration. Increase command supervision. Keep canteens full. Use flavored powdered drink mixes to encourage water consumption.	Generally tired and weak. Mouth, tongue, and throat become parched and dry, and swallowing becomes difficult. Darkening of urine. Constipation and painful urination. Loss of appetite. Rapid heartbeat. Headache, dizziness, and nausea with or without vomiting. Difficulty focusing eyes. Dehydration compounds the effects of cold and altitude.	Sufficient hydration to offset water loss. Rest. Severe cases may require an IV. Insulate as required and evacuate.

Table A-3. Giardiasis and Its Effects

GIARDIASIS (PARASITICAL ILLNESS)			
CAUSE	**PREVENTION**	**SYMPTOMS**	**TREATMENT**
Parasitical illness contracted from drinking unpurified water.	Drink only potable water. Boil water for 3 to 5 minutes. Use approved water purification tablets or filters. Keep water containers clean.	Abdominal pain. Weakness and nausea. Frequent diarrhea and intestinal gas. Loss of appetite.	Proper hydration with potable water. Evacuation and prescribed medications.

Table A-4. Hypoxia and Its Effects

HYPOXIA			
CAUSE	PREVENTION	SYMPTOMS	TREATMENT
Rapid ascent to high altitudes (above 3,000 to 4,000 meters or 10,000 to 13,000 feet).	Acclimatization. Slow ascent. Limited activities. Lonq rest periods.	Impaired judgment, perception, and higher mental functions increasing with altitude.	Evacuation to tower altitude.

Table A-5. Acute Mountain Sickness (AMS) and Its Effects

ACUTE MOUNTAIN SICKNESS (AMS)			
CAUSE	PREVENTION	SYMPTOMS	TREATMENT
Rapid ascent to high altitudes {2,400 meters or 8.000	Acclimatization. Staged and/or graded ascent. During stops, no strenuous activity and only mild activity with frequent rest periods. Increased carbohydrate intake (whole grains, vegetables. peas and beans, potatoes, fruits. honey, and refined sugar). Acetazol amide prescribed by a physician.	Headache and fatigue. Insomnia. irritability, and depression. Coughing and shortness of breath. Loss of appetite, nausea, and vomiting. Dizziness. Swelling ol the eyes and face.	Stop and rest Symptoms will normally subside in 3-7 days if soldiers do not continue to ascend. Observe tor the development of HAPE or HACE. If symptoms do not disappear, a rapid descent of 150 to 300 meters (500 to 1,000 feet) or greater is necessary. Re-ascent should take place only after symptoms are resolved.

Table A-6. High Altitude Pulmonary Edema (HAPE) and Its Effects

HIGH ALTITUDE PULMONARY EDEMA (HAPE)			
CAUSE	**PREVENTION**	**SYMPTOMS**	**TREATMENT**
Unacclimatized soldiers rapidly ascending to high altitudes (2,400 meters or 8,000 feet}*. Acclimatized soldiers ascending rapidly from a high to a higher altitude. Usually begins within the first 2 - 4 days after rapid ascent and generally appears during the second night of sleep at high or higher altitudes. Fluid accumulation in the lungs.	Acclimatization. Staged and/or graded ascent. Sleeping at the lowest altitude possible. Slow assumption of physical activity. Protection from the cofd.	Wheezing and coughing (possibly wkth pink sputum). Gurgling sound in chest. Difficulty breathing. Coma Death may occur if rapid descent is not initiated.	Rapid evacuation recommended. Observe for the development of HACE. Seek qualified medical assistance.
'HAI'fc most often <iol.^ -KH ';ur uni. .ibc-'A' 3 -300 ivehv-'-v I'I? 000 U-<>;\			

Table A-7. High Altitude Cerebral Edema (HACE) and Its Effects

HIGH ALTITUDE CEREBRAL EDEMA (HACE)			
CAUSE	**PREVENTION**	**SYMPTOMS**	**TREATMENT**
Unacclimatized soldiers rapidly ascending to high altitudes (2,400 meters or 8,000 feet)\ Acclimatized soldiers ascending rapidly from a high to a higher altitude. Excessive accumulation of fluid in the brain.	Acclimatization Staged and/or graded ascent. Slow assumption ot physical activity. Protection from the cold.	Most severe high altitude illness. Severe headache, nausea, and vomiting. Staggering walk/sway. Contusion, disorientation, and drowsiness. Coma, usually followed by death.	1 m m ed i ate evac uat ion; preferably by air evacuation Seek qualified medical assistance.
*HACE like haPE. most often does not occur unm above ^.000 meters [12.000 feel)			

Appendix B

Forecasting Weather in the Mountains

The Air Force provides the bulk of the weather support required by the Army; however, reports from other branches of the military service, our own National Weather Bureau, or a foreign country's weather service can also aid in developing accurate forecasts {see FM 2-33.2). Weather at different elevations and areas, even within the same general region, may differ significantly due to variations in cloud height, temperature, winds, and barometric pressure. Therefore, general reports and forecasts must be used in conjunction with the locally observed weather conditions to produce reliable weather forecasts for a particular mountain area of operations.

INDICATORS OF CHANGING WEATHER

MEASURABLE INDICATORS

B-1. In the mountains, a portable aneroid barometer, thermometer, wind meter, and hygrometer are useful to obtain measurements that will assist in forecasting the weather. Marked or abnormal changes within a 12-hour period in the indicators listed in Figure B-1 may suggest a potential change in the weather.

- Barometric Pressure
- Wind Velocity
- Wind Direction
- Temperature
- Moisture Content of the Air

Figure B-1. Measurable Weather Indicators

CLOUDS

B-2 Clouds are good indicators of approaching weather conditions. By reading doud shapes and patterns, observers can forecast weather even without additional equipment.

B-3. Shape and height are used to identify douds. Shape provides information about the stability of the atmosphere, and height above ground level provides an indication of the distance of an approaching storm. Taken together, both indicate the likelihood of predpitation (see Figure B-2). The heights shown in the figure are an estimate and may vary, based on geographical location.

Clouds by Shape

B-4. Clouds may be classified by shape as cumulus or stratus.

- Cumulus douds are often called "puffy" clouds, looking like tufts of cotton. Their thickness (bottom to top) is usually equal to or greater than their width. Cumulus douds are primarily composed of water droplets that cause them to have sharp, distinct edges. These clouds usually indicate instability at the altitude of the atmosphere where they are found. The stormy weather associated with cumuius clouds is usually violent with heavy rains or snow and strong, gusty winds. Precipitating cumulus douds are called cumulonimbus.

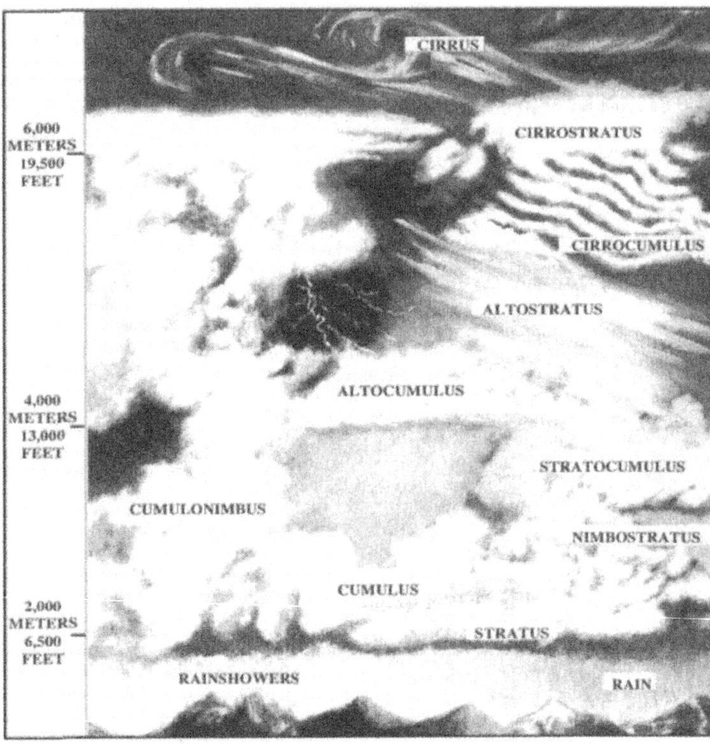

Figure B-2. Types of Clouds

- Stratus clouds are layered, often appearing flattened, with greater horizontal than vertical dimensions. They usually indicate a stable atmosphere, but can indicate the approach of a storm. Stormy weather assodated with stratus clouds usually does not normally include violent winds, and precipitation is usually light but steady, lasting up to 36 hours. Lightning is rarely associated with stratus douds, however, slee! may occur. Fog is also associated with the appearance of stratus clouds. Precipitating stratus clouds are called nimbostratus. and clouds that cannot be determined as stratus or cumulus are referred to as stratocumulus. These latter types may be evolving from one type to another, indicating a change in atmospheric stability.

Clouds by Height

B-5. Clouds are also classified by the height of their base above ground level into three categories- low. middle, and high.

- Low clouds, below 2,000 meters (6,500 feet), are either cumulus or stratus, or their predpitating counterparts. Low douds may be identified by their height above nearby surrounding relief of known elevation. Most precipitation originates from low douds because rain and snow from higher clouds usually evaporates before reaching the ground. As such, low douds usually indicate predpitation. especlally if they are more than 1.000 meters (3.000 feet) thick (douds that appear dark at the base usually are at least that thick).
- Middledouds, between 2.000 and 6,000 meters (6,500 and 19,500 feet) above ground, have a prefix of "alto", and are called either altostratus or altocumulus. Middle douds appear less distinct than low douds because of their height. Warm "alto" douds have sharper edges and are composed mainly of water droplets. Colder douds, composed mainly of ice crystals, have distinct edges that grade gradually into the surrounding sky. Middle clouds indicate potential storms, though usually hours away. Altocumulus clouds that are scattered in a blue sky are called "fair weather" cumulus and suggest the arrival of high pressure and clear skies. Lowering altostratus douds with winds from the south indicate warm front conditions, decreasing air pressure, and an approaching storm system within 12 to 24 hours.
- High douds, higher than 6,000 meters (19,500 feet), are drrus. drrostratus, and drrocumulus. They are usually frozen douds with a fibrous strudure and blurred outlines. The sky is often covered with a thin veil of drrus that partly obscures the sun or, at night, produces a ring of light around the moon. The arrival of drrus indicates moisture aloft and the approach of a storm system. Predpitation is often 24 to 36 hours away. As the storm approaches, the drrus thickens and lowers becoming altostratus and eventually stratus. Temperatures warm, humidity rises, and winds approach from the south or southeast.

Other Clouds

B-6 Somedoudsindicate serious weather ahead.

- Towering cumulus douds have bases below 2,000 meters (6,500 feet) and tops often over 6,000 meters (19.500 feet). They are the most dangerous of all types and usually do not occur when temperatures at the surface are below 32-degrees Fahrenheit. They indicate extreme instability in the atmosphere, with rapidly rising air currents caused by solar heating of the surface or air rising over a mountain barrier. Mature towering cumulus douds often exhibit frozen stratusdouds at their tops, producing an "anvil head" appearance. Towering cumulus clouds may be local in nature, or they may be associated with the cold front of an approaching storm. The iatter appears as an approaching line of thunderstorms or towering cumulus clouds. Towering cumulus clouds usually produce high, gusty winds, lightning, heavy showers, and occasionally hail and tornadoes (although tornadoes are rare in mountainous terrain). Such thunderstorms are usually short-lived and bring clear weather.

G Cloud caps often form above pinnacle and peaks, and usually indicate higher winds aloft. Cloud caps with a lens shape (similar to a "flying saucer") are called lenticular and indicate very high winds (over 40 knots). Cloud caps should always be watched for changes. If they grow and descend, bad weather can be expected.

APPLYING THE INDICATORS

B-7. Weather forecasts are simply educated estimations or deductions based on general scientific weather principles and meteorological evidence. Forecasts based on past results may or may not be accurate. However, even limited experience in a particular mountainous region and season may provide local indications of impending weather patterns and increased accuracy. Native weather lore, although sometimes greatly colored and surrounded in mystique, should not be discounted when developing forecasts, as it is normally based on the local inhabitants' long-term experience in the region.

BAD WEATHER

B-8. Signs of approaching bad weather (within 24 to48 hours) may include—

- A gradual lowering of the clouds. This may be the arrival or formation of new lower strata of clouds. It can also indicate the formation of a thunderhead.
- An increasinghaloaroundthesunorlhe moon.
- An increase in humidity and temperature.
- Cirrus clouds.
- A decrease in barometric pressure (registered as a gain in elevation on an altimeter).

STORM SYSTEMS

B-9. The approach of a storm system is indicated when—

- A thin veil of cirrus clouds spreads over the sky, thickening and lowering until altostratus clouds are formed. The same trend is shown at night when a halo forms around the moon and then darkens until only the glow of the moon is visible. When there is no moon, cirrus clouds only dim the stars, but altostratus clouds completely hide them.
- Low clouds, which have been persistent on lower slopes, begin to rise at the time upper clouds appear.

- Various layers of clouds move in at different heights and become abundant.
- Lenticular clouds accompanying strong winds lose their streamlined shape, and other doud types appear in increasing amounts,
- Achangeinthedi rection ofthewindis accompanied by a rapid rise in temperature not caused by solar radiation. This may also indicate a warm, damp period.
- A light green haze is observed shortly after sunrise in mountain regions above the timber line.

THUNDERSTORMS

B-10. Indications of local thunderstorms or squally weather are—

- An increase in size and rapid thickening of scattered cumulus clouds during the afternoon.
- The approach of a line of large cumulus or cumulonimbus clouds with an "advance guard" of altocumulus douds. At night, increasing lightning windward of the prevailing wind gives the same warning.
- Massive cumulus douds hanging over a ridge or summit (day or night).

STRONG WINDS

B-11. Indications of approaching strong winds may be—

- Plumes of blowing snow from the crests of ridges and peaks or ragged shreds of doud moving rapidly.
- Persistent lenticular douds, a band of douds over high peaks and ridges, or downwind from them.
- A turbulent and ragged banner doud that hangs to the lee of a peak.

PRECIPITATION

B-12. When there is predpitation and the sky cannot be seen—

- Small snowflakes or ice crystals i ndicate that thedouds above are thi n, and fair weather exists at high elevations.
- A steady fall of snowflakes or raindrops indicates that the predpitation has begun at high levels, and bad weather is likely to be encountered on ridges and peaks.

FAIR WEATHER

B-13. Continued fair weather may be assodated with—

- A doudless sky and shallow fog, or layers of haze at valley bottoms in early morning.
- A doudless sky that is blue down to the horizon or down to where a haze layer forms a secondary horizon.
- Conditions under which small cumulus douds appearing before noon do not increase, but instead decrease or vanish during the day.

- Clear skies except for a law cloud deck that does not rise or t hick en during the day.

B-14. Signs of approaching fair weather include—

- A gradual rising and diminishing of clouds.
- A decreasing halo around the sun or moon.
- Dew on theground in the morning.
- Small snowfiakes, ice crystals, or drizzle, which indicate that the clouds are thin and fair weather may exist at higher elevations.
- An increase in barometric pressure (registered as a loss in elevation on an altimeter).

GLOSSARY

ABN	airborne
acclimatization	the physiological changes that allow the body to adapt or get used to the effects of a new environment, especially low oxygen saturation at higher elevations
ACE	armored combat earth mover
acetazolamide	a pharmaceutical drug used to accelerate acclimatization
ADA	air defense artillery
ADAM	area denial artillery munitions
AH-64	attack helicopter also called the Apacbe
aid	in mountaineering, a climbing device, such as pitons, bolts, chocks, and stirrups, used for body support and upward progress; also used for artificial height in the absence of handholds and footholds
ALOC	air lines of communications
AM	amplitude modulation
ambient temperature	encompassing atmosphere
AMS	acute mountain sickness
anchor	a secure point (natural or artificial} to which a person or rope can be safely attached
aneroid	using no liquid
ANZAC	Australia and New Zealand Corps
AO	area of operations
apnea	temporary suspension of respiration
ARSOF	Army special operations forces
ART	Army tactical task
ARTEP	Army training and evaluation program
assault climber	military mountaineer possessing advanced (Level 2) skills, capable of leading small teams over class 4 and 5 terrain and supervising rigging/operation of all basic rope systems
AT4	a man-portable, lightweight, self-contained, antiarmor weapon
ATGM	anti-tank guided missile
basic mountaineer	a military mountaineer trained in fundamental (Level 1) travel/climbing skills necessary to move safely and efficiently in mountainous terrain

belay	a rope management technique used to ensure that a fall taken by a climber can be quickly arrested: belay techniques are also used for additional safetyfcontrol in rappelling. raising and lowering systems, and for mountain stream crossings
BFV	Bradley fighting vehicle
BSFV	Bradley Stinger fighting vehicle
BN	battalion
&.	command and control
CAFAD	combined arms for air defense
CFV	cavalry fighting vehicle
CHS	combat health support
CNR	combat net radio
continental climate	bitterly cold winters, extremely hot summers; annual rain and snowfall is minimal and often quite scarce for long periods
Cordillera	principal mountain ranges of the world, named after the Spanish word for rope
crampons	climbing irons, attached to the bottom of boots, used on ice or snow in mountaineering
crevice	a narrow opening resulting from a split or crack as in a cliff
CS	combat support
CSS	combat service support
DA	Department of the Army
defile	a narrow passage or gorge
DPI CM	dual-purpose improved conventional munition
DZ	drop zone
ECWCS	extended cold weather clothing system
edema	a local or general condition in which the body tissues contain an excessive amount of tissue fluid
evacuation team	a team trained to move casualties over steep slopes, cliffs, and other obstacles that would significantly impede the mobility of standard litter bearers
EW	electronic warfare
F	Fahrenheit
FARP	forward arming and refueling point
FASCAM	family of scatterable mines
FEBA	forward edge of the battle area

fixed alpine path	a mountain path created by any combination of aids, to include steps, stanchions, standoff ladders, suspended walkways, cableways, or other improvements made of materials available; normally an engineering task.
fixed rope	a rope, or series of ropes, anchored to the mountain at one or more points to aid soldiers over steep, exposed terrain; usually installed by lead climbing teams (normally Level 2 qualification)
flash defilade	to arrange fortifications to protect from fire
FM	field manual; frequency modulation
FSMC	forward support medical company
FOX system	a lightly-armored, wheeled laboratory that takes air, water, and ground samples and immediately analyzes them for signs of weapons of mass destruction
gabion	a wicker basket filled with earth and stones often use in building fortifications; can also be created out of similar materials, such as wire mesh/fence, lumber, plywood, or any suitable material that forms a stackable container for rocks, gravel, and soil
giardiasis	parasitical illness
glaciated	covered with glacial ice
GPS	global positioning system
GTA	graphictraining aid
guide	a soldier experienced in all aspects of mountaineering who has the skills and knowledge to identify obstacles and ways to overcome them; commander's advisor on technical mountaineering matters that could affect the tactical scheme of maneuver; primary function of mountain leaders (Level 3 qualification)
HACE	high altitude cerebral edema
HAPE	high altitude pulmonary edema
HE	high explosives
Hellfire	tank-killing missile carried by the Apache attack helicopter
high mountains	mountains that have a local relief usually exceeding 900 meters (3,000 feet)
HUMINT	human intelligence
HWY	highway
hygrometer	an instrument used to measure humidity or moisture content in the air
hypoxia	a deficiency of oxygen reaching the tissues of the body
ice fog trails	steam/smoke trails created by firing weapons
ID	infantry division

IFV	infantry fighting vehicle
IHFR	improved high frequency radio
IMINT	imagery intelligence
installation team	a team organized to construct and maintain rope installations used to facilitate unit movement; usually comprised of Level 1 and 2 mountaineers
interdiction	to stop or hamper
ionospheric	a part of the earth's atmosphere of which ionization of atmospheric gases affects the propagation of radio waves; starts at about 30 miles above ground
IPB	intelligence preparation of the battlefield
IV	intravenous
JSTARS	joint surveillance, target attack radar system
km	kilometer
lead climbing team	a roped dimbing team (usually Level 2 qualification) trained to lead on dass 4 and 5 terrain; establishes/prepares the entire route for the remainder of the unit
leeward	the side sheltered from the wind
lenticutar	having the shape of a double-convex lens
LOC	line of communication
local relief	the difference in elevation between valley floors and the surrounding summits
look-down angles	the angle from the aircraft to the target
low mountains	mountains that have a local relief of 300 to 900 meters (1,000 ta 3,000 feet)
LPT	logistics preparation of the theater
LRS	long-range surveillance
LRSU	long-range surveillance unit
LSDIS	light and spedal division interim sensor
LTC	lieutenant colonel
LZ	landing zone
MANPADS	man-portable air defense system
maritime climate	mild temperatures with large amountsof rain or snow
MBA	main battlearea
METT-TC	mission, enemy, terrain and weather, troops and support available, time available, dvil considerations
MK-19	40-mm grenade machine gun, MOD 3

MOPP	mission-oriented protective posture
motti	Finnish word meaning "a pile of logs ready to be sawed into lumber"; used in military terms to descri be setting the condit ions so a larger force can be defeated in detai l
mountain leader	a military mountaineer possessing the highest level (Level 3) of mountaineering skills with extensive experience in a variety of mountain environments in both winter and summer months
WISE	mobile subscriber equipment
MSRT	mobile subscriber radio terminal
MTF	manual terrain following
NBC	nuclear, biological, and chemical
OCOKA	observation and fields of fire, cover and concealment, obstacles, key terrain, and avenues of approach
OH-58D	a scout and attack helicopter known as the Kiowa Warrior
OP	observation post
OPORD	operation order
OPSEC	operations security
OR	operational readiness
orographic	pertaining to the physical geography of mountains and mountain ranges
PADS	Position Azimuth Determining System
POL	petroleum, oils, and lubricants
protection	in mountaineering, special anchor points established during a roped party dimb to limit potential fall distances, protecting dimbers from severe fall/ground-fall consequences
PSYOP	psychological operations
RAAMS	remote antiarmor mine system
rappel	method of controlled fridional descent down a rope
RCW	ration, cold weather
rockfai l	rockfall occurs on alt steep slopes. It is caused by other dimbers or by the continual erosion of the rock on a mountainside resulting from freezing, thawing, and heavy rain; grazing animals; or enemy action.
SATCOM	satellite communications
scree	small unconsolidated rocks or gravel, fist-**si2e** or smaller, located mostly below rock ridges and diffs
screen i ng crest	a hill or ridge located in front of a radar set to mask it from unwanted returns (dutter) at dose range, and to provide security

	against electronic detection or jamming; screening crest also prevents visual observation and attack by direct fires
SEE	small emplacement excavator
SHELREP	shelling report
SHORAD	short-range air defense
SINCGARS	Single-channel Ground and Airborne Radio System
SOP	special operations forces
squall	a sudden, violent wind
SR	special reconnaissance
talus	accumulated rock debris that is much larger than scree, usually basket ball-size or larger
TBP	to be published
TC	training circular
TCF	tactical combat force
TCP	traffic control point
temperature inversion	when the temperature is warmer at higher elevations than lower elevations
TM	technical manual
TOC	tactical operations center
TOW	tube-launched, optically tracked, wire-guided, heavy antitank missile system
TRADOC	United States Army Training and Doctrine Command
tundra	treeless, black, mucky soil with permanently frozen subsoil; located in mountainous regions above the timberline
tussocks	grassy clumps
UAV	unmanned aerial vehicle
UGR	unitized group ration
UHF	ultrahigh frequency
Venturi effect	as a fluid (such as air) flows through a constriction (like a mountain pass), the speed increases and the pressure drops
VFR	visual flight rules
VT	variable time
wind chill	the rate at which a man or object cools to the ambient temperature; wind increases the rate of cooling and adds to the risk of frostbite, hypothermia, and other cold-weather injuries
windward	being in or facing the direction from which the wind is blowing
WP	white phosphorous

Bibliography

The bibliography lists field manuals by new number followed by old number.

ARMY PUBLICATIONS

Most Army doctrinal publications are available online:
http://155.217.58.58/atdls.htm

AR 385-10. *The Army Safety Program*. 23 May 1988.

FM 1-0 (12-6). *Personnel Doctrine*. 09 September 1994.

FM 2-01.3 (34-130). *Intelligence Preparation of the Battlefield*. 08 July 1994.

FM 2-33.2 (34-81). *Weather Support for Army Tactical Operations*. 31 August 1989.

FM 2-33.201 (34-81-1). *Battlefield Weather Effects*. 23 December 1992.

FM 3-0 (100-5). *Operations*. 14 June 1993.

FM 3-01.8 (44-8). *Combined Arms for Air Defense*. 01 June 1999.

FM 3-01.43 (44-43). *Bradley Stinger Fighting Vehicle Platoon and Squad Operations*. 03 October 1995.

FM 3-01.44 (44-44). *Avenger Platoon, Section, and Squad Operations*. 03 October 1995.

FM 3-04.100 (1-100). *Army Aviation Operations*. 21 February 1997.

FM 3-04.203 (1-203). *Fundamentals of Flight*. 03 October 1988.

FM 3-05.27 (31-27). *Pack Animals in Support of Army Special Operations Forces*. 15 February 2000.

FM 3-09.4 (6-20-40). *Tactics, Techniques, and Procedures for Fire Support for Brigade Operations (Heavy)*. 05 January 1990.

FM 3-09.12 (6-121). *Tactics, Techniques, and Procedures for Field Artillery Target Acquisition*. 25 September 1990.

FM 3-09.40 (6-40). *Tactics, Techniques, and Procedures for Field Artillery Manual Cannon Gunnery*. 23 April 1996.

FM 3-09.42 (6-20-50). *Tactics, Techniques, and Procedures for Fire Support for Brigade Operations (Light)*. 05 January 1990.

FM 3-11.4 (3-4). *NBC Protection*. 29 May 1992.

FM 3-11.5 (3-5). *NBC Decontamination*. 28 July 2000.

FM 3-]] ,6 (3-6), *Field Behavior of NBC Agents (Including Snufke and Incendiaries).* 03 November (986.

FM 3-11.19 (3-19). *NBC Reconnaissance.* 19 November 1993.

F M 3-21.7 (7-7). *Tile Mechanized Infantry Platoon and Squad <APC).* 15 *Match* 1985.

FM 3-21,20(7-20), *The Infantry Battalion.* 06 April 1992,

FM 3-21.30 (7-30), *The Infantry Brigade.* 03 October 1995.

F M 3-21.38 (S7-38). *Pathfinder Operations.* 09 April 1993.

FM 3-23.10 (23-10). *SniperTraining.* 17 August 1994.

FM 3-23,30 (23-30), *Grenades and Pyrotechnic Signal).* 27 Decem be r 1988,

FM 3-24.3 (20-3). *Camouflage. Concealment, and Decoys.* 30 August 1999.

FM 3-24.32 (20-32). *Mine/Countermine Operations.* 29 May 1998.

FM 3-25,9(23-9), *MI6A) and M16A2 Rifle Marksmanship.* 03 July 1989.

FM 3-25.18(21-18). *Foot Marches.* 0 J June 1990.

FM 3-25.26 (21-26). *Map Reading and Land Navigation.* 07 May 1993.

FM 3-25,60(21-60), *Visual Signals.* 30 September 1987,

FM 3-25,76 (21-76). *Survival.* 05 June 1992.

FM 3-34.1 (90-7). *Combined Arms Obstacle Integration.* 29 September 1994.

FM i 34.2 (9(1 13 I 1. *Combined Arms Brea, lung Operations.* -N iilmi.iri I V) I .

FM 3-34.102 (5-102), *Cotmtermobility.* I4March 1985.

FM 3-34.112(5-103). *Survivability.* 10 June 1985.

FM 3-34.214 (5-250). *Explosives and Demolitions.* 30 July 1998.

FM 3-34,223 (5-7-30). *Brigade Engineer and Engineer Company Combat Operations I Airborne, Air Assault, Light}.* 28 December 1994.

FM 3-34.330 (5-33). *Terrain Analysis.* I 1 July 1990.

FM 3-60 (6-20-10). *Tactics. Techniques, and Procedures for the Targeting Process.* 08 May 1996.

FM 3-90.21 (6-20-1). *Tactics, Techniques, and Procedures for the Field Artillery Cannon Battalion.* 29 November 1990.

FM 3-91 -2 (7 I -2). *The Tank and Mechanized Infantry Battalion Task Force.* 21 September 1988.

FM 3-91,3 (71-3), *The Armored and Mechanized Infantry Brigade.* 08 January 1996,

FM 3-9], 323(71-123). *Tactics and Techniques for Combined Arms Heavy Forces: Armored Brigade, Battalion Task Force, and Company Team,* 30 September 1992,

FM 3-97.3 (90-3). *Desert Operations. 24* Augusl 1993.

FM 3-97.1 I (90-1 I). *Cold Weather Operations. TUP.*

FM 3-97,22 (90-22), *(Night) Multi-Service Night and Adverse Weather Combat Operations.* 31 January 1991.

FM 3-97,50 (3-50), *Smoke Operations,* 04 December 1990,

FM 3-100.14 (100-14). *Risk Management,* 23 April 1998.

FM 3-100.15 (100-15). *Corps Operations.* 29 Oelnber 1996.

FM 3-100,40(100-40), *Tactics,*TBP,

FM 3-100.55 (100-55). *Reconnaissance and Surveillance.* TBP,

FM 3-100.71 (71-100). *Division Operations.* 28 August 1996.

FM 4-02.2 (8-10-6). *Medictd Evacuation in a Theater of Operations Tactics. Techniques, and Procedures.* 14 April 2000.

FM 4-02,22 (22-51), *Leaders' Manual for Combat Stress Control.* 29 September 1994.

FM 4-02.92 (8-10-4). *Medical Planum Leaders' Handbook Tactics, Techniques, and Procedures.* l6November 1990,

FM 4-20.1 (10-27). *General Supply in Theaters of Operations.* 20 April 1993.

FM 4-20.2 (10-23). *Basic Dtfcrrine for Artnv Field Feedlng and Class I Operations Management.* lK April 1996.

FM 4-25,10 (21-10), *Field Hygiene and Sanitation.* 21 June 2000

FM 4-25.11(21-11). *First Aid for Soldiers.* 27 October 1988,

FM 4-30.32 (9-207). *Operations and Maintenance of Ordnance Materiel in Cold Weather.* 20 March 1998.

FM 5-0 (101-5). *Sniff Organization and Operations.* 31 May 1997.

FM 6-0 (100-34). *Command and Control.* TBP.

FM 6-02.1 I (24-1 I), *Tactical Satellite Communications.* 20 September 1990.

FM 6-02.18 (24-1 8). *Tactical Single-Channel Radio Communications Techniques.* 30 September 1987.

F'M 6-02.32 (I 1-32). *Combat Net Radio Operations.* 15 Otlciber 1990.

FM 6-02,43(1 1-43), *The Signal Leader's Guide,* 12 June 1995,

FM 6-02,55 (11-55). *Mobile Subscriber Equipment < MSB) Operations. 22* June 1999.

FM 6-22 (22-100), *Army Leadership Be, Know, Do.* 31 August 1999,

FM 7-0 (25-100). *Training die Farce.* 15 November 1988.

FM 7-10(25-101). *Battle Focused Truining.* 30 September 1990.

FM 7-I5(25-XX). *Army Universal Task List.* TBP,

FM 21-305. *Manna! far the Wheeled Vehicle Driver.* 27 August 1993.

GTA 8-5-60. *A Soldier's (hade to Staying Healthy at High Elevations.* 02 September 1996.

GTA 8-6-12, *Adverse Effects of Cold.* 01 August 1985.

TC3-10. *Commander's Tactical NBC Handbook.* 29 September 1994.

TC 24-20. *Tactical Wire and Cable Techniques.* 03 October 1 988.

TC 24-21. *Tactical Multichannel Radio Communications Techniques.* 03 October 1988.

TC 90-6-1. *Military Mountaineering.* 26 April 1989.

NONMILITARY PUBLICATIONS

Jalali. Ali Ahmad and Lester W. Grau. *The Other Side of the Mountain.* Ouanlko, Va.: Marine Corps Combat Development Command, USMC Studies and Analysis Division, 1999.

Ellis, Robert B. *See Naples and Die.* London: McFarland & Co. 1996.

Gawryeh, George W. "The Rock of Gallipoli" in *Studies in Battle Command.* Kurt Leavenworth. Kans.. US Army Command and General Staff College, Combat Studies Institute. 1995.

Cirau, LesLcr W. *The Bear Went Over the Mountain: Soviet Combat Tactics in Afghanistan.* Washington. D.C: National Defense University Press, 1996.

DOCUMENTS NEEDED

These documents should be available to the intended users of this manual.

DA Form 2028, Recommended Changes to Publications and Blank Forms. I February 1974,

FM 7-0 (25-100). *Training die Force.* 15 November 1988.

FM 7-10 (25-101). *Battle Focused Training.* 3(1 September 1990.

FM 3-0 (100-5), *Operations.* 14 June 1993,

FM 3-40(100-40). *Tactics.* TBP.

TC 90-6-1. *Military Mountaineering.* 26 April 1989.

Index

Entries are by paragraph number unless stated otherwise.

Made in United States
Orlando, FL
05 November 2026

90229002R00285